美国数学会经典影印系列

出版者的话

近年来，我国的科学技术取得了长足进步，特别是在数学等自然科学基础领域不断涌现出一流的研究成果。与此同时，国内的科研队伍与国外的交流合作也越来越密切，越来越多的科研工作者可以熟练地阅读英文文献，并在国际顶级期刊发表英文学术文章，在国外出版社出版英文学术著作。

然而，在国内阅读海外原版英文图书仍不是非常便捷。一方面，这些原版图书主要集中在科技、教育比较发达的大中城市的大型综合图书馆以及科研院所的资料室中，普通读者借阅不甚容易；另一方面，原版书价格昂贵，动辄上百美元，购买也很不方便。这极大地限制了科技工作者对于国外先进科学技术知识的获取，间接阻碍了我国科技的发展。

高等教育出版社本着植根教育、弘扬学术的宗旨服务我国广大科技和教育工作者，同美国数学会（American Mathematical Society）合作，在征求海内外众多专家学者意见的基础上，精选该学会近年出版的数十种专业著作，组织出版了"美国数学会经典影印系列"丛书。美国数学会创建于1888年，是国际上极具影响力的专业学术组织，目前拥有近30000会员和580余个机构成员，出版图书3500多种，冯·诺依曼、莱夫谢茨、陶哲轩等世界级数学大家都是其作者。本影印系列涵盖了代数、几何、分析、方程、拓扑、概率、动力系统等所有主要数学分支以及新近发展的数学主题。

我们希望这套书的出版，能够对国内的科研工作者、教育工作者以及青年学生起到重要的学术引领作用，也希望今后能有更多的海外优秀英文著作被介绍到中国。

高等教育出版社

2016年12月

美国数学会经典影印系列

Fine Regularity of Solutions of Elliptic Partial Differential Equations

椭圆偏微分方程的解的精细正则性

Jan Malý, William P. Ziemer

高等教育出版社·北京

To

Jaroslava and Suzanne

Contents

Preface	ix
Basic Notation	xiii

Chapter 1. Preliminaries		1
1.1	Basic results	1
	1.1.1 Covering theorems	7
	1.1.2 Densities of measures	9
	1.1.3 The maximal function and its applications	11
1.2	Potential estimates	14
1.3	Sobolev spaces	21
	1.3.1 Inequalities	28
	1.3.2 Imbeddings	31
	1.3.3 Pointwise differentiability of Sobolev functions	40
	1.3.4 Spaces $Y^{1,p}$	45
	1.3.5 Adams' inequality	50
	1.3.6 Bessel and Riesz potentials	57
1.4	Historical notes	61

Chapter 2. Potential Theory		63
2.1	Capacity	63
	2.1.1 Comparison of capacities; capacities of balls	66
	2.1.2 Polar sets	71
	2.1.3 Quasicontinuity	73
	2.1.4 Multipliers	77
	2.1.5 Capacity and energy minimizers	78
	2.1.6 Thinness	84
	2.1.7 Capacity and Hausdorff measure	86
	2.1.8 Lebesgue points for Sobolev functions	89
2.2	Laplace equation	92
	2.2.1 Green potentials	98
	2.2.2 Classical thinness	102
	2.2.3 Dirichlet problem and the Wiener criterion	105
2.3	Regularity of minimizers	109
	2.3.1 Abstract minimization	110
	2.3.2 Minimizers and weak solutions	110
	2.3.3 Higher regularity	115
	2.3.4 The De Giorgi method	118
	2.3.5 Moser's iteration technique	122

		2.3.6	Removable singularities	126
		2.3.7	Estimates of supersolutions	128
		2.3.8	Estimates of energy minimizers	131
		2.3.9	Dirichlet problem	135
		2.3.10	Application of thinness: the Wiener criterion	139
	2.4	Fine topology		143
	2.5	Fine Sobolev spaces		148
	2.6	Historical notes		157

Chapter 3. Quasilinear Equations — 161
  3.1 Basic properties of weak solutions — 161
      3.1.1 Upper bounds for weak solutions — 167
      3.1.2 Weak Harnack inequality — 174
      3.1.3 Removable sets for weak solutions — 178
  3.2 Higher regularity of equations with differentiable structure — 180
  3.3 Historical notes — 182

Chapter 4. Fine Regularity Theory — 185
  4.1 Basic energy estimates — 185
  4.2 Sufficiency of the Wiener condition for boundary regularity — 197
      4.2.1 The special case of harmonic functions — 204
  4.3 Necessity of the Wiener condition for boundary regularity — 210
      4.3.1 Main estimate — 211
      4.3.2 Necessity of the Wiener condition — 224
  4.4 Equations with measure data — 225
  4.5 Historical notes — 230

Chapter 5. Variational Inequalities – Regularity — 233
  5.1 Differential operators with measurable coefficients — 233
      5.1.1 Continuity in the presence of irregular obstacles — 234
      5.1.2 The modulus of continuity — 236
  5.2 Differential operators with differentiable structure — 238
  5.3 Historical notes — 251

Chapter 6. Existence Theory — 253
  6.1 Existence of solutions to variational inequalities — 253
      6.1.1 Pseudomonotone operators — 262
      6.1.2 Variational problems – existence of bounded solutions — 263
      6.1.3 Variational problems leading to unbounded solutions — 266
  6.2 The Dirichlet problem for equations with differentiable structure — 269
  6.3 Historical notes — 270

References — 273

Index — 283

Notation Index — 289

Preface

"An important problem ... is the question concerning existence of solutions of partial differential equations when the values on the boundary of the region are prescribed. ... Has not every regular variational problem a solution, provided certain assumptions regarding given boundary conditions are satisfied and provided also if need be that the notion of a solution shall be suitably extended?" These are the words that Hilbert used to conclude his question concerning the twentieth problem stated in an address delivered before the International Congress of Mathematicians in 1900.

Hilbert's twentieth problem and his related nineteenth problem have opened one of the most interesting chapters in analysis and have attracted the efforts of many mathematicians. Although there are still many open questions related to these problems of Hilbert, a great deal of progress has been made, some with dazzling success.

One of the components necessary to establish regularity of certain variational problems was the need to show that a weak solution of a linear equation in divergence form with bounded, measurable coefficients is Hölder continuous. This result resisted many attempts, but finally in 1957, De Giorgi and Nash, independently of each other, provided a proof of this crucial result. The De Giorgi–Nash result stimulated a great number of related investigations, one of the most important being that of Moser who, by entirely different methods, provided another proof of their result. A critical component in Moser's proof was the discovery that the logarithm of the solution is a function of bounded mean oscillation. He also proved the Harnack inequality, which states that locally, the supremum of the solution is bounded by its infimum.

Others generalized these results to weak solutions of quasilinear equations in divergence form with a very general structure. Ladyzhenskaya and Ural'tseva fully developed the De Giorgi method in this direction while Serrin and Trudinger extended the Moser method. In particular, Trudinger discovered that weak supersolutions possess a weak Harnack inequality. That is, the infimum of a nonnegative supersolution over a ball can be bounded below by its integral average.

The primary purpose of all of these results was to investigate the behavior of weak solutions in the interior of the domain. While these methods could be easily adapted to establish some regularity results of the solution at the boundary, a definitive one similar to the inspired result by Wiener to characterize continuity at the boundary for harmonic functions was lacking. The first step in this direction was made by Littman, Stampacchia, and Weinberger who proved that a point in the boundary of an arbitrary domain was simultaneously regular for harmonic functions and weak solutions of linear equations with bounded, measurable coefficients. In his work concerning the local behavior of solutions of quasilinear equations, Serrin discovered that a capacity, now known as the p-capacity, was the appropriate measurement for describing removable sets for weak solutions. Later, Maz'ya discovered a Wiener-type expression involving this capacity which provided a sufficient condition for continuity at the boundary of weak solutions of equations whose structure is similar to that of the p-Laplacian. Utilizing different techniques, it was shown in [**GZ3**] that Maz'ya's condition was also sufficient for boundary continuity for solutions of a large class of quasilinear equations in divergence form. The methods used in [**GZ3**] were of interest in that they did not employ any potential theoretic techniques, thus making a distinct departure from all previous attempts in this area. Indeed, a new proof of sufficiency of Wiener's theorem can be obtained by particularizing the results in [**GZ3**]. By carefully analyzing the methods leading to these results Evans and Jensen, [**EJ1**], [**EJ2**], established an even simpler proof of sufficiency and also included necessity, using essentially no more than Sobolev's inequality and the mean value property of superharmonic functions. Showing that the Wiener-type expression was also necessary for the general class of quasilinear equations proved to be difficult and was established only recently in [**KM4**]. Prior to this, Lindqvist and Martio, [**LM**], established necessity in the case $p > n - 1$.

The main thrust of this book is to give a comprehensive exposition of the results surrounding the work of [**GZ3**] and [**KM4**]. This includes a "fine" analysis of Sobolev spaces and a development of the associated nonlinear potential theory. The term "fine" refers to a topology on $\mathbf{R}^n$, actually the p-fine topology, which is induced in a natural way by the Wiener-type condition. It also includes an analysis of variational inequalities or obstacle problems where the obstacles are allowed to be irregular, even discontinuous. The exposition on variational inequalities includes the regularity theory associated with differential operators with a differentiable structure and $\mathcal{C}^{1,\alpha}$ obstacles.

Specifically, the book is organized in the following way. Chapter 1 contains preliminary material, including important results from measure and integration theory, the Besicovitch Covering theorem, the maximal function, basic estimates involving Riesz potentials, and Sobolev spaces. We find it convenient to introduce the space $Y^{1,p}$ which differs from the Sobolev space $W^{1,p}$ only in that its elements are required to lie in L^{p^*} ($p^* := np/(n-p)$), and not in L^p. The space $Y^{1,p}$ is associated with Riesz potentials whereas Bessel potentials are the counterparts to functions in $W^{1,p}$.

The second chapter begins with the development of p-capacity, the capacity which is used in the nonlinear version of the Wiener condition. Here it is shown that Sobolev functions possess Lebesgue points everywhere except for a set of p-capacity zero, that is, a p-polar set. The classical concepts of thinness and quasicontinuity are extended to this environment involving p-capacity. A result of Maz'ya and Shaposhnikova, [**MaSh**], relating to multipliers of Sobolev functions, provides a borderline case to the Sobolev-type imbedding theorem due to Adams, [**Ad1**], which is found in Chapter 1. Together, these two results allow a large class for the coefficients of the structure associated with our equations. The second section is focused on matters related to the classical Laplace equation, including Green potentials, classical thinness and the classical Wiener criterion. The subsection entitled "Regularity of minimizers" contains a short overview of the regularity theory associated with certain variational problems. The main purpose here is to indicate the important role that the De Giorgi–Nash result plays in this theory. We then give a brief treatment of De Giorgi's proof as well as the proof due to Moser involving his iteration method for solutions to equations with a simple structure. Although this material will be completely subsumed by the exposition in Chapters 3 and 4, we choose to include it for the benefit of the reader who wishes to understand the fundamental ideas of these proofs in a setting that is not encumbered with the additional complexities of a very general structure. A proof of the Wiener criterion for equations with a simple structure is presented, again for the purpose of preparing the reader for the complete development that appears in Chapter 4. The chapter is concluded with a treatment of the p-fine topology as well as fine Sobolev spaces. Many of the topics in this and the previous chapter have been exposed in other books that have recently appeared. The reader is recommended to consult the work of Adams and Hedberg, [**AH**], as well as that of Heinonen, Kilpeläinen, and Martio, [**HKM2**].

Chapter 3 begins the analysis of quasilinear equations with a general structure whose coefficients are permitted to lie in the multiplier spaces mentioned above. The results include local sup-norm estimates for weak subsolutions in terms of their integral averages, certain maximum principles, and the weak Harnack inequality for weak supersolutions. The techniques used here are due to Moser, [**Mos1**], [**Mos2**], Serrin [**Se1**], and Trudinger [**T1**]. The next topic in this chapter is removable sets for weak solutions, a topic first investigated by Serrin in [**Se3**] and [**Se4**]. The removability of a set is assured if its capacity is zero, thus indicating that one might suspect capacity to play a role in the criterion for boundary regularity. The chapter is concluded with a brief summary of regularity results of equations with differentiable structure. These results will be referenced in Chapters 5 and 6.

Regularity at the boundary is the main concern of Chapter 4. This includes basic energy estimates as well as a weak Harnack inequality for a class of functions termed "supersolutions up to level l." This Harnack inequality is crucial in establishing the Wiener criterion for boundary regularity as well as proving regularity for weak solutions of double obstacle problems. Sufficiency of the Wiener condition for boundary regularity is based on the work in [**GZ3**] while necessity of the condition comes from [**KM4**]. The methods used to establish the Wiener criterion are applied to the analysis of equations of the form

$$-\operatorname{div} \boldsymbol{A}(x,u,\nabla u) + \boldsymbol{B}(x,u,\nabla u) = \mu$$

where μ is a Radon measure satisfying suitable growth conditions.

The weak Harnack inequality for supersolutions up to level l is used in Chapter 5 to prove that weak solutions to the double obstacle problem are continuous provided the obstacles satisfy a regularity condition expressed in terms of the Wiener condition. In particular, it is shown that the solution is continuous at a point if both obstacles are continuous there. Moreover, an estimate for the modulus of continuity of the solution is obtained which implies that the solution inherits a Hölder continuity estimate from its obstacles. The last part of the chapter is devoted to the double obstacle problem for equations with a differentiable structure. In this context it is natural to assume that the obstacles are $C^{1,\alpha}$, and with this assumption, it is shown that the solution also possesses Hölder continuous derivatives.

The last chapter begins with the analysis of existence of solutions to variational inequalities. Assuming that the principle term of the differential operator satisfies a coercivity condition, existence is established with the help of pseudomonotone operators. The chapter concludes with the Dirichlet problem for equations with differentiable structure. The main result here is that if a continuous function is prescribed on the boundary of an arbitrary bounded domain, then there exists a solution which is locally $C^{1,\alpha}$ which continuously assumes the boundary values at all points satisfying the Wiener condition. By the Kellogg property, this will be satisfied at all points of the boundary except for a p-polar set.

There are several excellent sources related to this book that the reader may wish to consult. They include the recent monographs by Adams and Hedberg [**AH**] on nonlinear potential theory and function spaces, and by Heinonen, Kilpeläinen, and Martio [**HKM2**] concerning the potential theory related to p-Laplacian type equations. Books on Sobolev spaces that are related to our treatment on this subject are by Maz'ya [**Maz6**], Evans and Gariepy [**EG**], and [**Z5**]. Finally, the books by Ladyzhenskaya and Ural'tseva [**LU**], Morrey [**Mo4**], Gilbarg and Trudinger [**GT**] and Kinderlehrer and Stampacchia provide treatments of elliptic equations and related topics that are particularly relevant for our purposes.

This book is an outgrowth of lectures given by the authors at the Spring School on Potential Theory and Analysis which was held in Paseky, Czech Republic, in June, 1993. This conference was organized by the Faculty of Mathematics and Physics, Charles University, Prague. The authors wish to thank the organizers for the opportunity to participate in this conference.

In addition to support from Charles University and Indiana University, the authors gratefully acknowledge support from the following grants: 201/93/2174, 201/94/0474, and 201/96/0311 of Czech Grant Agency (GAČR), No. 354 of Charles University (GAUK), and the National Science Foundation (USA).

March, 1997

Jan Malý
William P. Ziemer

Basic Notation

$\mathbf{R}^n$	Euclidean n-space
E°	topological interior of E
$\overline{E}$	closure of E
∂E	topological boundary of E
χ_E	characteristic function of E
dist (A,B)	distance between sets A and B
diam (A)	diameter of the set A
$\mathrm{osc}_A\, u$	diameter of the set $\{u(x) : x \in A\}$
$\{u > \alpha\}$	$\{u(x) : x > \alpha\}$
$U \subset\subset \Omega$	$\overline{U}$ is compact and $\overline{U} \subset \Omega$ (for p-finely open sets see Definition 3.1)
$\mathcal{C}(E)$	the space of continuous functions on E
$\mathcal{C}_0(E)$	those functions $u \in \mathcal{C}(E)$ that satisfy $u(x_j) \to 0$ whenever $|x_j| \to \infty$ or $x_j \to x \notin E$
$\mathcal{C}^k(\Omega)$	k times continuously differentiable functions on Ω
$\mathcal{C}^k_c(\Omega)$	functions in $\mathcal{C}^k(\Omega)$ compactly supported in Ω
$\mathcal{C}^{0,\alpha}(\Omega)$	locally Hölder continuous functions on Ω
$\mathcal{C}^{k,\alpha}(\Omega)$	the functions whose derivatives up to order k are locally α-Hölder continuous on Ω
$\mathcal{C}^{k,\alpha}(\overline{\Omega})$	the functions whose derivatives up to order k are (globally) α-Hölder continuous on Ω
X^*	the dual of the space X
δ_z	Dirac measure supported at z
$|E|$	Lebesgue measure of $E \subset \mathbf{R}^n$
H^k	k-dimensional Hausdorff measure
$L^p(\Omega)$	p^{th}-power Lebesgue integrable functions defined on Ω
$L^p(\Omega; \mathbf{R}^n)$	p^{th}-power $\mathbf{R}^n$-valued Lebesgue integrable functions defined on Ω
$L^p_{\mathrm{loc}}(\Omega)$	locally p^{th}-power Lebesgue integrable functions defined on Ω
$\|u\|_{p;\Omega}$ or $\|u\|_p$	L^p norm of u on Ω
$B(x,r)$	$\{y : |x - y| < r\}$
$K(x,r)$	$\{y : |x - y| \leq r\}$
$\boldsymbol{\alpha}(n)$	$|B(0,1)|$
$u \llcorner E$	u restricted to the set E

$\int_E u(x)\,dx$	integration with respect to Lebesgue measure
$\overline{u}_E := \fint_E u(x)\,dx$	$\|E\|^{-1} \int_E u(x)\,dx$
$f * g$	convolution of f and g
$f \circ g$	composition $x \mapsto f(g(x))$
$a \cdot b$	scalar product in $\mathbf{R}^n$
spt u	closure of $\{x : u(x) \neq 0\}$
u^+	$\max\{u, 0\}$
u^-	$\max\{-u, 0\}$
$D_i u$	$\dfrac{\partial u}{\partial x_i}$
p'	$\dfrac{p}{p-1}$
p^*	$\begin{cases} \frac{np}{n-p} & \text{if } 1 \le p < n \\ \infty & \text{if } p = n \end{cases}$

CHAPTER 1

Preliminaries

1.1 Basic results

In this section, we use the notation Ω for an open subset of $\mathbf{R}^n$ and X for an abstract set endowed with a measure μ. The purpose here is to review some of the well-known results of analysis that will be used throughout this book.

1.1 LEMMA. *The inequality*

$$(1.1) \quad |a+b|^p \leq \begin{cases} (1+\varepsilon)^{p-1}|a|^p + (1+1/\varepsilon)^{p-1}|b|^p & \text{for } 1 \leq p < \infty \\ |a|^p + |b|^p & \text{for } 0 < p < 1 \end{cases}$$

holds for arbitrary $a, b \in \mathbf{R}$ and $\varepsilon > 0$.

PROOF. We may assume $a, b > 0$. For $p > 1$, inequality (1.1) follows from the fact that $t \mapsto t^p$ is a convex function and therefore

$$(a+b)^p = \left[\lambda\left(\frac{a}{\lambda}\right) + (1-\lambda)\left(\frac{b}{1-\lambda}\right)\right]^p$$
$$\leq \lambda\left(\frac{a}{\lambda}\right)^p + (1-\lambda)\left(\frac{b}{1-\lambda}\right)^p$$

whenever $0 < \lambda < 1$. Taking $\lambda = \frac{1}{1+\varepsilon}$ establishes the result in this case. If $0 < p < 1$ and $a, b \geq 0$, then the function $x \mapsto x^p$ dominates $x \mapsto \alpha x$ where $\alpha = (a+b)^{p-1}$. □

1.2 LEMMA (YOUNG'S INEQUALITY). *For $a, b \geq 0$ and $1 < p < \infty$, we have*

$$ab \leq \frac{a^p}{p} + \frac{b^{p'}}{p'}.$$

PROOF. Recall that $\ln(x)$ is an increasing, concave function on $(0, \infty)$ i.e.

$$\ln(\lambda x + (1-\lambda)y) \geq \lambda \ln(x) + (1-\lambda)\ln(y)$$

for $x, y \in (0, \infty)$ and $\lambda \in [0, 1]$.
Set $x = a^p$, $y = b^{p'}$, and $\lambda = \frac{1}{p}$ (thus $(1-\lambda) = \frac{1}{p'}$) to obtain

$$\ln(\frac{1}{p}a^p + \frac{1}{p'}b^{p'}) \geq \frac{1}{p}\ln(a^p) + \frac{1}{p'}\ln(b^{p'}) = \ln(ab). \quad \square$$

An important application of Young's inequality is *Hölder's inequality*.

1.3 THEOREM. *If $u \in L^p(X)$, and $v \in L^{p'}(X)$, $p \geq 1$, then*

(1.2) $$\int_X |uv|\, d\mu \leq \|u\|_p \|v\|_{p'}.$$

When $p = 1$, we take $p' = \infty$.

PROOF. The assertion is clear if $p = 1$. It is also clear when $1 < p < \infty$ unless $0 < \|u\|_p, \|g\|_{p'} < \infty$, in which case we define

$$\tilde{u} = \frac{u}{\|u\|_p} \quad \text{and} \quad \tilde{v} = \frac{v}{\|v\|_{p'}}.$$

Then, applying Young's inequality, we obtain

$$\frac{1}{\|u\|_p \|v\|_{p'}} \int_X |u|\,|v|\, d\mu = \int_X |\tilde{u}|\,|\tilde{v}|\, d\mu \leq \frac{1}{p} + \frac{1}{p'} = 1,$$

thus obtaining the desired result. □

1.4 REMARK. Hölder's inequality can be extended to the case of k functions,

$$u_1 \in L^{p_1}(X), \ldots, u_k \in L^{p_k}(X),$$

where

(1.3) $$\sum_{i=1}^{k} \frac{1}{p_i} = 1.$$

By an induction argument and (1.2) it follows that

(1.4) $$\int_X u_1 \ldots u_k\, d\mu \leq \|u_1\|_{p_1} \ldots \|u_k\|_{p_k}.$$

When proving that L^p spaces are normed linear spaces we need the following *Minkowski's inequality*.

1.5 THEOREM. *Let $u, v \in L^p(X)$. Then $u + v \in L^p(X)$ and*

(1.5) $$\|u + v\|_p \leq \|u\|_p + \|v\|_p.$$

PROOF. This is easy if $p = 1$ or $p = \infty$. If $1 < p < \infty$, we apply Hölder's inequality and get

$$\int_X |u+v|^p\, d\mu \leq \int_X |u+v|^{p-1}|u|\, d\mu + \int_X |u+v|^{p-1}|v|\, d\mu$$

$$\leq \left(\int_X |u+v|^p\, d\mu\right)^{1/p'} \left(\left(\int_X |u|^p\, d\mu\right)^{1/p} + \left(\int_X |v|^p\, d\mu\right)^{1/p}\right). \quad \square$$

1.6 COROLLARY. *Suppose that $1 \leq p \leq q$ and $\mu(X) < \infty$. Then*

(1.6) $$\left(\fint_X |u|^p d\mu\right)^{1/p} \leq \left(\fint_X |u|^q d\mu\right)^{1/q}.$$

The following is a simple interpolation result.

1.7 COROLLARY. *Suppose $p \leq q \leq r$, and $1/q = \lambda/p + (1-\lambda)/r$. If $u \in L^p(X) \cap L^r(X)$, then $u \in L^q(X)$ and*

(1.7) $$\|u\|_q \leq \|u\|_p^\lambda \|u\|_r^{1-\lambda}.$$

PROOF. Let $\alpha = \lambda q$, $\beta = (1-\lambda)q$ and apply Hölder's inequality to obtain

$$\int_X |u|^q \, d\mu = \int_X |u|^\alpha |u|^\beta \, d\mu \leq \left(\int_X |u|^{\alpha s} \, d\mu \right)^{1/s} \left(\int_X |u|^{\beta t} \, d\mu \right)^{1/t}$$

where $s = p/\lambda q$ and $t = r/(1-\lambda)q$. □

1.8 THEOREM. *If f is nonnegative and μ-measurable on X then*

(1.8) $$\int_X f \, d\mu = \int_{[0,\infty)} \mu\{x : f(x) > t\} \, dt.$$

PROOF. Write
$$E_t = \{x \in X : f(x) > t\}.$$
If $\mu(E_t) = \infty$ for some $t > 0$, then obviously both sides of (1.8) equal ∞. Otherwise $E_0 = \bigcup E_{1/k}$ with $\mu(E_{1/k}) < \infty$, so that the measure $\mu \llcorner E_0$ verifies assumptions needed for the product construction. Let λ denote the Lebesgue measure and ν the product measure $(\mu \llcorner E_0) \times \lambda$. Set
$$W = \{(x,t) : 0 < t < f(x)\} \subset X \times \mathbf{R}.$$
Since
$$W = \bigcup_{q \in \mathbf{Q}^+} \{E_q \times (0,q)\}$$
where $\mathbf{Q}^+$ is the set of all positive rationals, W is ν-measurable. Thus,

$$\int_{[0,\infty)} \mu(E_t) \, d\lambda(t) = \int_{\mathbf{R}} \left(\int_{E_0} \chi_W(x,t) \, d\mu(x) \right) d\lambda(t)$$
$$= \int_{E_0 \times \mathbf{R}} \chi_W(x,t) \, d\nu(x,t)$$
$$= \int_{E_0} \lambda(\{t : 0 < t < f(x)\}) \, d\mu(x)$$
$$= \int_X f \, d\mu. \quad \square$$

Thus, a nonnegative measurable function f is integrable over X with respect to μ if, and only if, its distribution function $t \mapsto \mu(E_t)$ is integrable over $[0,\infty)$ with respect to one-dimensional Lebesgue measure λ.

The simple idea behind the proof of Theorem 1.8 can readily be extended as in the following theorem.

1.9 Theorem. *If f is measurable and $1 \leq p < \infty$ then*

$$\int_X |f|^p \, d\mu = p \int_{[0,\infty)} t^{p-1} \mu(\{x : |f(x)| > t\}) \, dt$$
$$= \int_{[0,\infty)} \mu(\{x : |f(x)| > t\}) \, dt^p.$$

PROOF. Set
$$W = \{(x,t) : 0 < t < |f(x)|\}$$
and note that the function
$$(x,t) \mapsto pt^{p-1} \chi_W(x,t)$$
is ν-measurable. Hence,

$$\int_X |f|^p \, d\mu = \int_X \left(\int_{(0,|f(x)|)} pt^{p-1} \, d\lambda(t) \right) d\mu(x)$$
$$= \int_X \left(\int_{\mathbf{R}} pt^{p-1} \chi_W(x,t) \, d\lambda(t) \right) d\mu(x)$$
$$= \int_{\mathbf{R}} \left(\int_X pt^{p-1} \chi_W(x,t) \, d\mu(x) \right) d\lambda(t)$$
$$= p \int_{[0,\infty)} t^{p-1} \mu(\{x : |f(x)| > t\}) \, d\lambda(t). \qquad \square$$

1.10 Theorem. *Let $f \in L^p(\mathbf{R}^n)$, $1 \leq p \leq \infty$, and $g \in L^1(\mathbf{R}^n)$. Then*

(1.9) $$\|f * g\|_p \leq \|f\|_p \|g\|_1 .$$

PROOF. The easier parts for $p = 1$ or $p = \infty$ are left to the reader. Suppose that $1 < p < \infty$. Then using Hölder's inequality we obtain

$$|f * g(x)| = |\int_{\mathbf{R}^n} f(x-y) g(y) \, dy|$$
$$\leq \left(\int_{\mathbf{R}^n} |f(x-y)|^p |g(y)| \, dy \right)^{1/p} \left(\int_{\mathbf{R}^n} |g(y)| \, dy \right)^{1-1/p}.$$

Fubini's theorem yields

$$\int_{\mathbf{R}^n} |f * g(x)|^p \, dx \leq \|g\|_1^{p-1} \int_{\mathbf{R}^n} \left(\int_{\mathbf{R}^n} |f(x-y)|^p |g(y)| \, dy \right) dx$$
$$= \|g\|_1^{p-1} \int_{\mathbf{R}^n} \left(\int_{\mathbf{R}^n} |f(x-y)|^p |g(y)| \, dx \right) dy = \|f\|_p^p \|g\|_1^p . \qquad \square$$

1.11 DEFINITION. Let ϕ be a non-negative, real-valued function in $\mathcal{C}_c^\infty(\mathbf{R}^n)$ with the property that

(1.10) $$\int_{\mathbf{R}^n} \phi(x)\, dx = 1, \quad \operatorname{spt} \phi \subset K(0,1).$$

An example of such a function is given by

(1.11) $$\phi(x) = \begin{cases} C\exp(-1/(1-|x|^2)) & \text{if } |x| < 1 \\ 0 & \text{if } |x| \geq 1 \end{cases}$$

where C is chosen so that $\int_{\mathbf{R}^n} \phi = 1$. For $\varepsilon > 0$, the function $\phi_\varepsilon(x) := \varepsilon^{-n}\phi(x/\varepsilon)$ belongs to $\mathcal{C}_c^\infty(\mathbf{R}^n)$ and $\operatorname{spt} \phi_\varepsilon \subset K(0,\varepsilon)$. ϕ_ε is called a *regularizer* (or *mollifier*) and the convolution

(1.12) $$u_\varepsilon(x) := \phi_\varepsilon * u(x) := \int_{\mathbf{R}^n} \phi_\varepsilon(x-y) u(y)\, dy$$

defined for functions $u \in L^1_{\text{loc}}(\mathbf{R}^n)$ is called the *regularization* (*mollification*) of u.

Recall that a point $x_0 \in \mathbf{R}^n$ is said to be a *Lebesgue point* for a function u defined on a neighborhood of x_0 if

$$\lim_{r \to 0} \fint_{B(x_0,r)} |u(x) - u(x_0)|\, dx = 0.$$

1.12 THEOREM.
(i) *If $u \in L^1_{\text{loc}}(\mathbf{R}^n)$, then for every $\varepsilon > 0$, $u_\varepsilon \in \mathcal{C}^\infty(\mathbf{R}^n)$.*
(ii)
$$\lim_{\varepsilon \to 0} u_\varepsilon(x) = u(x)$$
whenever x is a Lebesgue point for u. In case u is continuous, then u_ε converges to u uniformly on compact subsets of $\mathbf{R}^n$.
(iii) *If $u \in L^p(\mathbf{R}^n)$, $1 \leq p < \infty$, then $u_\varepsilon \in L^p(\mathbf{R}^n)$, $\|u_\varepsilon\|_p \leq \|u\|_p$, and $\lim_{\varepsilon \to 0} \|u_\varepsilon - u\|_p = 0$.*

PROOF. Recall that $D_i u$ denotes the partial derivative of u with respect to the ith variable. The estimates

$$\sup_x |\phi_\varepsilon(x-y) u(y)| \leq C|u(y)|,$$
$$\sup_x |D_i \phi_\varepsilon(x-y) u(y)| \leq C|u(y)|,$$

verify that the function $\phi_\varepsilon * u$ is continuous and that we can differentiate under the integral sign to obtain

(1.13) $$D_i(\phi_\varepsilon * u) = (D_i \phi_\varepsilon) * u$$

for each $i = 1, 2, \ldots, n$. The right side of the equation is a continuous function thus showing that u_ε is $\mathcal{C}^1$. Then, if α denotes the n-tuple with $|\alpha| = 2$ and with 1 in the ith and j^{th} positions, it follows that

$$D^\alpha(\phi_\varepsilon * u) = D_i\big(D_j(\phi_\varepsilon * u)\big)$$
$$= D_i\big((D_j \phi_\varepsilon) * u\big)$$
$$= (D^\alpha \phi_\varepsilon) * u,$$

which proves that $u_\varepsilon \in \mathcal{C}^2$ since again the right side of the equation is continuous. Proceeding this way by induction, it follows that $u \in \mathcal{C}^\infty$. This establishes (i).

In case (ii), observe that

(1.14)
$$|u_\varepsilon(x) - u(x)| \leq \int_{\mathbf{R}^n} \phi_\varepsilon(x-y)\,|u(y) - u(x)|\,dy$$
$$\leq \max_{\mathbf{R}^n}\phi\,\varepsilon^{-n}\int_{B(x,\varepsilon)} |u(x) - u(y)|\,dy \to 0$$

as $\varepsilon \to 0$ whenever x is a Lebesgue point for u. Now consider the case where u is continuous and that $E \subset \mathbf{R}^n$ is a compact. Then u is uniformly continuous on E and also on the closure of the open set $U = \{x : \mathrm{dist}(x,E) < 1\}$. For each $\gamma > 0$, there exists $0 < \varepsilon < 1$ such that $|u(x) - u(y)| < \gamma$ whenever $|x - y| < \varepsilon$ and whenever $x, y \in U$; in particular when $x \in E$. Consequently, it follows from (1.14) that whenever $x \in E$,
$$|u_\varepsilon(x) - u(x)| \leq M\gamma,$$
where M is the product of $\max_{\mathbf{R}^n}\phi$ and the Lebesgue measure of the unit ball. Since γ is arbitrary, this shows that u_ε converges uniformly to u on E.

The first part of (iii) follows from Theorem 1.10 since $\|\phi_\varepsilon\|_1 = 1$.

Finally, addressing the second part of (iii), for each $\gamma > 0$, select a continuous function v with compact support on $\mathbf{R}^n$ such that

(1.15)
$$\|u - v\|_p < \gamma.$$

Because v has compact support, it follows from (ii) that $\|v - v_\varepsilon\|_p < \gamma$ for all ε sufficiently small. Now apply Theorem 1.12 (iii) and (1.15) to the difference $u - v$ to obtain
$$\|u - u_\varepsilon\|_p \leq \|u - v\|_p + \|v - v_\varepsilon\|_p + \|v_\varepsilon - u_\varepsilon\|_p \leq 3\gamma.$$
This shows that $\|u - u_\varepsilon\|_p \to 0$ as $\varepsilon \to 0$. $\square$

1.13 Theorem. *Suppose that $u \in L^p(\Omega)$ and*
$$\int_X u\psi\,dx \geq 0$$
for every nonnegative $\psi \in \mathcal{C}_c^\infty(\Omega)$. Then $u \geq 0$ a.e.

Proof. Fix $\eta \in \mathcal{C}_c^\infty(\Omega)$ and let v be a function on $\mathbf{R}^n$ which is $u\eta$ in Ω and vanishes outside Ω. Then, by the assumptions,
$$v * \phi_\varepsilon \geq 0$$
for every $\varepsilon > 0$. By Theorem 1.12 (ii), $v * \phi_\varepsilon$ approaches v in $L^p(\mathbf{R}^n)$ thus implying that $v \geq 0$ a.e. in $\mathbf{R}^n$. Observing that this holds for arbitrary $\eta \in \mathcal{C}_c^\infty(\Omega)$, we deduce that $u \geq 0$ a.e. in Ω. $\square$

1.14 Theorem. *Let $\Omega \subset \mathbf{R}^n$ be an open set and $K \subset \Omega$ be compact. Then there is a function $\eta \in \mathcal{C}_c^\infty(\Omega)$ such that $0 \leq \eta \leq 1$ and $\eta = 1$ on K.*

Proof. The set $K_\delta := \{x : \mathrm{dist}\,(x,K) \leq \delta\}$ is compact and contained in Ω if δ is sufficiently small. Now mollify the characteristic function χ of K_δ. If we take $\varepsilon \in (0, \delta)$ such that $K_{\delta+\varepsilon} \subset \Omega$, then $\phi_\varepsilon * \chi$ has the desired properties. $\square$

1.1.1 Covering theorems.

1.15 LEMMA. *Let $\mathcal{G}$ be a family of closed balls in $\mathbf{R}^n$ with*
$$R := \sup\{\operatorname{diam} K : K \in \mathcal{G}\} < \infty.$$
Then there is a countable subfamily $\mathcal{F} \subset \mathcal{G}$ of pairwise disjoint elements such that
$$\bigcup\{K : K \in \mathcal{G}\} \subset \bigcup\{5K : K \in \mathcal{F}\}.$$
In fact, for each $K \in \mathcal{G}$ there exists $K' \in \mathcal{F}$ such that $K \cap K' \neq \emptyset$ and $K \subset 5K'$.

PROOF. Let $a > 1$ be a constant which will be specified later. We determine $\mathcal{F}$ as follows. For $j = 1, 2, \ldots$ let
$$\mathcal{G}_j = \left\{K = K(x, r) \in \mathcal{G} : a^{-j+|x|} < \frac{r}{R} \leq a^{-j+1+|x|}\right\},$$
and observe that $\mathcal{G} = \bigcup_{j=1}^{\infty} \mathcal{G}_j$. For any $K = K(x, r) \in \mathcal{G}_j$ we have
$$|x| < j \quad \text{and} \quad r > a^{-j}R.$$
It follows that each disjoint subfamily of $\mathcal{G}_j$ is necessarily finite. We construct inductively a sequence $\mathcal{F}_j \subset \mathcal{G}_j$ of finite families of disjoint balls from $\mathcal{G}$. We set $\mathcal{F}_0 = \emptyset$. Assuming that $\mathcal{F}_{j-1}$ has been constructed, we will successively augment a disjoint family of balls from $\mathcal{G}_j$ which do not intersect any element of $\mathcal{F}_{j-1}$. We start with an empty family and adjoin elements of $\mathcal{G}_j$, one at a time, so that the augmented family remains disjoint and that each of its elements does not intersect any element of $\mathcal{F}_{j-1}$. This process can continue for only a finite number of steps and thus determines a family $\mathcal{H}_j$ to which no other elements of $\mathcal{G}_j$ can be adjoined. Then define $\mathcal{F}_j$ as $\mathcal{F}_{j-1} \cup \mathcal{H}_j$. The family $\mathcal{F} := \bigcup_j \mathcal{F}_j$ obviously has the required properties. Indeed, for each $K = K(x, r) \in \mathcal{G}_j$, $j \geq 1$, there exists $K' = K(x', r') \in \mathcal{F}_j$ such that $K \cap K' \neq \emptyset$. Moreover,
$$r \leq a^{1+|x|-|x'|}r' \leq a^{1+|x-x'|}r' \leq a^{1+2R}r' \leq 2r'$$
if a is small enough, which implies that $K \subset 5K'$, as required. □

1.16 BESICOVITCH COVERING LEMMA. *There is a positive number $N > 1$ depending only on n so that any family $\mathcal{B}$ of closed balls in $\mathbf{R}^n$ whose cardinality is no less than N and $R = \sup\{r : K(a, r) \in \mathcal{B}\} < \infty$ contains disjointed subfamilies $\mathcal{B}_1, \mathcal{B}_2, \ldots, \mathcal{B}_N$ such that if A is the set of centers of balls in $\mathcal{B}$, then*
$$A \subset \bigcup_{i=1}^{N} \bigcup_{K \in \mathcal{B}_i} K.$$

PROOF. Step I. Assume A is bounded. Choose $K_1 = K(a_1, r_1)$ with $r_1 > \frac{3}{4}R$. Assuming we have chosen $K_1, \ldots, K_{j-1}$ in $\mathcal{B}$ where $j \geq 2$ choose K_j inductively as follows. If $A_j := A - \cup_{i=1}^{j-1} K_i = \emptyset$, then the process stops and we set $J = j$. If $A_j \neq \emptyset$, continue by choosing $K_j = K(a_j, r_j) \in \mathcal{B}$ so that $a_j \in A_j$ and

(1.16) $$r_j > \frac{3}{4}\sup\{r : K(a, r) \in \mathcal{B}, a \in A_j\}.$$

If $A_j \neq \emptyset$ for all j, then we set $J = +\infty$. In this case $\lim_{j \to \infty} r_j = 0$ because A is bounded and the inequalities

$$|a_i - a_j| > r_i = \frac{r_i}{3} + \frac{2}{3}r_i > \frac{r_i}{3} + \frac{r_j}{2}, \text{ for } i < j,$$

imply that

(1.17) $\qquad \{K(a_j, r_j/3) : 1 \leq j \leq J\}$ is disjointed.

In case $J < \infty$, we clearly have the inclusion

(1.18) $\qquad A \subset \{\cup K_j : 1 \leq j \leq J\}.$

This is also true in case $J = +\infty$, for otherwise there would exist $K(a, r) \in \mathcal{B}$ with $a \in \cap_{j=1}^\infty A_j$ and an integer j with $r_j \leq 3r/4$, contradicting the choice of K_j.

Step II. We now prove there exists an integer M (depending only on n) such that for each k with $1 \leq k < J$, M exceeds the number of balls K_i with $1 \leq i \leq k$ and $K_i \cap K_k \neq \emptyset$.

First note that if $r_i < 10 r_k$, then

$$K(a_i, r_i/3) \subset K(a_k, 15 r_k)$$

because if $x \in K(a_i, r_i/3)$,

$$|x - a_k| \leq |x - a_i| + |a_i - a_k|$$
$$\leq 10 r_k/3 + r_i + r_k$$
$$\leq 43 r_k/3 < 15 r_k.$$

Hence, there are at most $(60)^n$ balls K_i with

$$1 \leq i \leq k, \ K_i \cap K_k \neq \emptyset, \text{ and } r_i \leq 10 r_k$$

because, for each such i,

$$K(a_i, r_i/3) \subset K(a_k, 15 r_k),$$

and by (1.16) and (1.17)

$$|K(a_i, r_i/3)| = |K_1| \cdot \left(\frac{r_i}{3}\right)^n > |K_1| \cdot \left(\frac{r_k}{4}\right)^n = \frac{1}{60^n}|K(a_k, 15 r_k)|.$$

To complete Step II, it remains to estimate the number of points in the set

$$I = \{i : 1 \leq i \leq k, K_i \cap K_k \neq \emptyset, r_i > 10 r_k\}.$$

For notational convenience, let us write $b_i = a_i - a_k$. An elementary mesh-like construction gives a family $\{Q_m\}$, $m = 1, \ldots, (22\,n)^n$, of closed cubes with edge length $1/(11n)$ (so that $\operatorname{diam} Q_m \leq 1/11$), which cover $[-1,1]^n$ and thus in particular the unit sphere. We claim that for each $m = 1, \ldots, (22\,n)^n$ there is at most one $i \in I$ such that $b_i/|b_i| \in Q_m$, which estimates the cardinality of I.

If the claim were not valid, then there would exist $i, j \in I$, $i < j$, such that

$$\left| \frac{b_i}{|b_i|} - \frac{b_j}{|b_j|} \right| < \frac{1}{11}.$$

Notice that

$$r_i < |b_i| \leq r_i + r_k,$$
$$r_j < |b_j| \leq r_j + r_k,$$

as the balls $K(a_i, r_i)$, $K(a_j, r_j)$ intersect $K(a_k, r_k)$ but do not contain a_k. Hence

$$\big| |b_i| - |b_j| \big| \leq |r_i - r_j| + r_k \leq |r_i - r_j| + \tfrac{r_j}{10}$$

and

$$|b_j| \leq r_j + r_k \leq r_j + \tfrac{1}{10} r_j = \tfrac{11}{10} r_j.$$

We have

$$|a_i - a_j| = |b_i - b_j| \leq \left| b_i - \frac{|b_j|}{|b_i|} b_i \right| + \left| \frac{|b_j|}{|b_i|} b_i - b_j \right| \leq \Big| |b_i| - |b_j| \Big| + \frac{|b_j|}{11}$$

$$\leq |r_i - r_j| + \frac{r_j}{10} + \frac{r_j}{10}$$

$$\leq \begin{cases} r_i - \tfrac{4}{5} r_j < r_i & \text{if } r_i > r_j, \\ -r_i + \tfrac{6}{5} r_j \leq -r_i + \tfrac{8}{5} r_i < r_i & \text{if } r_i \leq r_j. \end{cases}$$

In the last inequality we have used that $i < j$ and thus $r_j < \tfrac{4}{3} r_i$ by (1.16). We arrived at a contradiction, as $i < j$ and thus $a_j \notin K(a_i, r_i)$. Hence the number of points in I is estimated by $(22\,n)^n$.

Step III. Choice of $\mathcal{B}_1, \ldots, \mathcal{B}_M$ in case A is bounded.

With each positive integer j, we define an integer λ_j such that $\lambda_j = j$ whenever $1 \leq j \leq M$ and for $j > M$ we define λ_{j+1} inductively as follows. From Step II there is an integer $\lambda_{j+1} \in \{1, 2, \ldots, M\}$ such that

$$K_{j+1} \cap \{\cup K_i : 1 \leq i \leq j, \lambda_i = \lambda_{j+1}\} = \emptyset.$$

Now deduce from (1.18) that the unions of the disjointed families

$$\mathcal{B}_1 = \{K_i : \lambda_i = 1\}, \ldots, \mathcal{B}_M = \{K_i : \lambda_i = M\}$$

covers A.

Step IV. The case A is unbounded.

For each positive integer ℓ, apply Step III with A replaced by $E_\ell = A \cap \{x : 3(\ell - 1)R \leq |x| < 3\ell R\}$ and $\mathcal{B}$ replaced by the subfamily $\mathcal{C}_\ell$ of $\mathcal{B}$ of balls with centers in E_ℓ. We obtain disjointed subfamilies $\mathcal{B}_1^\ell, \ldots, \mathcal{B}_M^\ell$ of $\mathcal{C}_\ell$ such that

$$E_\ell \subset \bigcup_{i=1}^{M} \{\cup K : K \in \mathcal{B}_i^\ell\}.$$

Since $P \cap Q = \emptyset$ whenever $P \in \mathcal{B}^\ell$, $Q \in \mathcal{B}^m$ and $m \geq \ell + 2$, the theorem follows with

$$\mathcal{B}_1 = \bigcup_{\ell=1}^{\infty} \mathcal{B}_1^{2\ell-1}, \ldots, \mathcal{B}_M = \bigcup_{\ell=1}^{\infty} \mathcal{B}_M^{2\ell-1}$$

$$\mathcal{B}_{M+1} = \bigcup_{\ell=1}^{\infty} \mathcal{B}_1^{2\ell}, \ldots, \mathcal{B}_{2M} = \bigcup_{\ell=1}^{\infty} \mathcal{B}_M^{2\ell}$$

and $N = 2M$. □

1.1.2 Densities of measures.

Here some basic results concerning the densities of arbitrary measures are established that will be used later in the development of Lebesgue points for Sobolev functions.

1.17 Lemma. *Let $\mu \geq 0$ be a Radon measure on $\mathbf{R}^n$. Let $0 < \lambda < \infty$ and $0 < \alpha \leq n$. Suppose for an arbitrary Borel set $A \subset \mathbf{R}^n$ that*

(1.19) $$\limsup_{r \to 0} \frac{\mu(B(x,r))}{r^\alpha} > \lambda$$

for each $x \in A$. Then there is a constant $C = C(\alpha, n)$ such that

$$\mu(A) \geq C\lambda H^\alpha(A).$$

PROOF. Assume $\mu(A) < \infty$ and choose $\varepsilon > 0$. Let $U \supset A$ be an open set with $\mu(U) < \infty$. For each x and r, we have

$$\lim_{t \to r^-} \mu(K(x,t)) = \mu(B(x,r))$$

and therefore it is immaterial whether $B(x,r)$ or $K(x,r)$ is used in (1.19).

Let $\mathcal{G}$ be the family of all closed balls $K(x,r) \subset U$ such that

$$x \in A, \quad 0 < r < \varepsilon/2, \quad \frac{\mu(K(x,r))}{r^\alpha} > \lambda.$$

Now appeal to Lemma 1.15 to obtain a disjoint subfamily $\mathcal{F} \subset \mathcal{G}$ such that

$$A \subset \bigcup \{5K : K \in \mathcal{F}\}.$$

Thus, by the definition of Hausdorff measure,

$$H^\alpha_{5\varepsilon}(A) \leq C \sum_{K \in \mathcal{F}} 2^{-\alpha} (\delta(K))^\alpha$$

where $\delta(K)$ denotes the diameter of the ball K. Since $\mathcal{F} \subset \mathcal{G}$ and $\mathcal{F}$ is disjoint, we have

$$\sum_{K \in \mathcal{F}} 2^{-\alpha} (\delta(K))^\alpha \leq C\lambda^{-1} \sum_{K \in \mathcal{F}} \mu(K)$$
$$\leq C\lambda^{-1} \mu(U) < \infty.$$

We conclude
$$H^\alpha_{5\varepsilon}(A) \leq C\lambda^{-1} \mu(U).$$

Since μ is a Radon measure, we have that $\mu(A) = \inf\{\mu(U) : U \supset A, U \text{ open}\}$. Thus, letting $\varepsilon \to 0$, we obtain the desired result. $\square$

1.18 Lemma. *Let $\mu \geq 0$ be a Radon measure on $\mathbf{R}^n$ that is absolutely continuous with respect to Lebesgue measure. Let*

$$A = \left\{ x \in \mathbf{R}^n : \limsup_{r \to 0} \frac{\mu(B(x,r))}{r^\alpha} > 0 \right\}.$$

Then, $H^\alpha(A) = 0$ whenever $0 \leq \alpha < n$.

PROOF. The result is obvious for $\alpha = 0$, so choose $0 < \alpha < n$. For each positive integer i let
$$A_i = \left\{ x \in \mathbf{R}^n : |x| < i, \limsup_{r \to 0} \frac{\mu(B(x,r))}{r^\alpha} > i^{-1} \right\}$$
and conclude from the preceding lemma that
(1.20) $$\mu(A_i) \geq C i^{-1} H^\alpha(A_i).$$
Since A_i is bounded, $\mu(A_i) < \infty$ and therefore $H^\alpha(A_i) < \infty$. Since $\alpha < n$, $H^n(A_i) = 0$ which implies $|A_i| = 0$. The absolute continuity of μ implies $\mu(A_i) = 0$ and consequently $H^\alpha(A_i) = 0$ from (1.20). But $A = \cup_{i=1}^\infty A_i$, and the result follows. □

1.19 COROLLARY. *Suppose $u \in L^p(\mathbf{R}^n)$, $1 \leq p < \infty$, and let $0 \leq \alpha < n$. If E is defined by*
$$E = \left\{ x : \limsup_{r \to 0} r^{-\alpha} \int_{B(x,r)} |u(x)|^p dx > 0 \right\},$$
then $H^\alpha(E) = 0$.

PROOF. This follows directly from Lemma 1.18 by defining a measure μ as
$$\mu(A) = \int_A |u|^p dx. \qquad \square$$

1.1.3 The maximal function and its applications.

The use of the maximal function is a convenient way to establish the a.e. existence of Lebesgue points for L^1 functions. The concept of Lebesgue point will play an important role later in our discussion of the pointwise behavior of Sobolev functions.

1.20 DEFINITION. Let f be a locally integrable function defined on $\mathbf{R}^n$. The maximal function of f, $M(f)$, is defined by
$$M(f)(x) = \sup \left\{ \fint_{B(x,r)} |f(y)| dy : r > 0 \right\}.$$

1.21 THEOREM. *If $f \in L^1(\mathbf{R}^n)$, then*
$$|\{M(f) > t\}| \leq 5^n t^{-1} \|f\|_1.$$

PROOF. For each $t \in \mathbf{R}$, let $A_t = \{x : M(f)(x) > t\}$. From Definition 1.20 it follows that for each $x \in A_t$, there exists a ball B_x with center $x \in A_t$, such that
(1.21) $$\fint_{B_x} |f| dy > t.$$
If we let $\mathcal{F}$ be the family of n-balls defined by $\mathcal{F} = \{B_x : x \in A_t\}$, then Theorem 1.15 provides the existence of a disjoint subfamily $\{B_1, B_2, \ldots, B_k, \ldots\}$ such that
$$\sum_{k=1}^\infty |B_k| \geq 5^{-n} |A_t|$$

and therefore, from (1.21)

$$\|f\|_1 \geq \int_{\cup_{k=1}^\infty B_k} |f| dy > t \sum_{k=1}^\infty |B_k| \geq t 5^{-n} |A_t|,$$

or

(1.22) $$|A_t| \leq \frac{5^n}{t} \|f\|_1 \quad \text{whenever} \quad t \in \mathbf{R}. \qquad \square$$

1.22 THEOREM. *If $f \in L^p(\mathbf{R}^n)$, $1 < p \leq \infty$, then $M(f) \in L^p(\mathbf{R}^n)$ and there exists a constant $C = C(p,n)$ such that*

$$\|M(f)\|_p \leq C \|f\|_p.$$

PROOF. We may assume that $1 < p < \infty$, for the conclusion of the theorem obviously holds in case $p = \infty$. For each $t \in \mathbf{R}$, define

$$f_t(x) = \begin{cases} f(x) & \text{if } |f(x)| \geq t/2 \\ 0 & \text{if } |f(x)| < t/2. \end{cases}$$

Then, for all x,

$$|f(x)| \leq |f_t(x)| + t/2,$$
$$M(f)(x) \leq M(f_t)(x) + t/2$$

and thus,

$$\{x : M(f)(x) > t\} \subset \{M(f_t)(x) > t/2\}.$$

Applying Theorem 1.21 with f replaced by f_t yields

(1.23) $$|A_t| \leq |\{M(f_t)(x) > t/2\}| \leq \frac{2 \cdot 5^n}{t} \int_{\mathbf{R}^n} |f_t| dy = \frac{2 \cdot 5^n}{t} \int_{\{|f| \geq t/2\}} |f| dy.$$

Now, from Lemma 1.9, and (1.23)

$$\int_{\mathbf{R}^n} M(f)^p dy = p \int_0^\infty t^{p-1} |A_t| \, dt$$

$$\leq p 2 \cdot 5^n \int_0^\infty t^{p-2} \left(\int_{\{|f| > t/2\}} |f| dx \right) dt$$

$$= p 2^p \cdot 5^n \int_0^\infty t^{p-2} \left(\int_{\{|f| > t\}} |f| dx \right) dt$$

$$= p 2^p \cdot 5^n \int_0^\infty t^{p-2} \mu(\{|f| > t\}) \, dt$$

where μ is a measure defined by $\mu(E) = \int_E |f| dx$ for every Borel set E. Thus, appealing again to Lemma 1.9, we have

$$\int_{\mathbf{R}^n} M(f)^p dx = \frac{p 2^p \cdot 5^n}{p-1} \int_0^\infty \mu(\{|f| > t\}) \, dt^{p-1}$$

$$= \frac{p 2^p \cdot 5^n}{p-1} \int_{\mathbf{R}^n} |f|^{p-1} d\mu$$

$$= \frac{p 2^p \cdot 5^n}{p-1} \int_{\mathbf{R}^n} |f|^p dx < \infty.$$

Since $p > 1$, this establishes the theorem. $\square$

1.23 LEMMA. *Let $f \in L^1(\mathbf{R}^n)$ and $\varepsilon > 0$. Then there are an upper semicontinuous function $t \leq f$ and a lower semicontinuous function $s \geq f$ such that*

$$\int_{\mathbf{R}^n} (s-t)\, dx < \varepsilon.$$

PROOF. There is a sequence f_j of continuous functions such that $\|f_j - f\|_1 \to 0$. Passing to a subsequence, we may suppose that

$$\sum_{j=1}^\infty \|f_{j+1} - f_j\|_1 < \frac{1}{2}\varepsilon.$$

Then we may set

$$s := f_1 + \sum_{j=1}^\infty |f_{j+1} - f_j|,$$

$$t := f_1 - \sum_{j=1}^\infty |f_{j+1} - f_j|. \quad \square$$

1.24 THEOREM. *Let $f \in L^1_{\text{loc}}(\Omega)$. Then almost every $x \in \mathbf{R}^n$ is a Lebesgue point for f.*

PROOF. We may assume that $\Omega = \mathbf{R}^n$ and $f \in L^1(\mathbf{R}^n)$. Writing

$$E_\varepsilon = \{x \in \mathbf{R}^n : \limsup_{r \to 0} \fint_{B(x,r)} |f(y) - f(x)|\, dy > \varepsilon\},$$

it is enough to prove that $|E_\varepsilon| = 0$ for each $\varepsilon > 0$. Choose a $\delta > 0$. Appeal to Lemma 1.23 and find an upper semicontinuous function $t \leq f$ and a lower semicontinuous function $s \geq f$ such that

$$\int_{\mathbf{R}^n} (s-t)\, dx < \delta.$$

Choose $x \in \mathbf{R}^n$ and assume that $M(s-t)(x) < \frac{\varepsilon}{2}$. Since the function $s, -t$ are lower semicontinuous, there is $\rho > 0$ such that

$$s(y) > s(x) - \frac{\varepsilon}{2} \quad \text{and} \quad t(y) < t(x) + \frac{\varepsilon}{2} \quad \text{for all } y \in B(x, \rho).$$

If $0 < r < \rho$, then obviously

$$|f(y) - f(x)| \leq s(y) - t(y) + \frac{\varepsilon}{2} \quad \text{for all } y \in B(x, r).$$

Hence

$$\fint_{B(x,r)} |f(y) - f(x)|\, dy \leq M(s-t)(x) + \frac{\varepsilon}{2} \leq \varepsilon,$$

so that $x \notin E_\varepsilon$. By Theorem 1.21,

$$|E_\varepsilon| \leq |\{M(s-t) \geq \frac{\varepsilon}{2}\}| \leq \frac{C\|s-t\|_1}{\varepsilon} \leq C\frac{\delta}{\varepsilon}.$$

By letting $\delta \to 0$ we obtain $|E_\varepsilon| = 0$ which concludes the proof. $\square$

As a consequence we obtain the Lebesgue density theorem.

1.25 COROLLARY. *Let $E \subset \mathbf{R}^n$ be a measurable set. Then almost every point of E is a density point for E.*

PROOF. It is easy to see that any point of E which is a Lebesgue point for $\chi_E \in L^1_{\text{loc}}(\mathbf{R}^n)$ is a density point for E. □

1.2 Potential estimates

Inequalities involving Riesz potentials provide an important tool for estimating functions in terms of the norms of their derivatives.

1.26 DEFINITION. The Riesz kernel, I_α, $0 < \alpha < n$, is defined by
$$I_\alpha(x) = |x|^{\alpha-n}.$$

The Riesz potential of a function f is defined as the convolution
$$I_\alpha * f(x) = \int_{\mathbf{R}^n} \frac{f(y)\,dy}{|x-y|^{n-\alpha}}.$$

Sometimes we write $I_\alpha f = I_\alpha * f$. In the literature one frequently sees the normalized Riesz kernel, i.e.
$$\tilde{I}_\alpha(x) = \rho_\alpha^{-1} |x|^{\alpha-n}$$
where

(1.24) $$\rho_\alpha = \frac{\pi^{n/2} 2^\alpha \Gamma(\alpha/2)}{\Gamma(n/2 - \alpha/2)}.$$

The normalization is motivated by the role it plays in the Riesz composition formula:
$$\tilde{I}_\alpha * \tilde{I}_\beta = \tilde{I}_{\alpha+\beta}, \quad \alpha > 0,\ \beta > 0,\ \alpha + \beta < n$$

cf. [**St3**], p. 118.

Observe that $I_\alpha * f$ is lower semicontinuous whenever $f \geq 0$. Indeed, if $x_i \to x$, then $|x_i - y|^{\alpha-n} f(y) \to |x - y|^{\alpha-n} f(y)$ for all $y \in \mathbf{R}^n$, and lower semicontinuity thus follows from Fatou's lemma.

1.27 LEMMA. *Let μ be a signed Radon measure on $\mathbf{R}^n$, $x \in \mathbf{R}^n$ and $\alpha < n$. Then*
$$\int_{\mathbf{R}^n} \frac{d\mu(y)}{|x-y|^{n-\alpha}} = (n-\alpha) \int_0^\infty r^{\alpha-n-1} \mu(B(x,r))\,dr,$$
provided that either μ is nonnegative, or
$$\int_{\mathbf{R}^n} \frac{d|\mu|(y)}{|x-y|^{n-\alpha}} < \infty.$$

PROOF. Using Fubini's theorem as in the proof of Theorem 1.8 we obtain

$$\int_{\mathbf{R}^n} \frac{d\mu(y)}{|x-y|^{n-\alpha}} = (n-\alpha) \int_{\mathbf{R}^n} \left(\int_{r>|x-y|} r^{\alpha-n-1}\, dr \right) d\mu(y)$$

$$= (n-\alpha) \int_0^\infty \left(\int_{|y-x|<r} r^{\alpha-n-1} d\mu(y) \right) dr$$

$$= (n-\alpha) \int_0^\infty r^{\alpha-n-1} \mu(B(x,r))\, dr. \qquad \square$$

1.28 DEFINITION. A function $f \in L^1(\Omega)$ is said to belong to the Morrey space $\mathcal{M}^p(\Omega)$, $1 \leq p \leq \infty$, if there is a constant K such that

$$\int_{\Omega \cap B(x,r)} |f|\, dy \leq K r^{n/p'}$$

for all balls $B(x,r) \subset \mathbf{R}^n$. The norm is defined as the smallest constant K that satisfies this inequality and is denoted by $\|f\|_{\mathcal{M}^p(\Omega)}$. An application of Hölder's inequality shows that $L^p(\Omega) \subset \mathcal{M}^p(\Omega)$.

1.29 LEMMA. *Let* $f \in \mathcal{M}^p(\Omega)$ *and* $\alpha p > n$. *Then*

$$\int \frac{|f(y)|\, dy}{|x-y|^{n-\alpha}} \leq C K d^{\alpha - n/p}$$

where $d = \operatorname{diam} \Omega$ *and*

$$C = \begin{cases} n \frac{p-1}{\alpha p - n}, & \alpha < n, \\ 1, & \alpha \geq n. \end{cases}$$

PROOF. The case $\alpha \geq n$ is trivial since then $|x-y|^{\alpha-n} \leq d^{\alpha-n}$. Suppose now that $\alpha < n$. By Lemma 1.27 we have

$$\int \frac{|f(y)|\, dy}{|x-y|^{n-\alpha}} = \int_0^\infty \left(\int_{\Omega \cap B(x,\rho)} |f(y)|\, dy \right) \frac{d\rho}{\rho^{n-\alpha+1}}$$

$$\leq (n-\alpha) \int_0^d K\rho^{n/p'} \frac{d\rho}{\rho^{n-\alpha+1}} + (n-\alpha) \int_d^\infty K d^{n/p'} \frac{d\rho}{\rho^{n-\alpha+1}}$$

$$= CK d^{\alpha - n/p}. \qquad \square$$

1.30 LEMMA. *Let* $\Omega \subset \mathbf{R}^n$ *be a measurable set with* $0 < |\Omega| < \infty$, $x \in \mathbf{R}^n$ *and* $0 < \alpha < n$. *Then*

$$\int_\Omega |y-x|^{\alpha-n}\, dy \leq C\, |\Omega|^{\alpha/n},$$

where

(1.25) $$C = n\alpha^{-1} \boldsymbol{\alpha}(n)^{1-\alpha/n}.$$

PROOF. We may suppose that $x = 0$. We find $R > 0$ such $|B(0, R)| = |\Omega|$ and denote $B = B(0, R)$. Since

$$\int_{\Omega \setminus B} |y|^{\alpha-n} \, dy \leq R^{\alpha-n} |\Omega \setminus B| = R^{\alpha-n} |B \setminus \Omega| \leq \int_{B \setminus \Omega} |y|^{\alpha-n} \, dy,$$

we get

$$\int_{\Omega} |y|^{\alpha-n} \, dy \leq \int_{B} |y|^{\alpha-n} \, dy = n\alpha^{-1} \boldsymbol{\alpha}(n) R^{\alpha} = n\alpha^{-1} \boldsymbol{\alpha}(n)^{1-\alpha/n} |\Omega|^{\alpha/n}. \qquad \square$$

The following theorem is a particular case of Lemma 1.34, but we give a separate, simpler proof. It is an important ingredient in the proof of Poincaré's inequality (Theorem 1.51).

1.31 THEOREM. *Let $1 \leq p < \infty$. Suppose Ω is a domain with finite measure and $f \in L^p(\Omega)$. Then*

$$\|I_1 f\|_{p;\Omega} \leq C \|f\|_{p;\Omega}$$

where

(1.26) $$C = n\boldsymbol{\alpha}(n)^{1-1/n} |\Omega|^{1/n}.$$

PROOF. By Lemma 1.30,

(1.27) $$\int_{\Omega} |x - y|^{1-n} \leq C_1 := n\boldsymbol{\alpha}(n)^{1-1/n} |\Omega|^{1/n}.$$

If $p > 1$, we use Hölder's inequality and obtain

(1.28)
$$\int_{B(x,r)} |f(y)| |x - y|^{1-n} \, dy$$
$$\leq \left(\int_{\Omega} |f(y)|^p |x - y|^{1-n} \, dy \right)^{1/p} \left(\int_{\Omega} |x - y|^{1-n} \, dy \right)^{1-1/p}$$
$$\leq C_1^{1-1/p} \left(\int_{\Omega} |f(y)|^p |x - y|^{1-n} \, dy \right)^{1/p},$$

which holds trivially also for $p = 1$. Using (1.27) again, it follows

$$\int_{\Omega} |I_1 f|^p \, dx \leq C_1^{p-1} \int_{\Omega} \left(\int_{\Omega} |x - y|^{1-n} |f(y)|^p \, dy \right) dx$$
$$= C_1^{p-1} \int_{\Omega} |f(y)|^p \left(\int_{\Omega} |x - y|^{1-n} \, dx \right) dy$$
$$\leq C_1^p \int_{\Omega} |f(y)|^p \, dy. \qquad \square$$

Now, we will estimate the norm of Riesz potentials in more generality. The following theorem is an important tool in combination with Theorem 1.22.

1.32 Theorem. *If $0 < \alpha < n$, $\beta > 0$, and $\delta > 0$, then for each $x \in \mathbf{R}^n$,*

(i) $\displaystyle\int_{B(x,\delta)} \frac{|f(y)|\,dy}{|x-y|^{n-\alpha}} \leq C_\alpha \delta^\alpha M(f)(x),$

(ii) $\displaystyle\int_{\mathbf{R}^n - B(x,\delta)} \frac{|f(y)|\,dy}{|x-y|^{\beta+n}} \leq C_\beta \delta^{-\beta} M(f)(x),$

where
$$C_\alpha = \frac{n}{\alpha}\boldsymbol{\alpha}(n), \quad C_\beta = \left(1+\frac{n}{\beta}\right)\boldsymbol{\alpha}(n).$$

PROOF. For $x \in \mathbf{R}^n$ and $\delta > 0$, we use Lemma 1.27 and obtain

$$\int_{B(x,\delta)} \frac{|f(y)|\,dy}{|x-y|^{n-\alpha}} = (n-\alpha)\int_0^\infty \left(\int_{B(x,\rho)\cap B(x,\delta)} |f(y)|\,dy\right)\frac{d\rho}{\rho^{n-\alpha+1}}$$

$$\leq (n-\alpha)\left(\boldsymbol{\alpha}(n)\int_0^\delta Mf(x)\rho^n \frac{d\rho}{\rho^{n-\alpha+1}}\right.$$
$$\left.+ \boldsymbol{\alpha}(n)\int_\delta^\infty Mf(x)\delta^n \frac{d\rho}{\rho^{n-\alpha+1}}\right)$$

$$= \frac{n}{\alpha}\boldsymbol{\alpha}(n)\,Mf(x)\delta^\alpha.$$

Similarly,

$$\int_{\mathbf{R}^n\setminus B(x,\delta)} \frac{|f(y)|\,dy}{|x-y|^{n+\beta}} = (n-\beta)\int_0^\infty \left(\int_{B(x,\rho)\setminus B(x,\delta)} |f(y)|\,dy\right)\frac{d\rho}{\rho^{n+\beta+1}}$$

$$\leq (n+\beta)\boldsymbol{\alpha}(n)\int_\delta^\infty Mf(x)\rho^n \frac{d\rho}{\rho^{n+\beta+1}}$$

$$= \frac{n+\beta}{\beta}\boldsymbol{\alpha}(n)Mf(x)\delta^{-\beta}. \qquad \square$$

1.33 Theorem. *Let $\alpha > 0$, $1 < p < \infty$, and $\alpha p < n$. Then, there is a constant $C = C(n,p,\alpha)$ such that*

$$\|I_\alpha f\|_{p^\sharp} \leq C\|f\|_p, \qquad p^\sharp = \frac{np}{n-\alpha p},$$

whenever $f \in L^p(\mathbf{R}^n)$.

PROOF. For $\delta > 0$, Hölder's inequality implies that

$$\int_{\mathbf{R}^n\setminus B(x,\delta)} \frac{|f(y)|}{|x-y|^{n-\alpha}}\,dy \leq \|f\|_p \left(\int_{\mathbf{R}^n\setminus B(x,\delta)} |x-y|^{(\alpha-n)p'}\,dy\right)^{1/p'}$$

$$= \|f\|_p \left(n\boldsymbol{\alpha}(n)\int_\delta^\infty r^{n-1-p'(n-\alpha)}\,dr\right)^{1/p'}$$

$$= C\|f\|_p \delta^{\alpha-n/p}.$$

Therefore, by Lemma 1.32(i),

(1.29) $\qquad |I_\alpha f(x)| \leq C\left(\delta^\alpha M(f)(x) + \delta^{\alpha-n/p}\|f\|_p\right).$

If we choose

$$\delta = \left(\frac{M(f)(x)}{\|f\|_p}\right)^{-p/n},$$

then (1.29) becomes
$$|I_\alpha f(x)| \leq C\bigl(M(f)(x)\bigr)^{1-\alpha p/n}\|f\|_p^{\alpha p/n}$$
or,
$$|I_\alpha f(x)|^{p^\sharp} \leq C\bigl(M(f)(x)\bigr)^p \|f\|_p^{\alpha p p^\sharp/n}.$$
An application of Theorem 1.22 now yields
$$\int_{\mathbf{R}^n} |I_\alpha f(x)|^{p^\sharp}\,dx \leq C\|f\|_p^{\alpha p p^\sharp/n} \int_{\mathbf{R}^n} M(f)^p\,dx$$
$$\leq C\|f\|_p^{\alpha p p^\sharp/n} \int_{\mathbf{R}^n} |f|^p\,dx,$$
which is the desired conclusion. $\square$

Assuming that Ω has finite measure, we can obtain a bound on $\|I_\alpha f\|_q$ where $p \leq q < p^\sharp$ by a method that depends only on Hölder's inequality.

1.34 LEMMA. *Let $1 \leq p < \infty$ and $0 < \alpha \leq n/p$. Suppose Ω is a domain with finite measure and $f \in L^p(\Omega)$. Then for $p \leq q < p^\sharp := np/(n-\alpha p)$ ($p^\sharp = \infty$ when $\alpha p = n$), $\frac{1}{r} := 1 - (\frac{1}{p} - \frac{1}{q})$, and $\gamma := \frac{(\alpha-n)r}{n} + 1$,*
$$\|I_\alpha f\|_{q;\Omega} \leq C^{1/r}\|f\|_{p;\Omega}$$
where

(1.30) $$C = C(n,\alpha,p,q,|\Omega|) = \frac{\boldsymbol{\alpha}(n)}{\gamma}\left(\frac{|\Omega|}{\boldsymbol{\alpha}(n)}\right)^\gamma.$$

PROOF. Since $q < p^\sharp$,

(1.31) $$|x-y|^{\alpha-n} \in L^r(\Omega)$$

for each fixed $x \in \mathbf{R}^n$. By Lemma 1.30, for any $x \in \mathbf{R}^n$ we have

(1.32) $$\int_\Omega |x-y|^{(\alpha-n)r}\,dy = \int_\Omega |x-y|^{\gamma n - n}\,dy \leq C,$$

with C as declared in (1.30). Now, the constant C will be the same till the end of the proof. For each fixed x, observe that

(1.33) $$|x-y|^{(\alpha-n)}|f(y)| = \left(|x-y|^{(\alpha-n)r}|f(y)|^p\right)^{1/q}$$
$$\cdot \left(|x-y|^{(\alpha-n)r/p'}\right) \cdot |f(y)|^{p/r'}.$$

Because $\frac{1}{p'} + \frac{1}{q} + \frac{1}{r'} = 1$, we may apply Hölder's inequality to the three factors on the right side of (1.33) to obtain

$$|I_\alpha f(x)| \leq \left(\int_\Omega |x-y|^{(\alpha-n)r}|f(y)|^p\,dy\right)^{1/q}$$
$$\cdot \left(\int_\Omega |x-y|^{(\alpha-n)r}\,dy\right)^{1/p'} \left(\int_\Omega |f(y)|^p\,dy\right)^{1/r'}.$$

Therefore, by Fubini's theorem and (1.32),

$$\int_\Omega (I_\alpha f)^q\, dx \leq \int_\Omega \int_\Omega |x-y|^{(\alpha-n)r} |f(y)|^p\, dx\, dy$$
$$\cdot\, C^{\frac{q}{p'}} \cdot \|f\|_p^{pq/r'}$$
$$\leq C \cdot \|f\|_p^p \cdot C^{\frac{q}{p'}} \cdot \|f\|_p^{pq/r'}$$

Thus,

$$\|I_\alpha f\|_q \leq C^{1/q+1/p'} \|f\|_p$$
$$= C^{1/r} \|f\|_p. \qquad \square$$

1.35 THEOREM. *Suppose that* $|\Omega| < \infty$. *Let* $f \in L^p(\Omega)$, $p > 1$, *and define*

$$g = I_{n/p} * f.$$

Then there are constants C_1 *and* C_2 *depending only on* p *and* n *such that*

$$(1.34) \qquad \fint_\Omega \exp\left(\frac{g}{C_1 \|f\|_p}\right)^{p'} dx \leq C_2.$$

PROOF. Let $p \leq q < \infty$ and recall from Lemma 1.34 the estimate

$$(1.35) \qquad \|I_\alpha f\|_q \leq C_0^{1/r} \|f\|_p,$$

where $\frac{1}{r} = 1 - (\frac{1}{p} - \frac{1}{q})$,

$$C_0 = C(n,\alpha,p,q,|\Omega|) = \frac{\boldsymbol{\alpha}(n)}{\gamma} \left(\frac{|\Omega|}{\boldsymbol{\alpha}(n)}\right)^\gamma,$$

and

$$\gamma = \frac{(\alpha-n)r}{n} + 1.$$

In the present situation, $\alpha p = n$, and therefore

$$\gamma = r/q \in [1/q, 1/p].$$

Thus, we can write

$$C_0 \leq K |\Omega|^\gamma q$$

where K is a constant that depends only on p and n. Thus, since $\gamma q/r = 1$, from (1.35) we have

$$\int_\Omega |g|^q\, dx \leq C_0^{q/r} \|f\|_p^q$$
$$\leq (qK)^{q/r} |\Omega|\, \|f\|_p^q = (qK)^{1+q/p'} |\Omega|\, \|f\|_p^q.$$

Now replacement of q by $p'k$ (which requires $k > p-1$) yields

$$\int_\Omega |g|^{p'k}\, dx \leq (p'kK)^{1+k} |\Omega|\, \|f\|_p^{p'k}.$$

In preparation for an expression involving an infinite series, we consider a constant C to be specified later and rewrite the preceding inequality as

$$\fint_\Omega \left(\frac{|g|}{C\|f\|_p}\right)^{p'k} dx \leq a_k := p'Kk^{k+1}\left(\frac{Kp'}{C^{p'}}\right)^k.$$

For $k \leq p-1 < m$ we get from Hölder's inequality

$$\fint_\Omega \left(\frac{|g|}{C\|f\|_p}\right)^{p'k} dx \leq a_m^{k/m}.$$

Consequently,

$$\fint_\Omega \sum_{k=0}^\infty \frac{1}{k!}\left(\frac{|g|}{C\|f\|_p}\right)^{p'k} dx \leq \sum_{k<m} \frac{a_m^{k/m}}{k!} + \sum_{k\geq m} \frac{a_k}{k!}$$

where m is the smallest integer greater than $p-1$. The series on the right converges if $C^{p'} > eKp'$ and thus the result follows from the monotone convergence theorem. □

The following theorem is a potential-theoretic version of the John-Nirenberg inequality.

1.36 THEOREM. *Let $f \in \mathcal{M}^p(\Omega)$, $p > 1$, with $\|f\|_{\mathcal{M}^p(\Omega)} = K$, and $g = I_{n/p} * f$. Then there exist constants $C_1 = C_1(n,p)$ and $C_2 = C_2(n,p)$ such that*

$$\int_\Omega \exp\left(\frac{|g|}{C_1 K}\right) dx \leq C_2 d^n$$

where $d = \operatorname{diam} \Omega$.

PROOF. For any $q \geq 1$, write

$$g(x) = I_{n/p} * f(x) = \int \frac{f(y)\,dy}{|x-y|^{n-n/p}} = \int \frac{|f|^{1/q} \cdot |f|^{1/q'}\,dy}{|x-y|^{(n-\alpha_1)/q}|x-y|^{(n-\alpha_2)/q'}}$$

where

$$\alpha_1 = \frac{n}{pq} \quad \text{and} \quad \alpha_2 = \frac{n}{p}\left(1 + \frac{1}{q}\right).$$

Thus, by Hölder's inequality,

$$g(x) \leq \left(I_{\alpha_1} * |f|\,(x)\right)^{1/q} \left(I_{\alpha_2} * |f|\,(x)\right)^{1/q'}.$$

Referring to Lemmas 1.34 and 1.29, we find

$$\int_\Omega I_{\alpha_1} * |f|\,(x)\,dx \leq pq\boldsymbol{\alpha}(n)Kd^{n(1/p'+1/pq)},$$

$$I_{\alpha_2} * |f|\,(x) \leq pqKd^{n/pq}, \quad x \in \Omega.$$

Thus,

$$\int_\Omega |g|^q\,dx \leq (pqK)^q \boldsymbol{\alpha}(n)d^n$$

and therefore

$$\int_\Omega \sum_{m=0}^N \frac{|g|^m}{m!(c_1 K)^m}\, dx \le \alpha(n) d^n \sum_{m=0}^N \frac{(pm)^m}{C_1^m m!}$$
$$\le C_2 d^n$$

if $pe < C_1$. Letting $N \to \infty$, we obtain our result. $\square$

1.3 Sobolev spaces

In this section we collect some fundamental results on Sobolev spaces that serve as a foundation for further development.

Let $u \in L^1_{\text{loc}}(\Omega)$. For a given multi-index α, a function $v \in L^1_{\text{loc}}(\Omega)$ is called the αth *weak derivative* of u if

$$(1.36) \qquad \int_\Omega \varphi v\, dx = (-1)^{|\alpha|} \int_\Omega u D^\alpha \varphi\, dx$$

for all $\varphi \in \mathcal{C}_c^\infty(\Omega)$. The function v is also referred to as the *generalized derivative* of u and we write $v = D^\alpha u$. Notice also that weak derivatives are a special case of *distributional derivatives*. Clearly, $D^\alpha u$ is uniquely determined up to sets of Lebesgue measure zero. First order weak derivatives ($|\alpha| = 1$) are denoted by $D_1 u, \ldots, D_n u$.

1.37 DEFINITION. For $p \ge 1$ and k a non-negative integer, we define the Sobolev space

$$(1.37) \qquad W^{k,p}(\Omega) = \{u \in L^p(\Omega) : D^\alpha u \in L^p(\Omega), |\alpha| \le k\}.$$

Let $\nabla^k \in \mathbf{R}^{n^k}$ be the "vector" of all derivatives D^α with $|\alpha| = k$. The space $W^{k,p}(\Omega)$ is equipped with a norm

$$(1.38) \qquad \|u\|_{k,p;\Omega} = \begin{cases} \left(\int_\Omega (|\nabla^k u|^p + \cdots + |u|^p)\, dx\right)^{1/p} & \text{if } p < \infty, \\ \max\{\|\nabla^k u\|_\infty, \ldots, \|u\|_\infty\} & \text{if } p = \infty, \end{cases}$$

which is clearly equivalent to

$$(1.39) \qquad \sum_{|\alpha| \le k} \|D^\alpha u\|_{p;\Omega}.$$

The space $W_0^{k,p}(\Omega)$ is defined as the closure of $W_c^{1,p}(\Omega)$ relative to the norm (1.38), where $W_c^{1,p}(\Omega)$ is the set of all functions $u \in W^{1,p}(\Omega)$ with compact support in Ω.

Finally, $W_{\text{loc}}^{k,p}(\Omega)$ denotes the family of all functions u such that $u \llcorner \Omega' \in W^{k,p}(\Omega')$ for each $\Omega' \subset\subset \Omega$.

It is an easy matter to verify that $W^{k,p}(\Omega)$, $W_0^{1,p}(\Omega)$ are Banach spaces. If $u \in W_0^{k,p}(\Omega)$ then the zero extension of u is in $W^{k,p}(\mathbf{R}^n)$. In this sense, $W_0^{k,p}(\Omega)$ is a closed subspace of $W^{k,p}(\mathbf{R}^n)$.

1.38 REMARK. Observe that if $u \in W^{k,p}(\Omega)$, then u is determined only up to a set of Lebesgue measure zero. We agree to call these functions u continuous, bounded, etc., if there is a function $\bar{u}$ such that $\bar{u} = u$ a.e. and $\bar{u}$ has these properties.

We will show that elements in $W^{k,p}(\Omega)$ have representatives that permit us to regard them as generalizations of absolutely continuous functions on $\mathbf{R}$. First, we prove an important result concerning the convergence of regularizers of Sobolev functions.

1.39 LEMMA. *Suppose $u \in W^{1,p}(\Omega), 1 \leq p < \infty$. Then the mollifiers, u_ε, of u satisfy*
$$\lim_{\varepsilon \to 0} \|u_\varepsilon - u\|_{1,p;\Omega'} = 0$$
whenever $\Omega' \subset\subset \Omega$.

PROOF. Since Ω' is a bounded domain, there exists $\varepsilon_0 > 0$ such that
$$\varepsilon_0 < \operatorname{dist}(\Omega', \partial\Omega).$$

For $\varepsilon < \varepsilon_0$, we may differentiate under the integral sign to obtain for $x \in \Omega'$ and $1 \leq i \leq n$,

$$\begin{aligned}\frac{\partial u_\varepsilon}{\partial x_i}(x) &= \varepsilon^{-n} \int_\Omega \frac{\partial \varphi}{\partial x_i}\left(\frac{x-y}{\varepsilon}\right) u(y)\, dy \\ &= -\varepsilon^{-n} \int_\Omega \frac{\partial \varphi}{\partial y_i}\left(\frac{x-y}{\varepsilon}\right) u(y)\, dy \\ &= \varepsilon^{-n} \int_\Omega \varphi\left(\frac{x-y}{\varepsilon}\right) \frac{\partial u}{\partial y_i}\, dy \\ &= \left(\frac{\partial u}{\partial x_i}\right)_\varepsilon(x).\end{aligned}$$

Our result now follows from Theorem 1.12. $\square$

1.40 COROLLARY. *If $1 \leq p < \infty$, then $C_c^\infty(\Omega)$ is dense in $W_0^{1,p}(\Omega)$.*

Since the definition of a Sobolev function requires that its distributional derivatives belong to L^p, it is natural to inquire whether the function possesses any classical differentiability properties. To this end, we begin by showing that its partial derivatives exist almost everywhere. That is, in keeping with Remark 1.38, we will show that there is a function $\bar{u}$ such that $\bar{u} = u$ a.e. and that the partial derivatives of $\bar{u}$ exist almost everywhere.

1.41 THEOREM. *Suppose $u \in W^{1,p}(\Omega), p \geq 1$. Let $\Omega' \subset\subset \Omega$. Then u has a representative u^* that is absolutely continuous on almost all line segments of Ω' that are parallel to the coordinate axes and the classical partial derivatives of u^* agree almost everywhere with the weak derivatives of u. Conversely, if u has such a representative and the classical partial derivatives $D_1 u, \ldots, D_n u$ together with u are in $L^p(\Omega')$ then $u \in W^{1,p}(\Omega')$.*

1.3 SOBOLEV SPACES

PROOF. We may assume that $\Omega = \mathbf{R}^n$ and u has compact support.

(i) Let $\{u_k\}$ be a sequence of regularizations of u such that u_k has compact support independent of k and

$$\|u_k - u\|_1 + \|\nabla u_k - \nabla u\|_1 < 2^{-k-1}.$$

Let G be the set where the sequence u_k converges pointwise and let u^* denote the pointwise limit, $\lim_{k\to\infty} u_k$, on G and 0 elsewhere. We will fix a direction and assume without loss of generality that it is defined by the vector $(0,\ldots,0,1)$. Let

$$f_k(x_1,\ldots,x_{n-1}) = \int_{-\infty}^{+\infty} (|u_{k+1} - u_k| + |\nabla u_{k+1} - \nabla u_k|)(x_1,\ldots,x_n)\, dx_n$$

and

$$f(x_1,\ldots,x_{n-1}) = \sum_{k=1}^{\infty} f_k(x_1,\ldots,x_{n-1}).$$

Then, using the monotone convergence theorem and Fubini's theorem,

$$\int_{\mathbf{R}^{n-1}} f\, dx_1 \ldots dx_{n-1} = \sum_{k=1}^{\infty} \int_{\mathbf{R}^{n-1}} f_k\, dx_1 \ldots dx_{n-1}$$

$$= \sum_{k=1}^{\infty} \int_{\mathbf{R}^n} (|u_{k+1} - u_k| + |\nabla u_{k+1} - \nabla u_k|)\, dx < \infty.$$

Hence $f \in L^1(\mathbf{R}^{n-1})$ and, in particular, it is finite $(n-1)$-a.e.

Fix $\hat{x} = (x_1,\ldots,x_{n-1})$ satisfying $f(\hat{x}) < \infty$. Denote $g_k(t) = u_k(\hat{x},t)$ and $g(t) := u^*(\hat{x},t)$. We claim that g is absolutely continuous. The sum

$$g_1 + \sum_{k=1}^{\infty}(g_{k+1} - g_k)$$

converges uniformly on $\mathbf{R}$ since

$$|g_{k+1}(t) - g_k(t)| \leq \int_{-\infty}^{t} |(g_{k+1} - g_k)'(s)|\, ds \leq f_k(\hat{x}).$$

It follows that $\{\hat{x}\} \times \mathbf{R} \subset G$ and g is continuous on $\mathbf{R}$. Also, we see that the limit

$$\lim_{k\to\infty} g_k' = g_1' + \sum_{k=1}^{\infty}(g_{k+1} - g_k)'$$

converges in $L^1(\mathbf{R})$ to some L^1-function, call it Dg. Passing to the limit in

$$g_k(t) = \int_{-\infty}^{t} g_k'(s)\, ds$$

we see that g is the indefinite integral of Dg. Hence, g is absolutely continuous, differentiable a.e. and its derivative, denoted by g', equals Dg a.e. Now taking the limit in

$$\int_{\mathbf{R}} g_k \psi'\, dt = -\int_{\mathbf{R}} g_k' \psi\, dt$$

for an arbitrary $\psi \in C_0^\infty(\mathbf{R})$ implies that Dg is a weak derivative of g. Let us write $\frac{\partial u}{\partial x_n}$ for the pointwise partial derivative of u. Then we have shown that for given $\varphi \in C_c^\infty$,
$$\int_{\mathbf{R}} u(\hat{x}, x_n) \frac{\partial \varphi}{\partial x_n} \, dx_n = -\int_{\mathbf{R}} \frac{\partial u}{\partial x_n} \varphi(\hat{x}, x_n) \, dx_n,$$
and using Fubini's theorem we get
$$\int_{\mathbf{R}^n} u \frac{\partial \varphi}{\partial x_n} \, dx = -\int_{\mathbf{R}^n} \frac{\partial u}{\partial x_n} \varphi(x) \, dx.$$
Hence, we conclude that the pointwise partial derivatives of u exist a.e. and are the weak derivatives of u.

(ii) Suppose that u has a representative u^* as in the statement of the Theorem. For $\varphi \in C_c^\infty(\mathbf{R}^n)$, $u^*\varphi$ has the same absolutely continuous properties as u^*. Thus, we can apply the fundamental theorem of calculus to obtain
$$\int_{\mathbf{R}} \frac{\partial (u^* \varphi)}{\partial x_n}(\hat{x}, t) \, dt = 0$$
for $(n-1)$-almost all $\hat{x} \in \mathbf{R}^{n-1}$ and therefore,
$$\int_{\mathbf{R}} u^*(\hat{x}, t) \frac{\partial \varphi}{\partial x_n}(\hat{x}, t) \, dt = -\int_{\mathbf{R}} \frac{\partial u^*}{\partial x_n}(\hat{x}, t) \varphi(\hat{x}, t) \, dt.$$
Fubini's Theorem implies
$$-\int_{\mathbf{R}^n} u^* \frac{\partial \varphi}{\partial x_n} \, dx = \int_{\mathbf{R}^n} \frac{\partial u^*}{\partial x_n} \varphi \, dx$$
for all $\varphi \in C_c^\infty(\mathbf{R}^n)$ and similarly for the partial derivatives over $x_1, \ldots, x_{n-1}$. Since u^* is a representative of u, this states that the weak derivative of u is the function $\frac{\partial u^*}{\partial x_i}$. □

1.42 COROLLARY. *If Ω is connected, $u \in W^{1,p}(\Omega)$, $p \geq 1$, and $\nabla u = 0$ a.e. on Ω, then u is constant on Ω.*

1.43 COROLLARY. *Let $u, v \in W^{1,p}(\Omega)$, $p \geq 1$. Then $\max\{u, v\}, \min\{u, v\} \in W^{1,p}(\Omega)$ and*
$$\nabla \max\{u, v\} = \begin{cases} \nabla u & \text{a.e. on } \{u \geq v\}, \\ \nabla v & \text{a.e. on } \{v \geq u\}, \end{cases}$$
$$\nabla \min\{u, v\} = \begin{cases} \nabla u & \text{a.e. on } \{u \leq v\}, \\ \nabla v & \text{a.e. on } \{v \leq u\}. \end{cases}$$
In particular, $\nabla u = \nabla v$ a.e. on the set $\{u = v\}$.

PROOF. Because u, v have representatives that has the absolute continuity properties stated in Theorem 1.41 it follows immediately that $\max\{u, v\}, \min\{u, v\} \in W^{1,p}(\Omega)$. The second part, with the aid of Theorem 1.41, is reduced to one-dimensional observation, which is elementary. □

1.44 Theorem. *Let $\mathcal{G}$ be an open cover of a set $E \subset \mathbf{R}^n$. Then there exists a family $\mathcal{F}$ of functions $u \in \mathcal{C}_c^\infty(\mathbf{R}^n)$ such that $0 \leq u \leq 1$ and*

(i) *for each $u \in \mathcal{F}$, there exists $U \in \mathcal{G}$ such that $\operatorname{spt} u \subset U$,*

(ii) *if $F \subset E$ is compact, then $\operatorname{spt} u \cap F \neq 0$ for only finitely many $u \in \mathcal{F}$,*

(iii) $\sum_{u \in \mathcal{F}} u(x) = 1$ *for each $x \in E$.*

The family $\mathcal{F}$ is called a *smooth partition of unity* of E subordinate to the open covering $\mathcal{G}$.

PROOF. Any partition of unity for the open set $\cup\{U : U \in \mathcal{G}\}$ is also one for E. Hence we may assume that E is open. For each positive integer i define

$$E_i = E \cap K(0, i) \cap \{x : \operatorname{dist}(x, \partial E) \geq \frac{1}{i}\}.$$

Thus, E_i is compact, $E_i \subset \operatorname{int} E_{i+1}$, and $E = \cup_{i=1}^\infty E_i$. For each $x \in E_i$ we find an open ball B containing x such that $\operatorname{diam} B \leq 1/i$ and $\overline{B} \subset G$ for some $G \in \mathcal{G}$. We select a finite cover $\mathcal{G}_i$ of E_i by such balls. Let

$$\mathcal{G}_\infty = \bigcup_i \mathcal{G}_i.$$

Now, for each $d > 0$, there is only a finite number of balls in $\mathcal{G}_\infty$ whose diameter exceeds d. It follows that each ball in $\mathcal{G}_\infty$ is contained in some "maximal ball", i.e. such a ball $B \in \mathcal{G}_\infty$ which is not contained in any other $B' \in \mathcal{G}_\infty$. Let $\mathcal{V}$ be the collection of all such maximal balls. It is clear that $\mathcal{V}$ is a covering of E. We will show local finiteness of $\mathcal{V}$. Let $x \in E$ and $U \in \mathcal{V}$ is a neighborhood of x. We find r such that $B(x, 2r) \subset U$. Consider $B' \in \mathcal{V}$ such that $B(x, r) \cap \overline{B}' \neq \emptyset$. Since B' is maximal, B' is not a subset of U and thus the radius of B' is greater than r. As we already noticed, there are only finitely many such balls. If follows that for any compact set $F \subset E$, there is only a finite number of balls $B \in \mathcal{V}$ with $\overline{B} \cap F \neq \emptyset$.

With any $B \in \mathcal{V}$ we associate $v_B \in \mathcal{C}_c(\mathbf{R}^n)$ with $\operatorname{spt} v_B = \overline{B}$ such that $v_B > 0$ on B; we get an explicit formula for such a function as in (1.11). We set

$$s = \sum_{B \in \mathcal{V}} v_B.$$

The sum is well defined and smooth due to the local finiteness of $\mathcal{V}$. The required partition of unity is

$$\mathcal{F} = \{\frac{v_B}{s} : B \in \mathcal{V}\}. \qquad \square$$

Clearly, the set

$$S = \{u \in \mathcal{C}^\infty(\Omega) : \|u\|_{1,p;\Omega} < \infty\}$$

is contained in $W^{1,p}(\Omega)$. Moreover, the same is true of the closure of S in the Sobolev norm since $W^{1,p}(\Omega)$ is complete. The next result shows that $\overline{S} = W^{1,p}(\Omega)$.

1.45 Theorem. *If $1 \leq p < \infty$, then the space*

$$\{u \in \mathcal{C}^\infty(\Omega) : \|u\|_{1,p;\Omega} < \infty\}$$

is dense in $W^{1,p}(\Omega)$.

PROOF. Let Ω_i be subdomains of Ω such that
$$\overline{\Omega_i} \subset \Omega_{i+1} \quad \text{and} \quad \bigcup_{i=1}^{\infty} \Omega_i = \Omega.$$

Let $\mathcal{F}$ be a partition of unity of Ω subordinate to the covering $\{\Omega_{i+1} - \overline{\Omega_{i-1}}\}$, $i = 1, 2, \ldots$, where $\Omega_0 = \emptyset$. Let φ_i denote the sum of the finitely many $\varphi \in \mathcal{F}$ with spt $\varphi \subset \Omega_{i+1} - \overline{\Omega_{i-1}}$. Then $\varphi_i \in \mathcal{C}_c^{\infty}(\Omega_{i+1} - \overline{\Omega_{i-1}})$ and

$$\sum_{i=1}^{\infty} \varphi_i \equiv 1 \quad \text{on} \quad \Omega. \tag{1.40}$$

Choose $\varepsilon > 0$. By Lemma 1.39, for $u \in W^{1,p}(\Omega)$, there exists $\varepsilon_i > 0$ such that
$$\text{spt}\,((\varphi_i u)_{\varepsilon_i}) \subset \Omega_{i+1} - \overline{\Omega_{i-1}}, \tag{1.41}$$
$$\|(\varphi_i u)_{\varepsilon_i} - \varphi_i u\|_{1,p;\Omega} < \varepsilon 2^{-i}.$$

With $g_i := (\varphi_i u)_{\varepsilon_i}$, (1.41) implies that only finitely many of the g_i can fail to vanish on any given $\Omega' \subset\subset \Omega$. Therefore $g := \sum_{i=1}^{\infty} g_i$ is defined and is an element of $\mathcal{C}^{\infty}(\Omega)$. For $x \in \Omega_i$, we have
$$u(x) = \sum_{j=1}^{i} \varphi_j(x) u(x),$$
and by (1.41)
$$g(x) = \sum_{j=1}^{i} (\varphi_j u)_{\varepsilon_j}(x).$$

Consequently,
$$\|g - u\|_{1,p;\Omega_i} \leq \sum_{j=1}^{i} \|(\varphi_j u)_{\varepsilon_j} - \varphi_j u\|_{1,p;\Omega} < \varepsilon.$$

Now an application of the monotone convergence theorem establishes our desired result. $\square$

The previous result holds in particular when $\Omega = \mathbf{R}^n$ in which case we get the following apparently stronger result.

1.46 COROLLARY. *If $1 \leq p < \infty$, then $W_0^{1,p}(\mathbf{R}^n) = W^{1,p}(\mathbf{R}^n)$ and the space $\mathcal{C}_c^{\infty}(\mathbf{R}^n)$ is dense in $W^{1,p}(\mathbf{R}^n)$.*

PROOF. For each positive integer k, let $\varphi_k \in \mathcal{C}_c(\mathbf{R}^n)$ be such that $0 \leq \varphi_k \leq 1$, $\varphi_k = 1$ on $B(0, k)$, spt $\varphi_k \subset B(0, 3k)$ and $|\nabla \varphi_k| \leq 1/k$. Then
$$\|\nabla(u\varphi_k) - \nabla u\|_p \leq \|(\varphi_k - 1)\nabla u\|_p + \|u\nabla \varphi_k\|_p$$
$$\leq C\|u\|_{1,p;\mathbf{R}^n \setminus B(0,k)} \to 0.$$

Now, each function $u\varphi_k$ has a compact support and thus its mollifications belong to $\mathcal{C}_c^{\infty}(\mathbf{R}^n)$. By Lemma 1.39,
$$\lim_{\varepsilon \to 0} \|\phi_{\varepsilon} * (u\varphi_k) - u\varphi_k\|_{1,p} \to 0,$$
which concludes the proof. $\square$

1.47 REMARK. It is a natural question whether $\mathcal{C}^1(\overline{\Omega})$ is dense in $W^{1,p}(\Omega)$. This is true only for special domains. In particular, extension domains introduced in Definition 1.59 share the property. We now will prove that $\mathcal{C}^1(\overline{B})$ is dense in $W^{1,p}(B)$ as the first step in showing that balls are extension domains, which we will verify later.

1.48 LEMMA. *Let $B = B(x,r)$ be a ball in $\mathbf{R}^n$. Then $\mathcal{C}^1(\overline{B})$ is dense in $W^{1,p}(B)$.*

PROOF. We may assume that $B = B(0,1)$. By Theorem 1.45, it is enough to prove that $u \in \mathcal{C}^1(B) \cap W^{1,p}(B)$ can be approximated by functions smooth up to the boundary. We will show that
$$\|u_j - u\|_{1,p} \to 0$$
when u_j are defined by
$$u_j(x) = u(r_j x), \qquad r_j \uparrow 1.$$
We choose $\varepsilon > 0$ and find $\rho \in (0,1)$ such that
$$\int_{B(0,1) \setminus B(0,\rho^2)} \left(|\nabla u|^p + |u|^p\right) dy < \varepsilon.$$
Then for sufficiently large j we have $r_j > \rho$ and thus
$$\int_{B(0,1) \setminus B(0,\rho)} \left(|\nabla u_j(y)|^p + |u_j(y)|^p\right) dy$$
$$= \int_{B(0,r_j) \setminus B(0,r_j\rho)} \left(r_j^{p-n}|\nabla u(z)|^p + r_j^{-n}|u(z)|^p\right) dz < \rho^{-n}\varepsilon.$$
Due to the uniform continuity of u and ∇u on $B(0,\rho)$,
$$\|u_j - u\|_{1,p;B(0,\rho)} \to 0.$$
It follows that
$$\limsup_{j \to \infty} \int_B \left(|\nabla u_j(y) - \nabla u(y)|^p + |u_j(y) - u(y)|^p\right) dy$$
$$\leq 2^{p-1}(\varepsilon + \rho^{-n}\varepsilon) + \limsup_{j \to \infty} \int_{B(0,\rho)} \left(|\nabla u_j(y) - \nabla u(y)|^p + |u_j(y) - u(y)|^p\right) dy.$$
Letting $\varepsilon \to 0$ we get the assertion. □

Recall that if $u \in L^p(\mathbf{R}^n)$, then $\|u(x+h) - u(x)\|_p \to 0$ as $h \to 0$. A similar result provides a very useful characterization of $W^{1,p}$.

1.49 THEOREM. *Let $1 < p < \infty$ and suppose $\Omega \subset \mathbf{R}^n$ is an open set. If $u \in W^{1,p}(\Omega)$ and $\Omega' \subset\subset \Omega$, then $\left|h^{-1}\right| \|u(x+h) - u(x)\|_{p;\Omega'}$ remains bounded for all sufficiently small $h \in \mathbf{R}^n$. Conversely, if $u \in L^p(\Omega)$ and*
$$\left|h^{-1}\right| \|u(x+h) - u(x)\|_{p;\Omega'}$$
remains bounded for all sufficiently small h, then $u \in W^{1,p}(\Omega')$.

PROOF. Assume $u \in W^{1,p}(\Omega)$ and let $\Omega' \subset\subset \Omega$. By Theorem 1.45, there exists a sequence of $\mathcal{C}^\infty(\Omega)$ functions $\{u_k\}$ such that $\|u_k - u\|_{1,p;\Omega} \to 0$ as $k \to \infty$. For each $g \in \mathcal{C}^\infty(\Omega)$, we have

$$\frac{g(x+h) - g(x)}{|h|} = \frac{1}{|h|} \int_0^{|h|} \nabla g\left(x + t\frac{h}{|h|}\right) \cdot \frac{h}{|h|}\, dt,$$

so by Jensen's inequality,

$$\left|\frac{g(x+h) - g(x)}{h}\right|^p \leq \frac{1}{|h|} \int_0^{|h|} \left|\nabla g\left(x + t\frac{h}{|h|}\right)\right|^p dt$$

whenever $x \in \Omega'$ and $h < \delta := \operatorname{dist}(\partial\Omega', \partial\Omega)$. Therefore,

$$\|g(x+h) - g(x)\|_{p;\Omega'}^p \leq |h|^p \frac{1}{|h|} \int_0^{|h|} \int_{\Omega'} \left|\nabla g\left(x + t\frac{h}{|h|}\right)\right|^p dx\, dt$$

$$\leq |h|^{p-1} \int_0^{|h|} \int_\Omega |\nabla g(x)|^p\, dx\, dt,$$

or

$$\|g(x+h) - g(x)\|_{p;\Omega'} \leq |h|\, \|\nabla v\|_{p;\Omega}$$

for all $h < \delta$. As this inequality holds for each u_k, it also holds for u.

For the proof of the converse, let e_i denote the ith unit basis vector. By assumption, the sequence

$$\left\{\frac{u(x + e_i/k) - u(x)}{1/k}\right\}$$

is bounded in $L^p(\Omega')$ for all large k. Therefore, using reflexivity of L^p, there exist a subsequence (denoted by the full sequence) and $u_i \in L^p(\Omega')$ such that

$$\frac{u(x + e_i/k) - u(x)}{1/k} \to u_i$$

weakly in $L^p(\Omega')$. Thus, for $\varphi \in \mathcal{C}_0^\infty(\Omega')$,

$$\int_{\Omega'} u_i \varphi\, dx = \lim_{k \to \infty} \int_{\Omega'} \left(\frac{u(x + e_i/k) - u(x)}{1/k}\right) \varphi(x)\, dx$$

$$= \lim_{k \to \infty} \int_{\Omega'} u(x) \left(\frac{\varphi(x - e_i/k) - \varphi(x)}{1/k}\right) dx$$

$$= -\int_{\Omega'} u D_i \varphi\, dx.$$

This shows that $D_i u = u_i$ in the sense of distributions. Hence, $u \in W^{1,p}(\Omega')$. □

1.3.1 Inequalities.

We provide an elementary treatment of Poincaré's inequality and the important Morrey and Campanato imbedding results. We start with a lemma which links the theory of Sobolev spaces with Riesz potentials.

1.50 LEMMA. *Let $\Omega \subset \mathbf{R}^n$ be a bounded convex domain $\varphi \in L^\infty(\Omega)$,*
$$\overline{u}_\varphi := \int_\Omega u(y)\varphi(y)\,dy$$
and suppose that
$$\int_\Omega \varphi = 1 \quad \text{and} \quad \|\varphi\|_\infty \leq M.$$

(i) *Let $x \in \Omega \subset B(x, r)$ and $u \in W^{1,1}(\Omega) \cap \mathcal{C}^1(\Omega)$. Then*

(1.42) $$|u(x) - \overline{u}_\varphi| \leq C \int_\Omega \frac{|\nabla u(y)|}{|x-y|^{n-1}}\,dy$$

where

(1.43) $$C = \frac{Mr^n}{n}.$$

(ii) *Let $u \in W^{1,1}(\Omega)$. Then (1.42) holds for almost every $x \in \Omega$ with*

(1.44) $$C = \frac{M(\operatorname{diam}\Omega)^n}{n}.$$

PROOF. Let $S = \{z \colon |z| = 1\}$ be the unit sphere in $\mathbf{R}^n$ and σ be the surface measure on S (that is, $\sigma = H^{n-1} \mathbin{\llcorner} S$). We express $y \in \Omega$ in spherical coordinates
$$y = x + tz, \quad t > 0,\ z \in S.$$
Let $\delta(z) = \sup\{t \colon x + tz \in \Omega\}$. Since
$$|u(x) - u(y)| \leq \int_0^t |\nabla u(x + sz)|\,ds \leq \int_0^{\delta(z)} |\nabla u(x + sz)|\,ds,$$
we get
$$|u(x) - \overline{u}_\varphi| = \Big|\int_\Omega \big(u(x) - u(y)\big)\varphi(y)\,dy\Big|$$
$$\leq M \int_\Omega |u(x) - u(y)|\,dy$$
$$= M \int_S \Big(\int_0^{\delta(z)} t^{n-1}|u(x) - u(x+tz)|\,dt\Big)\,dz$$
$$\leq M \int_S \Big(\int_0^r t^{n-1}\,dt \int_0^{\delta(z)} \frac{|\nabla u(x+sz)|}{s^{n-1}} s^{n-1}\,ds\Big)\,dz$$
$$= \frac{Mr^n}{n}\int_\Omega \frac{|\nabla u(y)|}{|x-y|^{n-1}}\,dy.$$

This concludes the proof of (i). Part (ii) now follows using a standard approximation argument and Theorem 1.31. □

The proof of Poincaré's inequality below is based on the convexity of the domain and yields a clear description of the constant. Later, in Lemma 1.65, we give a more general version of Poincaré's inequality.

1.51 Theorem (Poincaré's inequality). *Let $\Omega \subset \mathbf{R}^n$ be convex and bounded with $d := \operatorname{diam} \Omega$, and $u \in W^{1,p}(\Omega)$. Let*

$$\bar{u}_\Omega = \fint_\Omega u \, dx.$$

Then

$$\int_\Omega |u(x) - \bar{u}_\Omega|^p \, dx \leq C d^p \int_\Omega |\nabla u(y)|^p \, dx,$$

where

$$C = \left(\frac{\boldsymbol{\alpha}(n) d^n}{|\Omega|}\right)^{p(1-1/n)}.$$

Proof. By Lemma 1.50 with $\varphi = |\Omega|^{-1} \chi_\Omega$, for a.e. $x \in \Omega$ we have

$$|u(x) - \bar{u}_\Omega| \leq C_0 (I_1 |\nabla u|)(x)$$

with

$$C_0 = \frac{d^n}{n |\Omega|}.$$

Using Lemma 1.31 we obtain

$$\int_\Omega |u(x) - \bar{u}_\Omega|^p \, dx \leq C_0^p \int_\Omega (I_1 |\nabla u|)^p \, dx \leq C_0^p C_1^p \int_\Omega |\nabla u|^p \, dx,$$

where

$$C_1 = n\boldsymbol{\alpha}(n)^{1-1/n} |\Omega|^{1/n}. \qquad \square$$

1.52 Theorem. *Let ϕ be a nonnegative bounded convolution kernel supported in $K(0,1)$ with $\|\phi\|_\infty \leq M$ and $\|\phi\|_1 = 1$. Then for each $u \in W^{1,p}(\mathbf{R}^n)$ and $\varepsilon > 0$ we have*

$$\|u - u * \phi_\varepsilon\|_p \leq \boldsymbol{\alpha}(n) M \varepsilon \|\nabla u\|_p.$$

Proof. We may suppose that $u \in \mathcal{C}_c^1(\mathbf{R}^n)$. By Lemma 1.50 with $\varphi(y) = \phi_\varepsilon(x-y)$ we have

$$|u(x) - u * \phi_\varepsilon(x)| \leq k * |\nabla u|(x),$$

where

$$k(y) = \frac{M}{n} |y|^{1-n} \chi_{B(0,\varepsilon)}.$$

Using Theorem 1.10 we obtain

$$\|u - u * \phi_\varepsilon\|_p \leq \|\nabla u\|_p \|k\|_1 = \boldsymbol{\alpha}(n) M \varepsilon \|\nabla u\|_p. \qquad \square$$

1.53 Theorem (Morrey). *Let $u \in W^{1,1}(\Omega)$ and $0 < \alpha \leq 1$. Suppose there is a positive constant M such that*

$$\int_{B(r)} |\nabla u| \, dx \leq M r^{n-1+\alpha}$$

for all balls $B(r) \subset \Omega$. Then $u \in \mathcal{C}^{0,\alpha}(\Omega)$ and for any ball $B(r) \subset \Omega$

$$\operatorname*{osc}_{B(x,r)} u \leq C M r^\alpha$$

where $C = C(n, \alpha)$.

PROOF. Let $B := B(z, r) \subset \Omega$ and recall from Lemma 1.50 that

$$|u(x) - \overline{u}_B| \leq C \int_{B(z,r)} \frac{|\nabla u(y)|}{|x-y|^{n-1}} \, dy$$

for a.e. $x \in B$. Using Lemma 1.29 we conclude

$$|u(x) - \overline{u}_B| \leq CMr^\alpha. \qquad \square$$

Theorem 1.53 can be proved by alternative methods using Poincaré's inequality (Theorem 1.51) together with the following estimate, which also has other important applications.

1.54 THEOREM (CAMPANATO). *Let $u \in L^1(\Omega)$ and $0 < \alpha \leq 1$. Suppose there is a positive constant M such that*

(1.45) $$\fint_B |u - \overline{u}_B| \, dx \leq Mr^\alpha$$

for all balls $B = B(r) \subset \Omega$. Then $u \in \mathcal{C}^{0,\alpha}(\Omega)$ and for any ball $B(R) \subset \Omega$,

$$\underset{B(R/2)}{\operatorname{osc}} u \leq CMR^\alpha$$

where $C = C(n, \alpha)$.

PROOF. Let x be a Lebesgue point for u. If $B(x, r/2) \subset B(z, r) \subset \Omega$, then

(1.46)
$$|\overline{u}_{B(x,r/2)} - \overline{u}_{B(z,r)}| \leq \fint_{B(x,r/2)} |u - \overline{u}_{B(z,r)}| \, dx$$
$$\leq 2^n \fint_{B(x,r)} |u - \overline{u}_{B(z,r)}| \, dx \leq 2^n M r^\alpha.$$

Iterating this inequality yields

$$|\overline{u}_{B(x,2^{-k}r)} - \overline{u}_{B(x,r)}| \leq 2^n M r^\alpha \sum_{i=1}^k 2^{-\alpha i} \leq CMr^\alpha.$$

Letting $k \to \infty$ we obtain

$$|u(x) - \overline{u}_{B(x,r)}| \leq CMr^\alpha.$$

Hence, using (1.46) again,

$$|u(x) - \overline{u}_{B(z,R)}| \leq |u(x) - \overline{u}_{B(x,R/2)}| - |\overline{u}_{B(x,R/2)} - \overline{u}_{B(z,R)}| \leq CMR^\alpha$$

holds for each $x \in B(z, R/2)$. This establishes the result. $\qquad \square$

1.55 REMARK. The previous theorem is in fact a characterization of Hölder spaces because also a converse is valid. Namely, if $u \in \mathcal{C}^{0,\alpha}$, then the inequality 1.45 holds locally in Ω. The proof of this assertion is easy.

1.3.2 Imbeddings.

The imbedding theorems of Sobolev and Kondrachov-Rellich are the main results of this section. They both play a critical part in the subsequent development.

1.56 Theorem (Sobolev Inequality). *Let $1 \leq p < n$ and let $\Omega \subset \mathbf{R}^n$ be an open set. Then for each $u \in W_0^{1,p}(\Omega)$,*

$$\|u\|_{p^*;\Omega} \leq C \|\nabla u\|_{p;\Omega}$$

holds with $C = p^*/1^*$.

PROOF. First, we will assume that $p = 1$ and $u \in \mathcal{C}_c^1(\Omega)$. We may assume that $\Omega = \mathbf{R}^n$. Appealing to the fundamental theorem of calculus and using the fact that u has compact support, it follows for each integer i, $1 \leq i \leq n$, that

$$u(x_1, \ldots, x_i, \ldots, x_n) = \int_{-\infty}^{x_i} \frac{\partial u}{\partial x_i}(x_1, \ldots, t_i, \ldots, x_n)\, dt_i,$$

and therefore,

$$|u(x)| = \int_{-\infty}^{\infty} \frac{\partial u}{\partial x_i}(x_1, \ldots, t_i, \ldots, x_n)\, dt_i.$$

Consequently,

$$|u(x)|^{\frac{n}{n-1}} \leq \prod_{i=1}^n \left(\int_{-\infty}^{\infty} |\nabla u(x_1, \ldots, t_i, \ldots, x_n)|\, dt_i \right)^{\frac{1}{n-1}}$$

This can be rewritten as

$$|u(x)|^{\frac{n}{n-1}}$$
$$\leq \left(\int_{-\infty}^{\infty} |\nabla u|\, dt_1 \right)^{\frac{1}{n-1}} \prod_{i=2}^n \left(\int_{-\infty}^{\infty} |\nabla u|\, dt_i \right)^{\frac{1}{n-1}}$$

where $\nabla u = \nabla u(x_1, \ldots, t_j, \ldots, x_n)$ and j is the index that reflects the integration. Only the first factor on the right is independent of x_1. Thus, when the inequality is integrated with respect to x_1 we obtain, with the help of generalized Hölder's inequality,

$$\int_{-\infty}^{\infty} |u(x)|^{\frac{1}{n-1}}\, dx_1$$
$$\leq \left(\int_{-\infty}^{\infty} |\nabla u|\, dt_1 \right)^{\frac{1}{n-1}} \int_{-\infty}^{\infty} \prod_{i=2}^n \left(\int_{-\infty}^{\infty} |\nabla u|\, dt_i \right)^{\frac{1}{n-1}} dx_1$$
$$\leq \left(\int_{-\infty}^{\infty} |\nabla u|\, dt_1 \right)^{\frac{1}{n-1}} \left(\prod_{i=2}^n \int_{-\infty}^{\infty} \int_{-\infty}^{\infty} |\nabla u|\, dx_1 dt_i \right)^{\frac{1}{n-1}}$$

Similarly, integration with respect to x_2 yields

$$\int_{-\infty}^{\infty} \int_{-\infty}^{\infty} |u(x)|^{\frac{n}{n-1}}\, dx_1 dx_2$$
$$\leq \left(\int_{-\infty}^{\infty} \int_{-\infty}^{\infty} |\nabla u|\, dx_1 dt_2 \right)^{\frac{1}{n-1}} \left(\int_{-\infty}^{\infty} \int_{-\infty}^{\infty} |\nabla u|\, dt_1 dx_2 \right)^{\frac{1}{n-1}}$$
$$\prod_{i=3}^n \left(\int_{-\infty}^{\infty} \int_{-\infty}^{\infty} \int_{-\infty}^{\infty} |\nabla u|\, dx_1 dx_2 dt_i \right)^{\frac{1}{n-1}}$$

Continuing this way for the remaining $n-2$ steps, we finally arrive at

$$\int_{\mathbf{R}^n} |u|^{\frac{n}{n-1}} \, dx \leq \prod_{i=1}^n \left(\int_{\mathbf{R}^n} |\nabla u| \, dx \right)^{\frac{1}{n-1}}$$

or

(1.47)
$$\|u\|_{\frac{n}{n-1}} \leq \int_{\mathbf{R}^n} |\nabla u| \, dx,$$

which is the desired result in the case $p = 1$. The case when $1 \leq p < n$ is treated by replacing u by positive powers of $|u|$. Assume still that $u \in \mathcal{C}_c^1(\Omega)$, denote $q = p^*/1^*$ and apply our previous step to $v := |u|^q$, which is, of course, also in $\mathcal{C}_c^1(\mathbf{R}^n)$. Then our result follows from

$$\|u\|_{p^*}^q = \|v\|_{1^*} \leq \|\nabla v\|_1 = q \int_{\mathbf{R}^n} |u|^{q-1} |\nabla u| \, dx$$
$$\leq q \left(\int_{\mathbf{R}^n} |u|^{p^*} \right)^{1/p'} \left(\int_{\mathbf{R}^n} |\nabla u|^p \right)^{1/p} = q \|u\|_{p^*}^{p^*/p'} \|\nabla u\|_p,$$

where we have used also Hölder's inequality.

Finally, if $u \in W_0^{1,p}(\Omega)$, let $\{u_i\}$ be a sequence of functions in $\mathcal{C}_c^1(\mathbf{R}^n)$ converging to u in the Sobolev norm. Then, with $p^* = np/(n-p)$, the previous step applied to $u_i - u_j$ implies

$$\|u_i - u_j\|_{p^*;\Omega} \leq C \|u_i - u_j\|_{1,p;\Omega}.$$

We leave it as an exercise for the reader to conclude that $u_i \to u$ in $L^{p^*}(\Omega)$. In view of the fact that $|\nabla u_i| \to |\nabla u|$ in $L^p(\Omega)$, our result now follows. $\square$

We sometimes refer to the following corollary as Sobolev's inequality while Poincaré inequality is used for Theorem 1.51, Corollary 1.57 and other similar estimates when $q = p$.

1.57 COROLLARY. *Suppose that $|\Omega| < \infty$, $p, q \geq 1$, $\alpha := \frac{1}{q} - \frac{1}{p} + \frac{1}{n} \geq 0$. Then there is $C = C(n,q)$ such that*

$$\|u\|_q \leq C \|\nabla u\|_p |\Omega|^\alpha$$

for each $u \in W_0^{1,p}(\Omega)$.

PROOF. Set

$$s = \begin{cases} 1, & \text{if } qn \leq q+n, \\ \frac{qn}{q+n}, & \text{if } qn > q+n. \end{cases}$$

Then

$$1 \leq s \leq p, \ s < n \quad \text{and} \quad s^* \geq q.$$

Using Theorem 1.56 and Hölder's inequality we get

$$\|u\|_q \leq \|u\|_{s^*} |\Omega|^{\frac{1}{q} - \frac{1}{s^*}} = \|u\|_{s^*} |\Omega|^{\frac{1}{q} - \frac{1}{s} + \frac{1}{n}} \leq C \|\nabla u\|_s |\Omega|^{\frac{1}{q} - \frac{1}{s} + \frac{1}{n}}$$
$$\leq C \|\nabla u\|_p |\Omega|^{\frac{1}{q} - \frac{1}{p} + \frac{1}{n}}$$

as required.

1.58 COROLLARY. *Suppose that* $1 \leq p \leq n$, $p \leq q < p^*$. *If* $u \in W^{1,p}(\mathbf{R}^n)$, *then* $u \in L^q(\mathbf{R}^n)$ *and*
$$\|u\|_q \leq C\|u\|_{1,p}$$
where $C = C(n,q)$.

PROOF. We may assume that $u \geq 0$ and $\|u\|_{1,p} = 1$. Then we decompose u as $u = u_1 + u_2$, where $u_1 = \min\{u, 1\}$. Since $0 \leq u_1 \leq 1$, we have

(1.48) $$\int_{\mathbf{R}^n} u_1^q \, dx \leq \int_{\mathbf{R}^n} u_1^p \, dx \leq 1.$$

Set $E = \{u_2 \neq 0\}$. Then

(1.49) $$|E| = |\{u_2 > 0\}| = |\{u > 1\}| \leq \int_{\{u>1\}} u^p \, dx \leq 1.$$

As in the proof of Corollary 1.57 we find an exponent s such that
$$1 \leq s \leq p, \; s < n \; \text{ and } \; s^* \geq q.$$
Using (1.49) and Hölder's inequality we obtain
$$\left(\int_{\mathbf{R}^n} |\nabla u_2|^s\right)^{1/s} = \left(\int_E |\nabla u|^s\right)^{1/s}$$
$$\leq \left(\int_E |\nabla u|^p\right)^{1/p} \leq 1$$
and thus, by Theorem 1.56 and Hölder's inequality

(1.50) $$\left(\int_{\mathbf{R}^n} |u_2|^q\right)^{1/q} = \left(\int_E |u_2|^q\right)^{1/q} \leq \left(\int_E |u_2|^{s^*}\right)^{1/s^*}$$
$$\leq C\left(\int_{\mathbf{R}^n} |\nabla u_2|^s\right)^{1/s} \leq C.$$

From (1.48) and (1.50) we obtain the required estimate. □

As an immediate consequence of Corollary 1.58 we obtain the imbedding $W^{1,p}(\Omega) \to L^q(\Omega)$ for the class of domains introduced in the following definition.

1.59 DEFINITION. We say that Ω is a $(1,p)$-*extension domain* if there exists a bounded linear operator
$$L : W^{1,p}(\Omega) \to W^{1,p}(\mathbf{R}^n)$$
such that $L(u) \, \llcorner \, \Omega = u$ for all $u \in W^{1,p}(\Omega)$. If the context makes it clear what index p is under consideration, for brevity we will use the term extension domain rather than $(1,p)$-extension domain.

1.60 REMARK. On several occasions (e.g. in connection with Dirichlet problem for elliptic equations) it will be useful to easily recognize those domains that are extension domains.

A fundamental result of Calderón–Stein states that every Lipschitz domain is an extension domain. An open set Ω is a Lipschitz domain if its boundary can be locally represented as the graph of a Lipschitz function defined on some open ball of $\mathbf{R}^{n-1}$. This result was proved by Calderón, [**Ca**], when $1 < p < n$ and Stein, [**St3**], extended Calderón's result to $p = 1, \infty$. Later, Jones, [**Jo**], introduced a class of domains that includes Lipschitz domains, called (ε, δ) domains, which he proved are extension domains for Sobolev functions. A domain Ω is called an (ε, δ) domain if whenever $x, y \in \mathbf{R}^n$ and $|x - y| < \delta$, there is a rectifiable arc $\gamma \subset \Omega$ joining x to y and satisfying

$$H^1(\gamma) \leq \varepsilon^{-1}|x - y|$$

and

$$\operatorname{dist}(z, \mathbf{R}^n - \Omega) \geq \frac{\varepsilon|x - z||y - z|}{|x - y|} \quad \text{for all } z \text{ on } \gamma.$$

One of the interesting results he obtained is the following: If $\Omega \subset \mathbf{R}^2$ is finitely connected, then Ω is an extension domain if and only if it is an (ε, δ) domain for some values of $\varepsilon, \delta > 0$.

The reader should also consult the book by Gol'dshtein and Reshetnyak, [**GR1**], for other treatments of extension domains. Here it is proved that a wide class of domains (class J) are extension domains.

Now we will show that the imbedding into $L^q(\Omega)$ possesses a compactness property if q is strictly less than the borderline exponent p^*. Specifically, we will show that the injection map from $W^{1,p}(\mathbf{R}^n)$ into $L^q(\Omega)$, $q < p^*$, has the property that the closure of the image of arbitrary bounded set in $W^{1,p}(\mathbf{R}^n)$ is compact in the range space. A mapping with this property is called *compact*.

1.61 THEOREM (KONDRACHOV-RELLICH). *Suppose $\Omega \subset \mathbf{R}^n$ is an open set with $|\Omega| < \infty$, $1 \leq p \leq n$ and $1 \leq q < p^*$. Then the restriction map*

$$u \mapsto u \, \llcorner \, \Omega$$

induces a compact mapping

$$W^{1,p}(\mathbf{R}^n) \to L^q(\Omega).$$

In particular, $W_0^{1,p}(\Omega)$ is compactly imbedded in $L^q(\Omega)$; if Ω is an extension domain, then $W^{1,p}(\Omega)$ is compactly imbedded in $L^q(\Omega)$.

PROOF. Clearly, it is sufficient to show that $\{u \, \llcorner \, \Omega : u \in A\}$ is compact in $L^q(\Omega)$, where

$$A := \{u \in W^{1,p}(\mathbf{R}^n) \colon \|u\|_{1,p;\mathbf{R}^n} \leq 1\}.$$

For $\varepsilon > 0$, we invoke our customary notation u_ε to denote the mollification of u, see (1.12). Notice that by Theorem 1.12 (iii), $u_\varepsilon \in A$ whenever $u \in A$. Then, for $u \in A$ and any $x \in \mathbf{R}^n$,

(1.51)
$$|u_\varepsilon(x)| \leq \int_{B(0,\varepsilon)} |u(x-y)| \, \phi_\varepsilon(y) \, dy$$
$$\leq \|u\|_p \|\phi_\varepsilon\|_{p'}.$$

Similarly,

(1.52)
$$|\nabla u_\varepsilon(x)| \leq \int_{B(0,\varepsilon)} |u(x-y)| \, |\nabla \phi_\varepsilon(y)| \, dy$$
$$\leq \|\nabla u\|_p \|\phi_\varepsilon\|_{p'}.$$

Thus, for arbitrary $u \in A$, u_ε is a Lipschitz function bounded by a constant depending only on n, p, ε and ϕ. Furthermore, its Lipschitz constant is bounded by $C(n,p,\varepsilon,\phi)$. Consequently, it follows that $A_\varepsilon := \{u_\varepsilon : u \in A\}$ is a bounded, equicontinuous subset of $\mathcal{C}(\mathbf{R}^n)$. Hence, the Arzelà-Ascoli compactness theorem implies that each sequence in A_ε has a subsequence that converges uniformly on compact subsets of $\mathbf{R}^n$.

Now, we will show that $\{u \llcorner \Omega : u \in A\}$ is compact in $L^1(\Omega)$. consider an exponent s such that $\min\{p,q\} < s < p^*$. Let $\{u_k\}$ be a sequence in A. Then, by Corollary 1.58, $\{u_k\}$ is bounded in $L^s(\mathbf{R}^n)$ and after passing to a subsequence, we conclude that there exist $u \in L^s(\mathbf{R}^n)$ such that

$$(1.53) \qquad u_k \to u \quad \text{weakly in} \quad L^s(\mathbf{R}^n).$$

The weak convergence of u_k to u implies that $\phi_\varepsilon * u_k(x) \to \phi_\varepsilon * u(x)$ for all $x \in \Omega$. Moreover, as we showed above, we may assume that $\{\phi_\varepsilon * u_k\}$ converges uniformly to $\phi_\varepsilon * u$ on compact subsets of $\mathbf{R}^n$. According to Theorem 1.52,

$$(1.54) \qquad \|u - u_\varepsilon\|_{1;\Omega} \leq \|u - u_\varepsilon\|_p |\Omega|^{1/p'} \leq C \varepsilon |\Omega|^{1/p'} \|\nabla u\|_p \leq C \varepsilon |\Omega|^{1/p'}.$$

We fix ε and denote

$$v_k = \phi_\varepsilon * u_k, \quad v = \phi_\varepsilon * u.$$

We claim that $v_k \to v$ in $L^1(\Omega)$. This is so because

$$(1.55) \quad \begin{aligned} \|v_k - v\|_{1;\Omega} &= \|v_k - v\|_{1;B(0,R)\cap\Omega} + \|v_k - v\|_{1;\Omega\setminus B(0,R)} \\ &\leq \|v_k - v\|_{1;B(0,R)\cap\Omega} + \|v_k - v\|_s |\Omega \setminus B(0,R)|^{1/s'} \\ &\leq \|v_k - v\|_{1;B(0,R)\cap\Omega} + C |\Omega \setminus B(0,R)|^{1/s'} \end{aligned}$$

for any ball $B(0,R)$. Since the first term on the right of (1.55) tends to 0 as $k \to \infty$, and the second one can be made arbitrarily small by choosing R sufficiently large, our claim is established.

By passing to another subsequence, we may assume that

$$\|\phi_{\varepsilon_k} * u_k - \phi_{\varepsilon_k} * u\|_{1;\Omega} < 1/k$$

where $\varepsilon_k := 1/k$. Thus, appealing to (1.54), we see that

$$\|u - u_k\|_{1;\Omega} \leq \|u - \phi_{\varepsilon_k} * u\|_{1;\Omega} + \|\phi_{\varepsilon_k} * u - \phi_{\varepsilon_k} * u_k\|_{1;\Omega} + \|\phi_{\varepsilon_k} * u_k - u_k\|_{1;\Omega}$$

tends to 0 as $k \to \infty$.

This establishes our result for $q = 1$. If $1 \leq q < p^*$ is arbitrary, with the exponent s as above, we use (1.7), (1.53) and the preceding part of the proof to obtain

$$(1.56) \qquad \|u_k - u\|_{q;\Omega} \leq \|u_k - u\|_{1;\Omega}^\alpha \|u_k - u\|_{s;\Omega}^{1-\alpha} \to 0.$$

Hence the imbedding $W_0^{1,p}(\Omega) \to L^q(\Omega)$ is compact. $\square$

1.62 THEOREM. Let $u \in W_0^{1,p}(\Omega)$ with $p > n$. Then $u \in C^{0,\alpha}(\overline{\Omega})$ where $\alpha = 1 - n/p$, and

(1.57) $$|u(x) - u(x')| \leq C|x - x'|^\alpha$$

for each $x, x' \in \Omega$ with $C = C(n,p)$. If $|\Omega| < \infty$, then $W_0^{1,p}(\Omega)$ is compactly imbedded in $C_0(\Omega)$ and there is $C = C(n,p)$ such that

$$\sup_\Omega |u| \leq C |\Omega|^{1/n - 1/p} \|\nabla u\|_{p;\Omega}$$

for any $u \in W_0^{1,p}(\Omega)$.

PROOF. We may extend the domain of u be setting $u = 0$ outside Ω. An application of Hölder's inequality shows that $|\nabla u|$ satisfies the condition of Theorem 1.53, which proves (1.57). Alternatively, by Lemma 1.50

$$|u(x) - \overline{u}_B| \leq C \int_B |x - y|^{1-n} |\nabla u(y)| \, dy$$

for any ball $B = B(z, r) \subset \mathbf{R}^n$ and almost every $x \in B$. Using Hölder's inequality we get

$$|u(x) - \overline{u}_B| \leq C \Big(\int_B |\nabla u|^p \, dy\Big)^{1/p} \Big(\int_B |x - y|^{(1-n)p'} \, dy\Big)^{1/p'} \leq C\|u\|_p \, r^\alpha.$$

Now, assume that $|\Omega| < \infty$. For the proof that $u \in C_0(\Omega)$, a standard approximation argument shows that this is enough to prove the supremum estimate and decay estimate for $|x| \to \infty$ when $u \in C_c^1(\Omega)$. We find r such that $|\Omega| = |B(0, r)|$. If $x \in \Omega$, we choose a point $x' \in \partial\Omega$ such that $|x - x'| = \text{dist}(x, \mathbf{R}^n \setminus \Omega)$. Then

$$\boldsymbol{\alpha}(n)|x - x'|^n = |B(x, |x - x'|)| \leq |B(x, r) \cap \Omega|.$$

Hence

$$|u(x)| = |u(x) - u(x')| \leq C|x - x'|^\alpha \leq C\big(\boldsymbol{\alpha}(n)^{-1} |B(x, r) \cap \Omega|\big)^{\alpha/n},$$

which is the decay estimate independent of u. In particular, we get the supremum bound

$$\sup_\Omega |u| \leq C |\Omega|^{\alpha/n} = C |\Omega|^{1/n - 1/p}.$$

Using all obtained estimates, the compactness of the imbedding follows easily from the Arzelà-Ascoli compactness theorem. □

In the following result, we give an elementary proof that an open ball is an extension domain.

1.63 THEOREM. Let $u \in W^{1,p}(B(x,r))$. Then there exists $v \in W^{1,p}(\mathbf{R}^n)$ such that $v = u$ on $B(x,r)$, $\text{spt}\, v \subset B(x, 3r/2)$ and

$$\int_{B(x, 3r/2)} |\nabla v|^p \, dy \leq C \int_{B(x,r)} \big(|\nabla u|^p + r^{-p}|u|^p\big) \, dy$$

where $C = C(n,p)$. The mapping $u \mapsto v$ is linear.

PROOF. We may assume that $x = 0$. Define w by

$$w(y) = \begin{cases} \left(\frac{2|x|}{r} - 1\right) u(x) & \text{for } |x| \geq r/2 \\ 0 & \text{for } |x| < r/2 \end{cases}$$

and observe that

$$|\nabla w(y)| \leq |\nabla u(y)| + \tfrac{2}{r} u(y)$$

for a.e. $y \in B(0, r) \setminus B(0, r/2)$. Hence,

(1.58) $$\int_{B(0,r)} |\nabla w|^p \leq C \int_{B(0,r)} \left(|\nabla u|^p + r^{-p} |u|^p\right) dy.$$

Now define $T: \{y : r \leq |y| \leq 3r/2\} \to \{y : r/2 \leq |y| \leq r\}$ by

$$T(y) = \left(\frac{2r}{|y|} - 1\right) y$$

and note that $\|\nabla T\| = 1$ and the Jacobian determinant JT satisfies $JT \geq 3^{1-n}$. Let

$$v(y) = \begin{cases} 0 & \text{for } |y| > 3r/2 \\ w \circ T(y) & \text{for } r \leq |y| \leq 3r/2 \\ u(y) & \text{for } |y| < r. \end{cases}$$

Then,

$$\int_{\{r \leq |y| \leq 3r/2\}} |\nabla v|^p \, dy \leq 3^{n-1} \int_{\{r \leq |y| \leq 3r/2\}} (|\nabla w \circ T|)^p \cdot JT$$

$$= 3^{n-1} \int_{\{r/2 \leq |y| \leq r\}} |\nabla w|^p \, dy.$$

If $u \in \mathcal{C}^1(\overline{B})$, then v is Lipschitz on $\mathbf{R}^n$ and a reference to (1.58) yields the result. In the general case we use approximation based on Lemma 1.48. □

1.64 COROLLARY (SOBOLEV-POINCARÉ'S INEQUALITY). *Let $B = B(x, r)$ and suppose that $1 \leq p \leq n$ and $1 \leq q \leq p^*$; if $p = n$, assume $1 \leq q < \infty$. Then, if $u \in W^{1,p}(B)$,*

(1.59) $$\left(\fint_B |u(y) - \overline{u}_B|^q \, dy\right)^{1/q} \leq Cr \left(\fint_B |\nabla u(y)|^p \, dy\right)^{1/p},$$

where $C = C(n, p, q)$.

PROOF. First we treat the case of $p < n$ and $q = p^*$. Without loss of generality, assume $\int_{B(x,r)} u \, dy = 0$. We know from Theorem 1.51 that

$$\int_B |u|^p \, dy = \int_B |u - \overline{u}_B|^p \, dy \leq Cr^p \int_B |\nabla u|^p \, dy.$$

Now refer to Theorem 1.63 to find a function $v \in W^{1,p}(\mathbf{R}^n)$ that agrees with u on B, $\operatorname{spt} v \subset 2B$ and

$$\int_{2B} |\nabla v|^p \, dy \leq C \int_B |\nabla u|^p \, dy + Cr^{-p} \int_B |u|^p \, dy.$$

By Sobolev's inequality

$$\left(\int_B |u|^{p^*}\,dy\right)^{1/p^*} \leq \left(\int_{2B} |v|^{p^*}\,dy\right)^{1/p^*} \leq C\left(\int_{2B} |\nabla v|^p\,dy\right)^{1/p}$$
$$\leq C\left(\int_B |\nabla u|^p\,dy + r^{-p}\int_B |u|^p\,dy\right)^{1/p} \leq C\left(\int_B |\nabla u|^p\,dy\right)^{1/p},$$

which is clearly equivalent to (1.59) with $q = p^*$. If $p = n$ or $q < p^*$, we use an exponent $s \in [1,p]$ such that $q \leq s^* < \infty$. Using Hölder's inequality and the preceding part we get

$$\left(\fint_B u^q\right)^{1/q} \leq \left(\fint_B u^{s^*}\right)^{1/s^*} \leq Cr\left(\fint_B |\nabla u|^s\right)^{1/s}$$
$$\leq Cr\left(\fint_B |\nabla u|^p\right)^{1/p}. \qquad \square$$

As an application of Theorem 1.61, we give another modification of Sobolev-Poincaré's Inequality (Corollary 1.64). This version holds for a more general domain, but doesn't yield an explicit estimate for the constant.

1.65 LEMMA. *Let $\Omega \subset \mathbf{R}^n$ be a connected extension domain with $|\Omega| < \infty$. Suppose that $1 \leq p \leq n$, $1 \leq q < p^*$ and $\gamma > 0$. Then there is a constant $C = C(n,p,q,\Omega,\gamma)$ such that*

(1.60)
$$\|u - \overline{u}_E\|_{q;\Omega} \leq C \|\nabla u\|_{p;\Omega}$$

for any $u \in W^{1,p}(\Omega)$ and any measurable set $E \subset \Omega$ with $|E| \geq \gamma$.

PROOF. The proof proceeds by contradiction. If (1.60) were false, then for each integer j there is a function $u_j \in W^{1,p}(\Omega)$ and a measurable set $E_j \subset \Omega$ with the property that
$$\|u_j - \overline{u_j}_{E_j}\|_{q;\Omega} \geq j \|\nabla u_j\|_{p;\Omega}.$$
Clearly, we may assume that
$$\fint_{E_j} u_j\,dx = 0 \quad \text{and} \quad \int_\Omega |u_j|^q\,dx = |\Omega|.$$
But then, there exist a subsequence (denoted by the full sequence) and $u \in W^{1,p}(\Omega)$ such that $\{u_j\}$ tends weakly to u in $W^{1,p}(\Omega)$. By the Kondrachov-Rellich compactness theorem, $\{u_j\}$ tends strongly to u in $L^1(\Omega)$, $L^q(\Omega)$ and in some $L^s(\Omega)$ with $s > 1$. Since $\|u_j\|_{q;\Omega} = |\Omega|^{1/q}$ it follows that $\|\nabla u_j\|_{p;\Omega} \to 0$ and therefore that $\|\nabla u\|_{p;\Omega} = 0$. Corollary 1.42 thus implies that u is constant on Ω. Since
$$\int_\Omega |u|^q\,dx = \lim_j \int_\Omega |u_j|^q\,dx = |\Omega|,$$
we get $u = 1$ on Ω. Now,

(1.61)
$$\left|\int_\Omega u_j\right| = \left|\int_{\Omega\setminus E_j} u_j\right| \leq \left(\int_\Omega |u_j|^s\,dx\right)^{1/s} |\Omega \setminus E_j|^{1-1/s}$$
$$\leq \left(\int_\Omega |u_j|^s\,dx\right)^{1/s} (|\Omega| - \gamma)^{1-1/s}.$$

Passing to limits in (1.61) we get

$$|\Omega| \leq |\Omega|^{1/s}(|\Omega| - \gamma)^{1-1/s},$$

which is a contradiction.

Lemma 1.50 and Theorem 1.36 immediately lead to the John-Nirenberg result, [**JN**].

1.66 THEOREM. *Let $u \in W^{1,1}(\Omega)$ where $\Omega \subset \mathbf{R}^n$ is convex. Suppose there is a constant M such that*

$$\int_{\Omega \cap B(r)} |\nabla u| \, dx \leq M r^{n-1}$$

for all balls $B(r)$. Then there exist positive constants σ_0 and C depending only on n such that

$$\int_{\Omega} \exp(\tfrac{\sigma}{M} |u - \overline{u}_\Omega|) \, dx \leq C(\operatorname{diam}\Omega)^n$$

whenever $\sigma < \sigma_0 |\Omega| (\operatorname{diam}\Omega)^{-n}$.

1.3.3 Pointwise differentiability of Sobolev functions.

Now we examine to what extent the existence of weak derivatives of Sobolev functions implies the existence of classical derivatives. A great deal more can be said about this than Theorem 1.71, but further exploration would take us too far afield.

We recall that the norm in $W^{1,\infty}(\mathbf{R}^n)$ is given by

$$\|u\|_{1,\infty} = \max\{\|u\|_\infty, \|\nabla u\|_\infty\}.$$

1.67 LEMMA. *Let $E \subset \mathbf{R}^n$ be a nonempty set and u be a function on E. Let a be a positive constant such that*

$$|u(y) - u(x)| \leq a|y - x| \quad \text{and} \quad |u(x)| \leq a \quad \text{for all } x, y \in E.$$

Then there is an extension v of u defined on $\mathbf{R}^n$ such that

$$|v(y) - v(x)| \leq a|y - x| \quad \text{and} \quad |v(x)| \leq a \quad \text{for all } x, y \in \mathbf{R}^n.$$

PROOF. The function

$$\tilde{v}(x) = \sup_{y \in E}\{u(y) - a|x - y|\}$$

is an extension of u defined on $\mathbf{R}^n$ and has the desired Lipschitz constant. Now, we set $v = \min\{a, \max\{-a, \tilde{v}\}\}$, i.e.

$$v(x) = \begin{cases} \tilde{v}(x), & \text{if } -a < \tilde{v}(x) < a, \\ a, & \text{if } \tilde{v}(x) \geq a, \\ -a, & \text{if } \tilde{v}(x) \leq -a. \end{cases}$$

□

1.68 LEMMA. *Let $u \in W^{1,1}(\mathbf{R}^n)$. Then for any $a > 0$, there is a Lipschitz function v on $\mathbf{R}^n$ such that*

(1.62) $$\|v\|_{1,\infty} \leq a,$$
(1.63) $$\|v\|_{1,1} \leq C\|u\|_{1,1},$$

and

(1.64) $$|\{u \neq v\}| \leq \frac{C\|u\|_{1,1}}{a}$$

with $C = C(n)$.

PROOF. Let
$$G_0 := \{x \in \mathbf{R}^n : |u(x)| > a\},$$
$$G_1 := \{x \in \mathbf{R}^n : M(\nabla u) > \theta a\},$$
where $\theta > 0$ will be specified later. Then
$$|G_0| \leq a^{-1} \int_{G_0} |u| \leq a^{-1} \|u\|_{1,1}$$
and by Theorem 1.21,
$$|G_1| \leq C \frac{\|\nabla u\|_1}{\theta a} \leq C \frac{\|u\|_{1,1}}{\theta a}.$$
Set
$$G = G_0 \cup G_2.$$
Let $B = B(z, r)$ be a ball in $\mathbf{R}^n$. By Lemma 1.50 and Theorem 1.32,
$$|u(x) - \overline{u}_B| \leq C \int_B \frac{|\nabla u(y)|}{|y-x|^{n-1}} \, dy$$
$$\leq C \int_{B(x,2r)} \frac{|\nabla u(y)|}{|y-x|^{n-1}} \, dy \leq CrM(\nabla u)(x)$$
holds for a.e. $x \in B$. It follows that chosen θ small enough, there is a set N of Lebesgue measure zero such that
$$|u(y) - u(x)| \leq C|y-x|\Big(M(\nabla u)(x) + M(\nabla u)(y)\Big) \leq C\theta a|y-x| \leq a|y-x|$$
holds for all $x, y \in \mathbf{R}^n \setminus (G \cup N)$. Now, by Lemma 1.67 there is a Lipschitz function v on $\mathbf{R}^n$ such that $v = u$ a.e. outside $G \cup N$ and
$$\|v\|_{1,\infty} \leq a.$$
Then
$$|\{u \neq v\}| \leq |G| < C\,a^{-1}\|u\|_{1,1}$$
and
$$\|v\|_{1,1} = \int_{\mathbf{R}^n \setminus G} (|u| + |\nabla u|) \, dx + \int_G (|v| + |\nabla v|) \, dx \leq \|u\|_{1,1} + 2a|G| \leq C\|u\|_{1,1}. \quad \square$$

The following theorem establishes a Luzin-type approximation for Sobolev functions.

1.69 Theorem. *Let $u \in W^{1,1}(\mathbf{R}^n)$. Then for any $a > 0$, there is a $\mathcal{C}^1$ function v on $\mathbf{R}^n$ such that*
$$\|v\|_{1,\infty} \leq a, \qquad \|v\|_{1,1} \leq C\|u\|_{1,1}$$
and
$$|\{u \neq v\}| \leq \frac{C\|u\|_{1,1}}{a}.$$

PROOF. By Lemma 1.68, there is a measurable set G_1 and a Lipschitz function w_1 such that
$$\|w_1\|_{1,1} \leq C\|u\|_{1,1}, \quad \|w_1\|_{1,\infty} \leq \frac{a}{2},$$
$$w_1 = u \text{ on } \mathbf{R}^n \setminus G_1 \quad \text{and} \quad |G_1| < \frac{2C}{a}\|u\|_{1,1}.$$

Let ϕ be a mollifier. It is easy to see that
$$\|\phi_\varepsilon * w\|_{1,\infty} \leq \|w\|_{1,\infty}$$
for any $w \in W^{1,\infty}(\mathbf{R}^n)$ and $\varepsilon > 0$. By mollifying of w_1 we find a function $v_1 \in \mathcal{C}^1(\mathbf{R}^n)$ such that
$$\|v_1\|_{1,\infty} \leq \|w_1\|_{1,\infty}, \quad \text{and} \quad \|v_1 - w_1\|_{1,1} < \frac{1}{4}\|u\|_{1,1}.$$

We proceed by induction. Consider $k > 1$ and suppose that $w_1, \ldots, w_{k-1} \in W^{1,\infty}(\mathbf{R}^n)$, $v_1, \ldots, v_{k-1} \in \mathcal{C}_1(\mathbf{R}^n)$ and $G_1, \ldots, G_{k-1} \subset \mathbf{R}^n$ satisfy

(1.65)
$$u = v_1 + \cdots + v_{j-1} + w_j \text{ on } \mathbf{R}^n \setminus (G_1 \cup \cdots \cup G_j),$$
$$\|w_j\|_{1,1} \leq C 4^{-j+1} \|u\|_{1,1},$$
$$|G_j| \leq C\, a^{-1} 2^{-j+2} \|u\|_{1,1},$$
$$\|v_j\|_{1,\infty} \leq \|w_j\|_{1,\infty} \leq 2^{-j} a, \quad \text{and}$$
$$\|v_j - w_j\|_{1,1} \leq 4^{-j} \|u\|_{1,1}$$

for each $j = 1, \ldots, k-1$. By Lemma 1.68, we construct $w_k \in W^{1,\infty}(\mathbf{R}^n)$ and a measurable set G_k such that

(1.66)
$$w_k = w_{k-1} - v_{k-1} \text{ on } \mathbf{R}^n \setminus G_k,$$
$$\|w_k\|_{1,1} \leq C\|w_{k-1} - v_{k-1}\|_{1,1},$$
$$\|w_k\|_{1,\infty} \leq a\, 2^{-k} \quad \text{and}$$
$$|G_k| \leq C \frac{\|w_{k-1} - v_{k-1}\|_{1,1}}{a\, 2^{-k}}.$$

By mollifying we obtain $v_k \in \mathcal{C}^1(\mathbf{R}^n)$ such that
$$\|v_k\|_{1,\infty} \leq \|w_k\|_{1,\infty} \quad \text{and} \quad \|v_k - w_k\|_{1,1} \leq 4^{-k}\|u\|_{1,1}.$$

It is easy to observe that the properties (1.65) are then satisfied also for $j = k$. Set
$$v = \sum_{j=1}^{\infty} v_j.$$

The sum is convergent both in $W^{1,1}(\mathbf{R}^n)$ and $W^{1,\infty}(\mathbf{R}^n)$. Due to uniform convergence of the sum of derivatives of v_j's, $v \in \mathcal{C}^1(\mathbf{R}^n)$. We have

$$\|v\|_{1,1} \leq \sum_j \|w_j\|_{1,1} + \sum_j \|v_j - w_j\|_{1,1} \leq C\|u\|_{1,1},$$

$$\|v\|_{1,\infty} \leq \sum_j \|v_j\|_{1,\infty} \leq a,$$

and

$$|\{v \neq u\}| \leq \sum_j |G_j| \leq C \frac{\|u\|_{1,1}}{a}. \qquad \square$$

1.70 DEFINITION. Recall that a point $x_0 \in E$ is said to be a (Lebesgue) *density point* of a measurable set $E \subset \mathbf{R}^n$ if

$$\lim_{r \to 0} \frac{|E \cap B(x_0, r)|}{|B(x_0, r)|} = 1.$$

We say that a function $u\colon \Omega \to \mathbf{R}$ is *approximately differentiable* at $x_0 \in \mathbf{R}^n$ if there exists a linear mapping $L\colon \mathbf{R}^n \to \mathbf{R}$ and a measurable set $E \subset \Omega$ such that x_0 is a density point of E and

$$\lim_{\substack{x \to x_0 \\ x \in E}} \frac{|u(x) - u(x_0) - L(x - x_0)|}{|x - x_0|} = 0.$$

1.71 THEOREM. *Any weakly differentiable function u on Ω is approximately differentiable a.e. in Ω and the approximate derivative coincides a.e. with the weak derivative.*

PROOF. Since the properties in consideration are local, we may suppose that $\Omega = \mathbf{R}^n$ and $u \in W^{1,1}(\mathbf{R}^n)$. Let Z be the set of all points where u is not approximately differentiable. Choose $\varepsilon > 0$. By Theorem 1.69, there is a measurable set E and a function $v \in \mathcal{C}^1(\mathbf{R}^n)$ such that $v = u$ on E and $|\mathbf{R}^n \setminus E| < \varepsilon$. Clearly, $\nabla u = \nabla v$ a.e. in $\{u = v\}$. If $x \in E$ is a point of density of E such that $\nabla u(x) = \nabla v(x)$, then obviously $\nabla u(x)$ is an approximate derivative of u at x. This shows that $|Z \cap E| = 0$, and thus $|Z| < \varepsilon$. $\qquad \square$

1.72 THEOREM. *Let $u \in W^{1,p}_{\mathrm{loc}}(\Omega)$ with $n < p \leq \infty$. Then u is differentiable a.e. in Ω and the pointwise derivative coincide a.e. with the weak derivative.*

PROOF. We may suppose that $p < \infty$. Let us consider $x_0 \in \Omega$ such that $\nabla u(x_0)$ is an approximate derivative at x_0 and

$$a = a(x_0) := \left(M(|\nabla u|^p)(x_0)\right)^{1/p} < \infty.$$

By Theorem 1.71 and Theorem 1.21, this is the case at almost every $x_0 \in \Omega$. Let E be a measurable set such that x_0 is a density point for E and

$$\lim_{\substack{x \to x_0 \\ x \in E}} \frac{u(x) - u(x_0) - \nabla u(x_0) \cdot (x - x_0)}{|x - x_0|} = 0.$$

Choose $\varepsilon \in (0, 1/2)$. Find $\delta > 0$ such that

(1.67) $\quad |u(x) - u(x_0) - \nabla u(x_0) \cdot (x - x_0)| \le \varepsilon |x - x_0| \quad$ for all $x \in B(x_0, 2\delta)$,

and
$$|B(x_0, r) \setminus E| < \varepsilon^n |B(x_0, r)| \quad \text{for each } r \in (0, 2\delta).$$
Let $x \in B(x_0, \delta)$ and $r := 2|x - x_0|$. Notice that $B(x, \varepsilon r) \subset B(x_0, r)$. Using Lemma 1.50, we obtain

(1.68)
$$\begin{aligned}
|u(z) - \overline{u}_{B(x,\varepsilon r)}| &\le C \int_{B(x,\varepsilon r)} \frac{|\nabla u(y)|}{|z-y|^{n-1}} \, dy \\
&\le C \Big(\int_{B(x,\varepsilon r)} |\nabla u(y)|^p \, dy \Big)^{1/p} \Big(\int_{B(x,\varepsilon r)} |y-z|^{-p'(n-1)} \, dy \Big)^{1/p'} \\
&\le C \Big(\int_{B(x_0, r)} |\nabla u(y)|^p \, dy \Big)^{1/p} (\varepsilon r)^{1-\frac{n}{p}} \\
&\le C \left(a^p r^n \right)^{1/p} \varepsilon^{1-\frac{n}{p}} r^{1-\frac{n}{p}} \\
&\le C a \varepsilon^{1-\frac{n}{p}} |x - x_0|
\end{aligned}$$

for each $z \in B(x, \varepsilon r)$. Since
$$|B(x, \varepsilon r)| = \varepsilon^n |B(x_0, r)| > |B(x_0, r) \setminus E|,$$
there exists $z_0 \in E \cap B(x, \varepsilon r)$. From (1.68) we get

(1.69) $\qquad\qquad |u(x) - u(z_0)| \le C a \varepsilon^{1-\frac{n}{p}} |x - x_0|.$

By (1.67) and (1.69),
$$\begin{aligned}
|u(x) - u(x_0) - \nabla u(x_0) \cdot (x - x_0)| &\le |u(x) - u(z_0)| \\
&+ |u(z_0) - u(x_0) - \nabla u(x_0) \cdot (x_0 - z_0)| + |\nabla u(x_0)||x - z_0| \\
&\le C \left(a\varepsilon^{1-\frac{n}{p}} + 2\varepsilon + 2\varepsilon |\nabla u(x_0)| \right) |x - x_0|.
\end{aligned}$$

It follows that u is differentiable at x_0. $\qquad\square$

1.73 COROLLARY (RADEMACHER'S THEOREM). *Let u be a Lipschitz function on Ω. Then u is differentiable a.e. in Ω and the pointwise derivative coincide a.e. with the weak derivative.*

The following theorem is called a *chain rule*. It is useful for checking validity of test functions in the theory of weak solutions.

1.74 THEOREM. *Let f be a Lipschitz function on $\mathbf{R}$ such that $f(0) = 0$.*
(i) *If $u \in W^{1,1}_{\text{loc}}(\Omega)$, then $f \circ u \in W^{1,1}_{\text{loc}}(\Omega)$. Moreover, for a.e. $x \in \Omega$ we have either*

(1.70) $\qquad\qquad \nabla(f \circ u)(x) = f'(u(x)) \, \nabla u(x)$

or

(1.71) $\qquad\qquad \nabla(f \circ u)(x) = \nabla u(x) = 0.$

(ii) *If $u \in W^{1,p}_0(\Omega)$, then $f \circ u \in W^{1,p}_0(\Omega)$ and*
$$\|f \circ u\|_{1,p} \le \|f'\|_\infty \|u\|_{1,p}.$$

PROOF. (i) As a first step in the proof, we suppose in addition that $u \in \mathcal{C}^1(\mathbf{R}^n)$. Then $u \circ f$ is locally Lipschitz and by Rademacher's theorem (Theorem 1.73), $u \circ f$ is differentiable a.e. and the derivative a.e. coincides with the weak derivative. Therefore, it is easily seen that

$$\nabla(f \circ u)(x) = f'(u(x))\, \nabla u(x)$$

holds a.e. on the set where $f'(u(x))$ exists. By Theorem 1.73, there is a Borel measurable set $N \subset \mathbf{R}$ such that $|N| = 0$ and f' exists outside N. Denote

$$E = \{x : u(x) \in N\},$$
$$U_i = \{x : D_i u \neq 0\}, \quad i = 1, \ldots, n,$$
$$Z = \{x : \nabla u(x) = 0\}.$$

The set E is measurable. We have $\nabla(f \circ u) = \nabla u = 0$ a.e. in $E \cap Z$. Let $x \in U_i$. Then there is $\delta > 0$ such that $x + t e_i \in U_i$ for each $t \in [-\delta, \delta]$ and

$$t \mapsto u(x + t e_i)$$

is strictly monotone. From the classical one-dimensional change of variable formula we obtain

$$\int_{-\delta}^{\delta} |D_i u| \chi_{\{x+te_i \in E\}}\, dt = \int_{u(x-\delta e_i)}^{u(x+\delta e_i)} \chi_N\, dy = 0$$

and therefore,

(1.72) $$|\{t : x + t e_i \in U_i \cap E\}| = 0$$

for each $i = 1, \ldots, n$ and $x \in \mathbf{R}^n$. We use Fubini theorem to integrate (1.72) with respect to all variables different from i and obtain that $|U_i \cap E| = 0$. This concludes the proof of the particular case when $u \in \mathcal{C}^1(\mathbf{R}^n)$. Now, if u is in $W^{1,1}_{\mathrm{loc}}(\Omega)$, then by Theorem 1.41 we may suppose that u is absolutely continuous along almost all lines parallel to the coordinates axes. It is easily seen that $f \circ u$ has then the same property and from the obvious pointwise inequality $|D_i(f \circ u)| \leq \|f'\|_\infty |D_i u|$, it follows that $D_i(f \circ u)$ is locally integrable. It remains to prove that the either (1.70) or (1.71) holds for almost every $x \in \Omega$. This property is local, so we may assume that $u \in W^{1,1}(\mathbf{R}^n)$. By Theorem 1.69, given $\varepsilon > 0$, there is a function $v \in \mathcal{C}^1(\mathbf{R}^n)$ and a measurable set G with $|G| < \varepsilon$ such that

$$v = u \quad \text{outside} \quad G,$$

and

$$\|v\|_{1,1} \leq C \|u\|_{1,1}.$$

By the first part of the proof, the alternative (1.70) or (1.71) holds for v, so that by Lemma 1.43 the measure of the set where this does not hold for u is less that ε.

(ii) is an easy consequence of (i). □

1.3.4 Spaces $Y^{1,p}$.

Here we introduce spaces which correspond better to Riesz potentials (case $\alpha = 1$) than Sobolev spaces. They will be also useful later, in our development of capacities.

1.75 DEFINITION. With each open set $\Omega \subset \mathbf{R}^n$ and $1 \leq p < n$, the space $Y^{1,p}(\Omega)$ is defined as all the family of all weakly differentiable functions $u \in L^{p^*}(\Omega)$ whose weak derivatives are functions in $L^p(\Omega)$. The space $Y^{1,p}(\Omega)$ is endowed with the norm
$$\|u\|_Y := \|u\|_{p^*} + \|\nabla u\|_p.$$
Next, we define $Y_0^{1,p}(\Omega)$ as the closure of $\mathcal{C}_c^\infty(\Omega)$ in $Y^{1,p}(\Omega)$.

From Theorem 1.56 it follows that $W_0^{1,p}(\Omega) \subset Y_0^{1,p}(\Omega)$ and $W_0^{1,p}(\Omega) = Y_0^{1,p}(\Omega)$ when Ω has finite Lebesgue measure.

1.76 LEMMA. *Let $1 \leq p < n$. Then $Y_0^{1,p}(\mathbf{R}^n) = Y^{1,p}(\mathbf{R}^n)$.*

PROOF. Let $u \in Y^{1,p}(\mathbf{R}^n)$. For each positive integer k, let $\varphi_k \in \mathcal{C}_c^\infty(\mathbf{R}^n)$ be such that $0 \leq \varphi_k \leq 1$, $\varphi_k = 1$ on $B(0,k)$, $\operatorname{spt} \varphi_k \subset B(0,3k)$ and $|\nabla \varphi_k| \leq 1/k$. Then
$$\begin{aligned}\|\nabla(u\varphi_k) - \nabla u\|_p &\leq \|(\varphi_k - 1)\nabla u\|_p + \|u \nabla \varphi_k\|_p \\ &\leq \|\nabla u\|_{p; \mathbf{R}^n \setminus B(0,k)} \\ &\quad + \frac{1}{k} \|u\|_{p^*; \mathbf{R}^n \setminus B(0,k)} |B(0,3k)|^{1/p - 1/p^*} \to 0.\end{aligned}$$

Also
$$\|u\varphi_k - u\|_{p^*} \leq \|u\|_{p^*; \mathbf{R}^n \setminus B(0,k)} \to 0.$$

Now, each function $u\varphi_k$ belongs to $W_0^{1,p}(\mathbf{R}^n)$ and thus it can be approximated by functions from $\mathcal{C}_c^\infty(\Omega)$. □

1.77 COROLLARY. *Let $1 \leq p < n$. If $u \in Y^{1,p}(\mathbf{R}^n)$, then*
$$\|u\|_{p^*} \leq C \|\nabla u\|_p$$
where $C = C(n,p)$.

PROOF. This follows immediately from Theorem 1.56 and the previous result. □

1.78 THEOREM. *Suppose that $1 \leq p < n$. Let u be a weakly differentiable function on $\mathbf{R}^n$ such that $\nabla u \in L^p$. Then there is a unique $\xi \in \mathbf{R}$ such that $u - \xi \in Y^{1,p}(\mathbf{R}^n)$.*

PROOF. Suppose first that $u - \xi$ and $u - \xi'$ are in $Y^{1,p}(\mathbf{R}^n)$. Then $\xi - \xi' \in L^{p^*}(\mathbf{R}^n)$ and consequently $\xi = \xi'$. Thus the uniqueness is done.

For $j = 1, 2, \ldots$ we denote
$$B_j = B(0, 2^j), \quad \xi_j = \fint_{B_j} u \quad \text{and} \quad u_j = (u - \xi_j)\chi_{B_j}.$$

By Corollary 1.64,
$$(1.73) \qquad \int_{\mathbf{R}^n} u_j^{p^*} dx = \int_{B_j} |u - \xi_j|^{p^*} dx \leq C \left(\int_{B_j} |\nabla u|^p dx \right)^{p^*/p} \leq C \|\nabla u\|_p^{p^*}$$

and

$$|\xi_{j+1} - \xi_j| = \left|\fint_{B_j} (u - \xi_{j+1})\, dx\right|$$
$$\leq 2^n \fint_{B_{j+1}} |u - \xi_{j+1}|\, dx \leq 2^n \left(\fint_{B_{j+1}} |u - \xi_{j+1}|^{p^*}\, dx\right)^{1/p^*}$$
$$\leq C\, 2^{j(1-n/p)} \|u\|_{1,p}.$$

It follows that
$$\sum_j |\xi_{j+1} - \xi_j| < \infty$$
and thus there exists a limit
$$\xi = \lim_j \xi_j.$$
Now, using Fatou's lemma, we get from (1.73) that
$$\int_{\mathbf{R}^n} |u - \xi|^{p^*}\, dx \leq C \|\nabla u\|_p^{p^*}.$$

This shows that $u - \xi$ belongs to $L^{p^*}(\mathbf{R}^n)$ and thus also to $Y^{1,p}(\mathbf{R}^n)$. □

1.79 THEOREM. *Let $u \in Y^{1,p}(\mathbf{R}^n)$, $1 \leq p < n$. Then*

(1.74) $$|u(x)| \leq \frac{1}{n\boldsymbol{\alpha}(n)} I_1 * |\nabla u|\,(x)$$

for a.e. $x \in \mathbf{R}^n$.

PROOF. Suppose first that $u \in \mathcal{C}_c^\infty(\mathbf{R}^n)$. Then by Lemma 1.50,
$$|u(x)| \leq |u(x) - \overline{u}_B| + |\overline{u}_B| \leq \frac{1}{n\boldsymbol{\alpha}(n)} I_1 * |\nabla u|\,(x) + |\overline{u}_B|$$
where $B = B(x,k)$ and k is so large that $B \supset \operatorname{spt} u$. We have
$$\left|\fint_{B(x,k)} u\, dy\right| \leq \left(\fint_{B(x,k)} u^{p^*}\, dy\right)^{1/p^*} \leq |B(x,k)|^{-1/p^*} \|u\|_{p^*} \to 0$$
which establishes the result for smooth functions. In general case, we approximate u by a sequence $\{u_k\}$ of functions from $\mathcal{C}_c^\infty(\mathbf{R}^n)$ which tends to u in $Y^{1,p}(\mathbf{R}^n)$. By Theorem 1.33, $I_1 * |\nabla u_k - \nabla u| \to 0$ in $L^{p^*}(\mathbf{R}^n)$ and thus for a subsequence a.e. At the points of pointwise convergence we may pass to limit and obtain (1.74). □

Now we will treat the case $p = n$ which differs from $p < n$ in many ways.

1.80 DEFINITION. Suppose that $n > 1$. An open set $\Omega \subset \mathbf{R}^n$ is said to be an *n-Green domain* if $\{u \llcorner \Omega : u \in \mathcal{C}_c^\infty(\mathbf{R}^n)\} \not\subset W_0^{1,n}(\Omega)$.

For an arbitrary open set $\Omega \subset \mathbf{R}^n$, we define $Y_0^{1,n}(\Omega)$ as the set of all weakly differentiable functions on Ω such that $\nabla u \in L^n(\Omega)$ and $u\eta \in W_0^{1,n}(\Omega)$ for any $\eta \in \mathcal{C}_c^\infty(\mathbf{R}^n)$.

If Ω is an n-Green domain, we introduce a norm on $Y_0^{1,n}(\Omega)$ by

(1.75) $$u \mapsto \|\nabla u\|_n.$$

We will show later (Theorem 1.85) that this assignment actually defines a norm on $Y_0^{1,n}(\Omega)$ and that $Y_0^{1,n}(\Omega)$ is a Banach space.

If Ω is not an n-Green domain (in particular, if $\Omega = \mathbf{R}^n$), then all constants belong to $Y_0^{1,n}(\Omega)$ and the assignment (1.75) defines only a seminorm on $Y_0^{1,n}(\Omega)$.

Similarly as in the case $p < n$, $W_0^{1,n}(\Omega) \subset Y_0^{1,n}(\Omega)$ for any open set Ω.

The next theorem shows that at least domains of finite measure are n-Green domains. Later in Theorem 2.16 we will see that there is an abundance of n-Green domains. The second assertion of the following theorem, namely that $Y_0^{1,n}(\Omega) = W_0^{1,n}(\Omega)$, need not be valid for n-Green domains of infinite measure.

1.81 THEOREM. *Suppose that $|\Omega| < \infty$. Then Ω is a n-Green domain and $Y_0^{1,n}(\Omega) = W_0^{1,n}(\Omega)$.*

PROOF. By Poincaré's inequality (Corollary 1.57)

$$\|u\|_n \leq C \|\nabla u\|_n |\Omega|^{1/n} \tag{1.76}$$

for all $u \in W_0^{1,n}(\Omega)$. Let η_j, $j = 1, 2, \ldots$, be functions from $\mathcal{C}_c^\infty(\mathbf{R}^n)$ such that $0 \leq \eta_j \leq 1$, $\eta_j = 1$ on $B(0,j)$ and $|\nabla \eta_j| < 1/j$. Then

$$\|\nabla \eta_j\|_{n;\Omega} \to 0,$$

whereas

$$\|\eta_j\|_{n;\Omega} \to |\Omega|^{1/n}.$$

Appealing to (1.76), there must be j such that $\eta_j \llcorner \Omega \notin W_0^{1,n}(\Omega)$. This shows that Ω is an n-Green domain.

Obviously, $W_0^{1,n}(\Omega) \subset Y_0^{1,n}(\Omega)$. Conversely, suppose that $u \in Y_0^{1,n}(\Omega)$. Using (1.76), we derive that $\{u\eta_j\}$ is a fundamental sequence in $W_0^{1,n}(\Omega)$. Thus u, as its limit, belongs to $W_0^{1,n}(\Omega)$. □

1.82 LEMMA. *Suppose that $n > 1$ and $0 < r < R < \infty$. Let*

$$\eta(x) = \eta(x;r,R) = \begin{cases} \frac{\ln|x| - \ln R}{\ln r - \ln R}, & r < |x| < R, \\ 0, & |x| \geq R, \\ 1, & |x| \leq r. \end{cases} \tag{1.77}$$

Then η is Lipschitz on $\mathbf{R}^n$, $0 \leq \eta \leq 1$, $\eta = 1$ on $B(0,r)$, $\eta = 0$ outside $B(0,R)$ and

$$\int_{\mathbf{R}^n} |\nabla \eta|^n \, dx = n\boldsymbol{\alpha}(n) \ln^{1-n}(r/R).$$

PROOF. All assertions are easy to observe or compute. □

1.83 LEMMA. *Suppose that $n > 1$. Let $u \in Y_0^{1,p}(\Omega)$. Then there are $u_j \in \mathcal{C}_c^\infty(\Omega)$ such that*

$$\|\nabla u_j - \nabla u\|_n \to 0$$

and

$$\|u_j - u\|_{n;B \cap \Omega} \to 0$$

for each ball $B \subset \mathbf{R}^n$.

PROOF. By mollifying functions $\eta(\cdot; k^2, k^3)$ from Lemma 1.82 we get functions $\varphi_k \in \mathcal{C}_c^\infty(\mathbf{R}^n)$ such that $0 \leq \varphi_k \leq 1$, $\varphi_k = 1$ on $B(0, k)$ and
$$\|\nabla \varphi_k\|_n \to 0.$$
Let u satisfy the assumptions. If t_k denotes the truncated function
$$t_k(x) = \begin{cases} k, & u(x) > k, \\ u(x), & -k \leq u(x) \leq k, \\ -k & u(x) < -k, \end{cases}$$
then obviously
$$\|\nabla t_k - \nabla u\|_{n;\Omega} \to 0$$
and
$$\|t_k - u\|_{n;B \cap \Omega} \to 0$$
for any ball $B \subset \mathbf{R}^n$. Hence we may assume that u is bounded. We estimate
$$\|\nabla(u\varphi_k) - \nabla u\|_{n;\Omega} \leq \|(\varphi_k - 1)\nabla u\|_{n;\Omega} + \|u\nabla \varphi_k\|_{n;\Omega}$$
$$\leq \|\nabla u\|_{n;\Omega \setminus B(0,k)} + \|u\|_{\infty;\Omega} \|\nabla \varphi_k\|_n \to 0.$$
It is easily seen that also
$$\|u\varphi_k - u\|_{n;B \cap \Omega} \to 0$$
for any ball $B \subset \mathbf{R}^n$. Now, each function $u\varphi_k$ belongs to $W_0^{1,n}(\Omega)$ and thus it can be approximated by functions from $\mathcal{C}_c^\infty(\Omega)$. □

1.84 LEMMA. *Let Ω be an n-Green domain and $B \subset \mathbf{R}^n$ be a ball. Then there is a constant $C = C(\Omega, B)$ such that*

(1.78) $$\int_{\Omega \cap B} |u|^n \leq C \int_\Omega |\nabla u|^n$$

for all $u \in Y_0^{1,n}(\Omega)$.

PROOF. The proof is similar as that of Theorem 1.65. A routine approximation argument using Lemma 1.83 shows that it is enough to prove the inequality for functions in $\mathcal{C}_c^\infty(\Omega)$. We may assume that $|\Omega \cap B| > 0$. If (1.78) were false, then for each integer j there is a function $u_j \in \mathcal{C}_c^\infty(\Omega)$, with the property that

(1.79) $$\|u_j\|_{n;\Omega \cap B} \geq j \|\nabla u_j\|_{n;\Omega}.$$

Clearly, we may assume that

(1.80) $$\int_{\Omega \cap B} |u_j|^n \, dx = |\Omega \cap B|.$$

We set $u_j = 0$ outside Ω to get functions from $\mathcal{C}_c^\infty(\mathbf{R}^n)$. By (1.79),

(1.81) $$\int_{\mathbf{R}^n} \|\nabla u_j\|^n \to 0.$$

From Lemma 1.65 and (1.80), it follows that the sequence $\{u_j\}$ is bounded in $W^{1,n}(B')$ for each ball $B' \subset \mathbf{R}^n$. Since balls are extension domains, using the Kondrachov-Rellich compactness theorem (Theorem 1.61) and the idea of diagonalization (this means that if we define recurrently sequences $\{u_j^{(k)}\}$, where each

$\{u_j^{(k)}\}$ is a subsequence of $\{u_j^{(k-1)}\}$, then the diagonal sequence $\{u_j^{(j)}\}$ is a subsequence of each $\{u_j^{(k)}\}$ after neglecting finitely many terms) we obtain a subsequence (denoted as the full sequence) and a weakly differentiable function u on $\mathbf{R}^n$ such that
$$\|u_j - u\|_{n;B'} \to 0$$
for each ball $B' \subset \mathbf{R}^n$. Passing to limit in (1.81) and (1.80) we get that $\nabla u = 0$ in $\mathbf{R}^n$ and
$$\int_{\Omega \cap B} |u|^n \, dx = |\Omega \cap B|.$$
Appealing to Corollary 1.42, we deduce that u is constant and the constant value of u cannot be anything else than 1.

Now, since Ω is a n-Green domain, there is $\eta \in \mathcal{C}_c^\infty(\mathbf{R}^n) \setminus W_0^{1,n}(\Omega)$. From the above considerations it easily follows that $u_j \eta \to \eta$ in $W_0^{1,n}(\Omega)$, which is a contradiction. □

1.85 THEOREM. *Let $\Omega \subset \mathbf{R}^n$ be an n-Green domain. Then $Y_0^{1,n}(\Omega)$ is a Banach space when endowed with the norm*
$$u \mapsto \|\nabla u\|_{n;\Omega}.$$

PROOF. Let $u \in Y_0^{1,n}(\Omega)$ with $\|\nabla u\|_n = 0$. Then by Lemma 1.84, $\|u\|_{n;B \cap \Omega} = 0$ for any ball $B \subset \mathbf{R}^n$ and thus $u = 0$ a.e.

Now, let $\{u_j\}$ be a fundamental sequence in $Y_0^{1,n}(\Omega)$. Then by Lemma 1.84, $\{u_j\}$ is also a fundamental sequence in $L^n(B \cap \Omega)$ for each ball $B \subset \mathbf{R}^n$. It follows that there are $u \in L_{\text{loc}}^n(\Omega)$ and $g \in L^n(\Omega; \mathbf{R}^n)$ such that $u_j \to u$ in $L^n(\Omega \cap B)$ for each ball $B \subset \mathbf{R}^n$ and $\nabla u_j \to g$ in $L^n(\Omega; \mathbf{R}^n)$. Standard arguments show that $u \in Y_0^{1,n}(\Omega)$, $g = \nabla u$ and $u_j \to u$ in $Y_0^{1,n}(\Omega)$. □

1.3.5 Adams' inequality.

We establish here an extension of the Sobolev inequality established by D. Adams [Ad1]. It depends on a special case of the Marcinkiewicz interpolation theorem which is proved in the first part of the subsection.

As a first step we need the following result which, in the setting of a finite measure space, would run counter to an application of Hölder's inequality.

1.86 LEMMA. *If $f \geq 0$ is a non-increasing function on $(0, \infty)$, $0 < p \leq q \leq \infty$ and $\alpha \in \mathbf{R}$, then*
$$\left(\int_0^\infty (t^\alpha f(t))^q \frac{dt}{t} \right)^{1/q} \leq C \left(\int_0^\infty (t^\alpha f(t))^p \frac{dt}{t} \right)^{1/p},$$
where $C = C(p, q, \alpha)$.

PROOF. Since f is non-increasing, we have for any $t > 0$
$$t^\alpha f(t) \leq C \left(\int_{t/2}^t (s^\alpha f(s))^p \frac{ds}{s} \right)^{1/p},$$

which implies the desired result when $q = \infty$. The general result follows by writing
$$\int_0^\infty (t^\alpha f(t))^q \frac{dt}{t} \le \sup_{t>0}(t^\alpha f(t))^{q-p} \int_0^\infty (t^\alpha f(t))^p \frac{dt}{t}. \qquad \square$$

1.87 DEFINITION. If f is a measurable function defined on a measure space (X, μ), recall that its distribution function $\lambda_{f;\mu}(t)$ is defined by
$$\lambda_{f;\mu}(t) = \mu(\{x : |f(x)| > t\}), \quad t \ge 0.$$

1.88 LEMMA. Let $f \in L^p(X)$. For any $t > 0$, denote
$$f^t = f \chi_{\{|f|>ct\}}$$
and decompose f as $f = f^t + f_t$, where $f_t = f - f^t$. Then
$$\|f^t\|_p^p = t^p \lambda_{f;\mu}(t) + p \int_t^\infty \lambda_{f;\mu}(s) s^{p-1} ds,$$
$$\|f_t\|_p^p = p \int_0^t (\lambda_{f;\mu}(s) - \lambda_{f;\mu}(t)) s^{p-1} ds.$$

PROOF. Obviously
$$\{|f^t| > s\} = \begin{cases} \{f > s\}, & s \ge t, \\ \{f > t\}, & s < t \end{cases}$$
and
$$\{|f_t| > s\} = \begin{cases} \emptyset, & s \ge t, \\ \{f > s\} \setminus \{f > t\}, & s < t. \end{cases}$$
Hence the result follows from Theorem 1.9. $\qquad \square$

1.89 DEFINITION. Let (X, μ), (Y, ν) be measure spaces. Let T be a linear (or more generally, subadditive) functional defined on $L^p(Y)$ whose values are measurable functions on (X, μ). We will write $Tf := T(f)$. The operator T is said to be of *weak type* (p, q) if there is a constant C^T such that for any $f \in L^p(Y)$ and $t > 0$,
$$(1.82) \qquad \mu(\{x : |(Tf)(x)| > t\}) \le (t^{-1} C^T \|f\|_p)^q.$$

1.90 LEMMA. If $f = f_1 + f_2$ is a decomposition of $f \in L^p(Y)$ and $T : L^p(Y) \to L^p(X)$ is a subadditive operator, then for any $t > 0$,
$$\lambda_{Tf;\mu}(2t) \le \lambda_{Tf_1;\mu}(t) + \lambda_{Tf_2;\mu}(t).$$

PROOF. By subadditivity, $|Tf(x)| \le |Tf_1(x)| + |Tf_2(x)|$ for almost every x. If $|Tf(x)| > 2t$, then either $|Tf_1(x)| > t$ or $|Tf_2(x)| > t$. The result follows now from subadditivity of the measure. $\qquad \square$

We present here the Marcinkiewicz interpolation theorem. We will suppose that (p_0, q_0) and (p_1, q_1) are pairs of numbers such that
$$(1.83) \qquad \begin{aligned} & 1 \le p_i \le q_i < \infty, \quad i = 0, 1, \\ & p_0 < p_1, \quad \text{and} \quad q_0 \ne q_1. \end{aligned}$$

In the proof we restrict to the case

(1.84)
$$\frac{q_0}{p_0} = \frac{q_1}{p_1}$$

which is sufficient for our applications. For full strength of the Marcinkiewicz theorem we refer to [**St3**], Appendix B.

1.91 THEOREM. *Let T be a linear (or, more generally, subadditive) operator defined on $L^{p_0}(Y) + L^{p_1}(Y)$ whose values are measurable functions on (X, μ). Suppose T is simultaneously of weak types (p_0, q_0) and (p_1, q_1) with constants C_0^T, C_1^T. If $0 < \theta < 1$, and*

(1.85)
$$1/p = \frac{1-\theta}{p_0} + \frac{\theta}{p_1},$$
$$1/q = \frac{1-\theta}{q_0} + \frac{\theta}{q_1},$$

then T is of strong type (p, q); that is,

$$\|Tf\|_{q;\mu} \leq C \|f\|_p, \quad f \in L^p(Y),$$

where $C = C(C_0^T, C_1^T, p_0, q_0, p_1, q_1, \theta)$.

PROOF. As announced above, we will prove the theorem only under the assumption (1.84). Let $f \in L^p(Y)$. For each $t > 0$ we decompose $f = f^t + f_t$ as in Lemma 1.88. We first observe that $f^t \in L^{p_0}(Y)$ and $f_t \in L^{p_1}(Y)$, so that $f \in L^{p_0}(Y) + L^{p_1}(Y)$ and Tf is defined. By Theorem 1.9, Lemma 1.90 and (1.82) we have

$$\begin{aligned}
\|Tf\|_q &= \left(q \int_0^\infty s^{q-1} \lambda_{Tf;\mu}(s)\, ds \right)^{1/q} \\
&\leq C \left(\int_0^\infty t^{q-1} \lambda_{Tf;\mu}(2t)\, dt \right)^{1/q} \\
&\leq C \left(\int_0^\infty t^{q-1} \lambda_{Tf^t;\mu}(t)\, dt \right)^{1/q} \\
&\quad + C \left(\int_0^\infty t^{q-1} \lambda_{Tf_t;\mu}(t)\, dt \right)^{1/q} \\
&\leq C \left(\int_0^\infty t^{q-1-q_0} \|f^t\|_{p_0}^{q_0}\, dt \right)^{1/q} + C \left(\int_0^\infty t^{q-1-q_1} \|f_t\|_{p_1}^{q_1}\, dt \right)^{1/q} \\
&= J_1 + J_2.
\end{aligned}$$

Using Theorem 1.9, Lemma 1.86, Lemma 1.88 and Fubini's theorem we obtain

$$J_1 = C\Big(\int_0^\infty t^{q-q_0}\|f^t\|_{p_0}^{q_0}\frac{dt}{t}\Big)^{1/q} \leq C\Big(\int_0^\infty t^{p-p_0}\|f^t\|_{p_0}^{p_0}\frac{dt}{t}\Big)^{1/p}$$

$$= C\Big(\int_0^\infty t^{p-p_0}\Big[t^{p_0}\lambda_{f;\nu}(t) + p_0\int_t^\infty s^{p_0-1}\lambda_{f;\nu}(s)\,ds\Big]\frac{dt}{t}\Big)^{1/p}$$

$$\leq C\Big(\int_0^\infty t^p\lambda_{f;\nu}(t)\frac{dt}{t}\Big)^{1/p}$$

$$+ C\Big(\int_0^\infty \Big[\int_t^\infty t^{p-p_0-1}s^{p_0-1}\lambda_{f;\nu}(s)\,ds\Big]dt\Big)^{1/p}$$

$$\leq C\Big(\int_0^\infty s^{p-1}\lambda_{f;\nu}(s)\,ds\Big)^{1/p}$$

$$+ C\Big(\int_0^\infty \Big[\int_0^s t^{p-p_0-1}\,dt\Big]s^{p_0-1}\lambda_{f;\nu}(s)\,ds\Big)^{1/p}$$

$$\leq C\Big(\int_0^\infty s^{p-1}\lambda_{f;\nu}(s)\,ds\Big)^{1/p}$$

$$\leq C\|f\|_p.$$

Similarly,

$$J_1 = C\Big(\int_0^\infty t^{q-q_1}\|f_t\|_{p_1}^{q_1}\frac{dt}{t}\Big)^{1/q} \leq C\Big(\int_0^\infty t^{p-p_1}\|f_t\|_{p_1}^{p_1}\frac{dt}{t}\Big)^{1/p}$$

$$\leq C\Big(\int_0^\infty \Big[t^{p-p_1}\int_0^t s^{p_1-1}\lambda_{f;\nu}(s)\,ds\Big]\frac{dt}{t}\Big)^{1/p}$$

$$= C\Big(\int_0^\infty \Big[\int_s^\infty t^{p-p_1-1}\,dt\Big]s^{p_1-1}\lambda_{f;\nu}(s)\,ds\Big)^{1/p}$$

$$= C\Big(\int_0^\infty s^{p-1}\lambda_{f;\nu}(s)\,ds\Big)^{1/p}$$

$$= C\|f\|_p$$

and thus, our result is established. □

We are now in a position to prove the basic estimate of this subsection, the Adams inequality.

1.92 THEOREM. *Let μ be a Radon measure on $\mathbf{R}^n$ such that for all $x \in \mathbf{R}^n$ and $0 < r < \infty$, there is a constant M with the property that*

$$\mu(B(x,r)) \leq Mr^a$$

where $a = q(n-kp)/p$, $k > 0$, $1 < p < q < \infty$, and $kp < n$. If $f \in L^p(\mathbf{R}^n)$, then

$$\Big(\int |I_k * f|^q d\mu\Big)^{1/q} \leq CM^{1/q}\|f\|_p$$

where $C = C(k,p,q,n)$.

PROOF. Assume first that $M = 1$. For $t > 0$ let
$$A_t = \{y : I_k * |f|(y) > t\}.$$
Our objective is to estimate $\mu(A_t)$ in terms of $\|f\|_p$. Let $\mu_t = \mu \mathop{\llcorner} A_t$. Then

(1.86)
$$\begin{aligned} t\mu(A_t) &\leq \int_{A_t} I_k * |f| \, d\mu = \int_{\mathbf{R}^n} I_k * |f| \, d\mu_t \\ &= \int_{\mathbf{R}^n} I_k * \mu_t(x) \, |f(x)| \, dx \end{aligned}$$

where the last equality is a consequence of Fubini's theorem. Referring to Lemma 1.27 it follows that
$$I_k * \mu_t(x) = (n-k) \int_0^\infty \mu_t(B(x,r)) r^{k-n-1} \, dr.$$

For $R > 0$ which will be specified later, (1.86) becomes

(1.87)
$$\begin{aligned} t\mu(A_t) &\leq (n-k) \int_0^R \int_{\mathbf{R}^n} |f(x)| \mu_t(B(x,r)) r^{k-n-1} \, dx \, dr \\ &\quad + (n-k) \int_R^\infty \int_{\mathbf{R}^n} |f(x)| \mu_t(B(x,r)) r^{k-n-1} \, dx \, dr \\ &= J_1 + J_2. \end{aligned}$$

Since $\mu(B(x,r)) \leq r^a$ by hypothesis, the first integral, J_1, is estimated by observing that
$$\mu_t(B(x,r)) \leq r^{a/p} \mu_t(B(x,r))^{1/p'}$$
and then applying Hölder's inequality to obtain

(1.88) $$J_1 \leq (n-k)\|f\|_p \int_0^R \left(\int_{\mathbf{R}^n} \mu_t(B(x,r)) \, dx \right)^{1/p'} r^{k-n-1+a/p} \, dr.$$

We now will evaluate
$$\int_{\mathbf{R}^n} \mu_t(B(x,r)) \, dx.$$
For this purpose, let
$$D_r := \{(x,y) \in \mathbf{R}^n \times \mathbf{R}^n : |x-y| < r\}, \quad \text{and} \quad F = \chi_{D_r}.$$
Then, by Fubini's theorem
$$\begin{aligned} \int_{\mathbf{R}^n} \mu_t(B(x,r)) \, dx &= \int_{\mathbf{R}^n} \left(\int_{\{y : |y-x|<r\}} d\mu_t(y) \right) dx \\ &= \iint_{\mathbf{R}^n \times \mathbf{R}^n} F(x,y) \, dx \, d\mu_t(y) \\ &= \int_{\mathbf{R}^n} \left(\int_{\{x : |y-x|<r\}} dx \right) d\mu_t(y) \\ &= \int_{\mathbf{R}^n} |B(y,r)| \, d\mu_t(y) \\ &= \alpha(n) r^n \mu(A_t). \end{aligned}$$

Therefore (1.88) yields

$$J_1 \leq \frac{p(n-k)}{kp-(n-a)}\|f\|_p \boldsymbol{\alpha}(n)^{1/p'}\mu(A_t)^{1/p'}R^{k-(n-a)/p}.$$

Similarly, by employing the elementary estimate

$$\mu_t(B(x,r)) \leq \mu_t(B(x,r))^{1/p'}\mu(A_t)^{1/p},$$

we have

$$J_2 \leq (n-k)\|f\|_p \mu(A_t)^{1/p} \int_R^\infty \left(\int_{\mathbf{R}^n} \mu_t(B(x,r))dx\right)^{1/p'} r^{k-n-1}dr$$

$$\leq \frac{p(n-k)}{n-kp}\|f\|_p \mu(A_t)\boldsymbol{\alpha}(n)^{1/p'}R^{k-n/p}.$$

Hence

$$J_1 + J_2 \leq p(n-k)\boldsymbol{\alpha}(n)^{1/p'}\|f\|_p \left(\frac{\mu(A_t)^{1/p'}R^{k-(n-a)/p}}{kp-(n-a)} + \frac{\mu(A_t)R^{k-n/p}}{n-kp}\right).$$

In order for this inequality to achieve its maximum effectiveness, we seek that value of R for which the right-hand side attains a minimum. An elementary calculation shows that

$$R = \mu(A_t)^{1/a},$$

and the value of the right-hand side for this value of R is

$$\frac{p(n-k)a}{(n-kp)(kp-n+a)}\boldsymbol{\alpha}(n)^{1/p'}\mu(A_t)^{1-1/q}\|f\|_p.$$

Consequently, from (1.87)

(1.89)
$$\begin{aligned}
t\mu(A_t)^{1/q} &= t\mu(A_t)\mu(A_t)^{1/q-1} \\
&\leq \left(\frac{p(n-k)a}{(n-kp)(kp-n+a)}\alpha(n)^{1/p'}\right) \\
&\quad \left(\mu(A_t)^{1-1/q}\mu(A_t)^{1/q-1}\|f\|_p\right) \\
&= \frac{(n-k)pq}{(n-kp)(q-p)}\alpha(n)^{1/p'}\|f\|_p.
\end{aligned}$$

Expression (1.89) states that the Riesz potential operator I_k is of weak type (p,q) whenever p and q are numbers such that

(1.90)
$$1 < p < q < \infty, \quad kp < n.$$

Given p,q satisfying (1.90), we find $p_0, q_0, p_1, q_1, \theta$ such that (1.83), (1.85), and (1.84) are satisfied, and $1 < p_0 < p_1 < n/k$. Then I_k is both of weak type (p_0, q_0) and (p_1, q_1) and the Marcinkiewicz Interpolation Theorem 1.91 states that I_k is of type (p,q). This establishes our assertion if $M=1$. In the general case we apply the preceding part to the measure μ/M. □

Of special interest to us is the following imbedding theorem which now results from Theorem 1.79.

1.93 COROLLARY (ADAMS). *Let μ satisfy the conditions of the previous theorem with $k = 1$. If $u \in Y^{1,p}(\mathbf{R}^n)$, then*

$$\left(\int_{\mathbf{R}^n} |u|^q \, d\mu\right)^{1/q} \leq C M^{1/q} \|\nabla u\|_p.$$

The previous result does not include the case $p = n$ and as usual, it requires special treatment. For this we state the following result without proof for it requires techniques that go beyond the scope of this book. The reader may consult [**AH**], Theorem 7.2.2 for a proof.

1.94 THEOREM. *Let $q > n = p$, and assume μ satisfies*

$$\mu(B(x,r)) \leq \begin{cases} M \ln^{-q(1-1/n)}(1/r) & \text{if } r < 1/2, \\ Mr^q & \text{if } r \geq 1/2 \end{cases}$$

for some constant M and all balls $B(x,r)$. Then there is a constant $C = C(q,n,M)$ such that

$$\left(\int_{\mathbf{R}^n} |u|^q \, d\mu\right)^{1/q} \leq C \|\nabla u\|_n$$

whenever $u \in W^{1,n}(\mathbf{R}^n)$.

We conclude this subsection by an important consequence of Corollary 1.93.

1.95 COROLLARY. *Suppose that $0 < \varepsilon \leq \min\{p, n(p-1)\}$. Let μ be a nonnegative Radon measure on $\mathbf{R}^n$. Suppose that*

$$\mu(B(x,\rho)) \leq M\rho^{n-p+\varepsilon}$$

for each $B(x,\rho) \subset \mathbf{R}^n$. Then

$$\int_{\mathbf{R}^n} |u|^p \, d\mu \leq CM\left(\delta \int_{\mathbf{R}^n} |\nabla u|^p \, dx + \delta^{1-p/\varepsilon} \int_{\mathbf{R}^n} |u|^p \, dx\right)$$

for any $u \in W^{1,p}(\mathbf{R}^n)$ and any $\delta > 0$. The constant C depends on n, p, ε.

PROOF. Let $s = \frac{np}{n+\varepsilon} \geq 1$. Choose a ball $B(r) \subset \mathbf{R}^n$. Let η be a cut-off function such that $\chi_{B(r)} \leq \eta \leq \chi_{B(2r)}$ and $|\nabla \eta| \leq 2r^{-1}$. Write

$$u_0 = \fint_{B(2r)} u.$$

By Corollary 1.93,

$$\int_{B(2r)} |(u-u_0)\eta|^p \, d\mu \leq CM\left(\int_{B(r)} |\nabla([u-u_0]\eta)|^s \, dx\right)^{p/s}.$$

Hence, using Hölder's inequality and Poincaré's inequality

(1.91)
$$\int_{B(r)} |u-u_0|^p \, d\mu \le C \int_{B(2r)} |u-u_0|^p \eta^p \, d\mu$$
$$\le C \Big(\int_{B(2r)} (|\nabla u|^s \eta^s + |u-u_0|^s |\nabla \eta|^s) \, dx \Big)^{p/s}$$
$$\le C \Big(\int_{B(2r)} |\nabla u|^s \, dx + r^{-s} \int_{B(2r)} |u-u_0|^s \, dx \Big)^{p/s}$$
$$\le C \Big(\int_{B(2r)} |\nabla u|^s \, dx \Big)^{p/s}$$
$$\le C |B(2r)|^{p/s-1} \int_{B(2r)} |\nabla u|^p \, dx$$
$$\le C r^\varepsilon \int_{B(2r)} |\nabla u|^p \, dx.$$

Also, using Hölder's inequality we have

(1.92)
$$\int_{B(r)} |u_0|^p \, d\mu = \mu(B(r)) \Big[\fint_{B(2r)} u \, dx \Big]^p \le M r^{n-p+\varepsilon} \fint_{B(2r)} |u|^p \, dx$$
$$\le C M r^{\varepsilon-p} \int_{B(2r)} |u|^p \, dx.$$

Adding (1.91) and (1.92) we arrive at

$$\int_{B(r)} |u|^p \, d\mu \le 2^{p-1} \Big(\int_{B(r)} |u-u_0|^p \, d\mu + \int_{B(r)} |u_0|^p \, d\mu \Big)$$
$$\le C M \Big(r^\varepsilon \int_{B(2r)} |\nabla u|^p \, dx + r^{\varepsilon-p} \int_{B(2r)} |u|^p \, dx \Big).$$

Given $\delta > 0$, set $r = \delta^{1/\varepsilon}$ so that $\delta^{1-p/\varepsilon} = r^{\varepsilon-p}$. We cover $\mathbf{R}^n$ with balls $B(z_j, r)$ where $\{z_j\}$ is a sequence consisting of all points such that nz_j/r have integer coordinates. It is obvious that each point of $\mathbf{R}^n$ belongs to at most N balls $B(z_j, 2r)$ where N depends on n. Hence,

$$\int_{\mathbf{R}^n} |u|^p \, d\mu \le \sum_j \int_{B(z_j, r)} |u|^p \, d\mu$$
$$\le C M \sum_j \Big(\delta \int_{B(z_j, 2r)} |\nabla u|^p \, dx + \delta^{1-p/\varepsilon} \int_{B(z_j, r)} |u|^p \, dx \Big)$$
$$\le C M \Big(\delta \int_{\mathbf{R}^n} |\nabla u|^p \, dx + \delta^{1-p/\varepsilon} \int_{\mathbf{R}^n} |u|^p \, dx \Big)$$

which concludes the proof. □

1.3.6 Bessel and Riesz potentials.

In this section we establish relation between Riesz potentials and spaces $Y^{1,p}(\mathbf{R}^n)$. We also introduce Bessel potentials which are in a similar way related to Sobolev functions. We will not attempt a thorough development of this topic.

Instead, we will introduce the basic properties of Bessel and Riesz potentials and refer the interested reader to other sources for a complete development cf. [**St3**].

We will use Fourier transform in the form formally expressed as

$$\hat{f}(x) = (2\pi)^{-n/2} \int_{\mathbf{R}^n} e^{-ix\cdot y} f(y) dy. \tag{1.93}$$

The n Riesz transforms are defined for f in $L^p(\mathbf{R}^n)$ by

$$R_i(f)(x) = \lim_{\varepsilon \to 0} \frac{n-1}{\rho_1} \int_{|y|\geq \varepsilon} \frac{y_i}{|y|^{n+1}} f(x-y)\, dy, \qquad i=1,\ldots,n,$$

where

$$\rho_1 = (n-1)\, \frac{\pi^{(n+1)/2}}{\Gamma((n+1)/2)},$$

cf. (1.24). It can be shown that

$$\widehat{R_i f}(x) = i\frac{x_i}{|x|} \hat{f}(x), \qquad i=1,\ldots,n.$$

We will utilize two basic properties of the Riesz transform. First, if

$$u = \tfrac{1}{\rho_1} I_1 f, \quad \text{then} \quad \frac{\partial u}{\partial x_i} = -R_i(f) \tag{1.94}$$

whenever $f \in \mathcal{S}$, where $\mathcal{S}$ denotes the Schwartz class of rapidly decreasing $\mathcal{C}^\infty$ functions on $\mathbf{R}^n$. Second, a smooth function u can be expressed as the Riesz potential of Riesz transforms of its derivatives; that is,

$$u = \tfrac{1}{\rho_1} I_1 * \left(\sum_{i=1}^n R_i D_i u \right).$$

The following fundamental result of Calderón and Zygmund [**CZ**] states that the Riesz transform is a bounded operator on $L^p(\mathbf{R}^n)$.

1.96 THEOREM. *Let $f \in L^p$, $1 \leq p < \infty$. There is a constant $C = C(n,p)$ such that*

(i) *if $p=1$, then*

$$|\{x : R_i(x) > t\}| \leq \frac{C}{t} \|f\|_1 \, ;$$

(ii) *if $1 < p < \infty$, then*

$$\|R_i(f)\|_p \leq C \|f\|_p \, .$$

1.97 THEOREM. *Let $1 < p < n$. The space of all Riesz potentials $\{I_1 f : f \in L^p(\mathbf{R}^n)\}$ is equal to $Y^{1,p}(\mathbf{R}^n)$ with equivalent norms; that is, there is a constant $C = C(n,p)$ such that*

$$C^{-1} \|f\|_p \leq \|I_1 f\|_Y \leq C \|f\|_p \, .$$

PROOF. Assume $u = \frac{1}{\rho_1} I_1 f$, $1 \leq p < n$ and let $\{f_k\}$ be a sequence in $\mathcal{S}$ such that $f_k \to f$ in $L^p(\mathbf{R}^n)$. Then with $u_k := \frac{1}{\rho_1} I_1 f_k$, we have $D_i u_k = -R_i(f_k)$, $i = 1, \cdots, n$, and by Theorem 1.96,

$$\|D_i u_k - D_i u_j\|_p \leq C \|f_k - f_j\|_p$$

for all positive integers k, j. Thus, $D_i u_k \to g_i$ for some $g_i \in L^p(\mathbf{R}^n)$ while Theorem 1.33 implies that $u_k \to u$ in $L^{p^*}(\mathbf{R}^n)$. Therefore, $D_i u_k \to D_i u$ in the space of Schwartz distributions, thus implying $D_i u = g_i$. This proves that $u \in Y^{1,p}(\mathbf{R}^n)$.

To prove the opposite inclusion, assume $u \in Y^{1,p}(\mathbf{R}^n)$ and let $u_k \in \mathcal{S}$ be a sequence such that $u_k \to u$ in $Y^{1,p}(\mathbf{R}^n)$. Since

$$u_k = \frac{1}{\rho_1} I_1 * \left(\sum_{i=1}^n R_i D_i u_k \right)$$

and

$$R_i D_i u_k \to R_i D_i u \quad \text{in} \quad L^p(\mathbf{R}^n),$$

it follows that

$$\frac{1}{\rho_1} I_1 * \left(\sum_{i=1}^n R_i D_i u_k \right) \to \frac{1}{\rho_1} I_1 * \left(\sum_{i=1}^n R_i D_i u \right) \quad \text{in} \quad L^{p^*}(\mathbf{R}^n).$$

This implies

(1.95) $$u = \frac{1}{\rho_1} I_1 * \left(\sum_{i=1}^n R_i D_i u \right),$$

which states that u is the potential of an $L^p(\mathbf{R}^n)$ function, as required.

To show the equivalence of the norms, suppose $u = \frac{1}{\rho_1} I_1 f$, $f \in L^p(\mathbf{R}^n)$. Then by Theorem 1.33 and (1.94),

$$\|u\|_Y \leq C \|\nabla u\|_p \leq C \|f\|_p.$$

On the other hand, if $u \in Y^{1,p}$, then (1.95) implies

$$u = \frac{1}{\rho_1} I_1 * f$$

where

$$f = \left(\sum_{i=1}^n R_i D_i u \right).$$

Therefore, $\|f\|_p \leq C \|\nabla u\|_p$, by Theorem 1.96. $\square$

The Riesz potential, which was introduced in Definition 1.26, leads to many important applications, but for the purpose of investigating Sobolev functions, the Bessel potential is more suitable. For an analysis of the Bessel kernel, we refer the reader to [**St3**], Chapter 5 and quote here without proof the facts relevant to our development.

The Bessel kernel, g_α, $\alpha > 0$, is defined as that function whose Fourier transform is

$$\hat{g}_\alpha(x) = (2\pi)^{-n/2} (1 + |x|^2)^{-\alpha/2}.$$

It is known that g_α is a positive, integrable function which is analytic except at $x = 0$. Similar to the Riesz kernel, we have

(1.96) $$g_\alpha * g_\beta = g_{\alpha+\beta}, \quad \alpha, \beta \geq 0.$$

There is an intimate connection between Bessel and Riesz potentials which is exhibited by g_α near the origin and infinity. Indeed, an analysis shows that for some $C > 0$,
$$g_\alpha(x) \sim C|x|^{(1/2)(\alpha-n-1)} e^{-|x|} \quad \text{as} \quad |x| \to \infty.$$
Here, $a(x) \sim b(x)$ means that $a(x)/b(x)$ is bounded above and below for all large $|x|$. Moreover, it can be shown that
$$g_\alpha(x) = \rho_\alpha^{-1} |x|^{\alpha-n} + o(|x|^{\alpha-n}) \quad \text{as} \quad |x| \to 0$$
if $0 < \alpha < n$. Thus, it follows for some constants C_1 and C_2, that

(1.97) $$g_\alpha(x) \leq \frac{C_1}{|x|^{n-\alpha}} e^{-C_2|x|}$$

for all $x \in \mathbf{R}^n$. Moreover, it also can be shown that

(1.98) $$|\nabla g_\alpha(x)| \leq \frac{C_1}{|x|^{n-\alpha+1}} e^{-C_2|x|}.$$

From our point of view, one of the most interesting facts concerning Bessel potentials is that they can be employed to characterize the Sobolev spaces $W^{k,p}(\mathbf{R}^n)$. This is expressed in the following theorem where we employ the notation
$$L^{\alpha,p}(\mathbf{R}^n), \quad \alpha > 0, \ 1 \leq p \leq \infty$$
to denote all functions u such that
$$u = g_\alpha * f$$
for some $f \in L^p(\mathbf{R}^n)$.

1.98 THEOREM. *If k is a positive integer and $1 < p < \infty$, then*
$$L^{k,p}(\mathbf{R}^n) = W^{k,p}(\mathbf{R}^n).$$
Moreover, if $u \in L^{k,p}(\mathbf{R}^n)$ with $u = g_\alpha * f$, then
$$C^{-1} \|f\|_p \leq \|u\|_{k,p} \leq C\|f\|_p$$
where $C = C(\alpha, p, n)$.

1.99 REMARK. The equivalence of the spaces $L^{k,p}$ and $W^{k,p}$ fails when $p = 1$ or $p = \infty$.

It is also interesting to observe the following dissimilarity between Bessel and Riesz potentials. In view of the fact that $\|g_\alpha\|_1 \leq C$, Young's inequality for convolutions (Theorem 1.10) implies

(1.99) $$\|g_\alpha * f\|_p \leq C\|f\|_p, \quad 1 \leq p \leq \infty.$$

On the other hand, we saw in Theorem 1.33 that the Riesz potential satisfies

(1.100) $$\|I_\alpha * f\|_q \leq C\|f\|_p, \quad p > 1$$

where $q = np/(n - \alpha p)$. However, an inequality of type (1.100) is possible for only such q, thus disallowing an inequality of type (1.99) for I_α and for every $f \in L^p$.

1.4 Historical notes

The Besicovitch Covering Lemma, Lemma 1.16, is the keystone in the theory of differentiation of measures as developed in [**Be1**] and [**Be2**]. The proof of the Covering Lemma 1.15 we present is a slight variation of a proof due to Robert Hardt. The original version of this lemma is due to Vitali [**Vi**] who employed closed cubes and Lebesgue measure.

Lebesgue [**Le1**] was the first to prove that the derivative of a measure (with respect to Lebesgue measure) exists almost everywhere. Some consequences of this are the fundamental results Theorem 1.24 and Corollary 1.25. His result is valid if cubes are replaced by general sets that are "regular" when compared to cubes. A sequence of sets $\{E_k\}$ is called regular at a point x_0 if $x_0 \in \cap_{k=1}^\infty E_k$, $\operatorname{diam}(E_k) \to 0$ and $\liminf_{k \to 0} \rho(E_k) > 0$ where $\rho(E_k)$ is defined as the infimum of the numbers $|C|/|E_k|$ with C ranging over all cubes containing E_k. In particular, one is allowed to consider coverings by nested cubes or balls that are not necessarily concentric. However, when Lebesgue measure is replaced by a Radon measure, Lemma 1.16 no longer remains valid if the balls in the covering are allowed to become too non-concentric. At about the time that Besicovitch made his contributions, Morse [**Mor1**], [**Mor2**], developed a theory which allowed coverings by a general class of sets rather than by concentric closed balls.

The maximal function estimate, Theorem 1.22, was first proved by Hardy-Littlewood for $n = 1$, [**HL**], and by Wiener for arbitrary n, [**Wie3**]. Riesz potentials were introduced in [**Rim**] and Theorem 1.33 was proved by Hardy and Littlewood [**HL1**], [**HL2**]. Our proof, based on Theorem 1.32, is due to Hedberg [**He2**]. Theorem 1.66, which is due to John and Nirenberg [**JN**], plays a critical role in the development of quasilinear equations with measurable coefficients. The optimal companion result, the borderline case of the Sobolev embedding theorem, was proved by Adams, [**Ad3**]. It states that if $f \in L^p(\mathbf{R}^n)$, $\alpha p = n$, $\operatorname{spt} f \subset B(0, r)$ and $\|f\|_p = 1$, then there is a constant $C = C(n, p)$ such that

$$\int_{B(0,r)} \exp(\beta |I_\alpha * f|^{p'}) \, dx \leq C r^n$$

with $\beta = n/\omega(n-1)$. Others made contributions to this inequality, but without achieving the best constant β, cf. Trudinger [**T2**], Hempel, Morris, and Trudinger, [**HMT**], Strichartz, [**Str**], and Moser, [**Mos3**].

There is a long history of authors employing functions that have weak derivatives in some sense. Beppo Levi [**BL**], Tonelli [**Ton1**] and Nikodým [**Nik**] used the class of functions that are absolutely continuous on almost all lines parallel to the coordinate axes, the property that essentially characterizes Sobolev functions (Theorem 1.41). Along with work of Sobolev, [**So1**], [**So2**], [**So3**], Calkin, [**Cal**], and Morrey, [**Mo2**], [**Mo4**], developed many of the properties of Sobolev functions that are used today. Theorem 1.45 was proved by Deny and Lions [**DL**], and the general version that applies to the spaces $W^{k,p}$ is due to Meyers and Serrin, [**MS**]. Lemma 1.50 was used by Sobolev [**So1**], [**So2**].

Inequalities estimating $\|u\|_2$ in terms of $\|\nabla u\|_2$ (in fact, imbedding of $W^{1,2}$ to L^2) were known already to Poincaré [**Po**]. Utilizing an idea introduced by Meyers [**Me4**], a general and comprehensive form of Poincaré-type inequalities was developed in [**Z5**].

The concept of functions satisfying the conditions of Definition 1.28 was introduced by Morrey in [**Mo1**]. Other major contributions to the analysis related to Morrey spaces were made by Campanato, [**Cam1**], [**Cam2**]. Theorem 1.56, the Sobolev inequality, is due to Sobolev [**So2**] ($p > 1$) and Gagliardo [**Ga**] ($p = 1$). This inequality was also developed by Morrey [**Mo2**], and Nirenberg [**Ni2**]. The proof of Theorem 1.56 for the case $p < n$ is due to Nirenberg [**Ni2**]. The case of $p > n$ is due to Tonelli [**Ton2**] and Morrey [**Mo2**]. Theorem 1.61 originated in a paper by Rellich [**Re**] in the case $p = 2$ and by Kondrachov [**Ko**] when $p \neq 2$. Calderón and Zygmund [**CZ**] proved the approximation property of Sobolev functions stated in Theorem 1.69. Later, Liu [**Liu**] proved that the approximating function v also has the property that it is close to u in the Sobolev norm. Further results in this direction were established in [**MZ1**]. The exposition here is based on pointwise inequalities involving maximal functions, as they were recently employed by Bojarski and Hajłasz in their study of Sobolev functions, [**Bo**], [**BH**], [**Ha**]. The pointwise differentiability a.e. of Sobolev function is due to Cesari [**Ce**] and the $W^{1,\infty}$ case to Rademacher [**Ra**]. The chain rule, Theorem 1.74, is due to Marcus and Mizel [**MM**].

Bessel potentials and their corresponding fractional order Sobolev spaces were introduced by Aronszajn and Smith [**AS2**], and Calderón [**Ca**]. The intimate relation between Bessel and Riesz potentials was observed by Stein [**St1**], [**St2**]. The interpolation result, Theorem 1.91, was established by Marcinkiewicz in [**Mar**].

The general version of the Sobolev inequality, Theorem 1.92, is due to Adams [**Ad1**].

CHAPTER 2

Potential Theory

2.1 Capacity

The capacity that we introduce in this section is fundamental to the analysis of pointwise behavior of Sobolev functions. We will see in later chapters that it is also used in a critical way to describe the boundary behavior of solutions of certain elliptic equations.

2.1 DEFINITION. Let $\mathcal{O} \subset \mathbf{R}^n$ be an open set and $1 < p \leq n$. Let $r \in (0, \infty]$ and let us agree that $r^{-p} = 0$ if $r = \infty$. We consider the space

$$\mathcal{U} = \mathcal{U}_{p;r}(\mathcal{O}) = \begin{cases} Y_0^{1,p}(\mathcal{O}) & \text{if } r = \infty, \\ W_0^{1,p}(\mathcal{O}) & \text{if } r < \infty. \end{cases}$$

Unless the case when $p = n$, $r = \infty$ and $\mathcal{O}$ is not an n-Green domain, $\mathcal{U}_{p;r}$ is a reflexive Banach space endowed with the norm

$$\|u\|_{\mathcal{U}} = \left[\int_{\mathcal{O}} \left(|\nabla u|^p + r^{-p}|u|^p\right) dx\right]^{1/p}.$$

If $E \subset \mathbf{R}^n$, the $(p;r)$-capacity of E relative to $\mathcal{O}$ is defined by

$$(2.1) \qquad \gamma_{p;r}(E;\mathcal{O}) := \inf\left\{\int_{\mathcal{O}} \left(|\nabla u|^p + r^{-p}|u|^p\right) dx : u \in \mathcal{Y}_{p;r}(E;\mathcal{O})\right\},$$

where

$$\mathcal{Y}_{p;r}(E;\mathcal{O}) := \{u \in \mathcal{U}_{p;r}(\mathcal{O}),\ E \subset \{u \geq 1\}^\circ\},$$

and $\{u \geq 1\}^\circ$ denotes the interior of the set $\{x \in \mathcal{O}: u(x) \geq 1\}$. The set $\mathcal{O}$ is called a *reference domain*; we simplify the notation as

$$\gamma(E) = \gamma_{p;r}(E) = \gamma_{p;r}(E;\mathcal{O}),$$
$$\mathcal{Y}(E) = \mathcal{Y}_{p;r}(E) = \mathcal{Y}_{p;r}(E;\mathcal{O})$$
$$\mathcal{U} = \mathcal{U}_{p;r} = \mathcal{U}_{p;r}(\mathcal{O}).$$

We write

$$\mathbf{c}_p(E) = \mathbf{c}_p(E;\mathcal{O}) = \gamma_{p;\infty}(E;\mathcal{O}),$$
$$\mathbf{C}_p(E) = \mathbf{C}_p(E;\mathcal{O}) = \gamma_{p;1}(E;\mathcal{O}).$$

In the sequel, when using symbols like γ instead of $\gamma_{p;r}$, we assume that p, r and $\mathcal{O}$ are given, and if not stated otherwise they will be no more specific than above.

2.2 Remarks.

(i) The definition of capacity cannot be "simplified" to

$$(2.2) \qquad E \mapsto \inf\Big\{ \int_{\mathcal{O}} \big(|\nabla u|^p + r^{-p}|u|^p\big)\, dx : u \in \mathcal{U}_{p;r},\ u \geq 1 \text{ on } E \Big\}.$$

This formula gives another set function, perhaps one not so important. However, the formula 2.2 turns out to be correct if we admit only special representatives of Sobolev functions. This problem will be considered later, see Corollary 2.25.

(ii) In general, the infimum in (2.1) is not attained. Nevertheless, minimizing sequences for (2.1) converge to the so-called capacitary extremals, which will be studied later in Theorem 2.31.

(iii) The capacity $\mathbf{c}_n(\cdot, \mathbf{R}^n)$ is not useful as it vanishes identically on all sets. Indeed, the function $1 \in Y_0^{1,n}(\mathbf{R}^n)$ is an admissible function in the definition of $\mathbf{c}_n(\cdot, \mathbf{R}^n)$. This is the reason why we use either the capacity relative to an n-Green domain or the capacity $\gamma_{n;r}$ in case $p = n$.

2.3 Theorem.
Let $\mathcal{O}$ be a reference domain, $1 < p \leq n$ and $r \in (0, \infty]$.

(i) If $E \subset \mathcal{O}$, then $\gamma(E) = \inf\{\gamma(U) : U \subset \mathcal{O} \text{ is open and } E \subset U\}$,

(ii) If $E_1 \supset E_2 \supset \cdots$ are compact subsets on $\mathcal{O}$, then

$$\gamma\Big(\bigcap_{j=1}^{\infty} E_j\Big) = \lim_{j \to \infty} \gamma(E_j),$$

(iii) If $E \subset \mathcal{O}$ is compact, then

$$\gamma(E) = \inf\Big\{ \int_{\mathcal{O}} \big(|\nabla u|^p + r^{-p}|u|^p\big)\, dx : u \in C_c^{\infty}(\mathcal{O}),\ u \geq 1 \text{ on } E \Big\},$$

(iv) If $E_1 \subset E_2 \subset \cdots \subset \mathcal{O}$ are arbitrary sets, then

$$\gamma\Big(\bigcup_{j=1}^{\infty} E_j\Big) = \lim_{j \to \infty} \gamma(E_j),$$

(v) $\gamma\Big(\bigcup_{j=1}^{\infty} E_j\Big) \leq \sum_{j=1}^{\infty} \gamma(E_j)$ whenever $E_1, E_2, \cdots$ are arbitrary subsets of $\mathcal{O}$.

PROOF. Notice that either $\mathcal{U}$ is a Banach space or the capacity is trivial. We need only consider the former case.

Property (i) is evident.

If $E = \bigcap_j E_j$ and E_j are compact, then

$$\lim_{j \to \infty} \gamma(E_j) \leq \gamma(U)$$

for any open set $U \supset E$, and thus by (i)

$$\lim_{j \to \infty} \gamma(E_j) \leq \gamma(E).$$

The converse inequality of (ii) is obvious.

To prove (iii), we choose $E \subset \mathcal{O}$ compact. Suppose $u \in \mathcal{U}$, $u \geq 1$ on an open set $U \supset E$. Let $\eta \in \mathcal{C}_c^\infty(\mathcal{O})$ be a cutoff function such that $0 \leq \eta \leq 1$ and $\eta = 1$ on E. Let v_j be a sequence of functions from $\mathcal{C}_c^\infty(\mathcal{O})$ such that $v_j \to u$ in $\mathcal{U}$. We set

$$u_j = (\phi_{\varepsilon_j} * u)\eta + v_j(1-\eta),$$

where $\varepsilon_j \to 0$, $0 < \varepsilon_j < \operatorname{dist}(E, \partial U)$. Then $u_j \in \mathcal{C}_c^\infty(\mathcal{O})$, $u_j \geq 1$ on E and $u_j \to u$ in $\mathcal{U}$. This proves that

$$\gamma(K) \geq \inf \left\{ \int_\mathcal{O} \left(|\nabla u|^p + r^{-p}|u|^p\right) dx : u \in C_c^\infty(\mathcal{O}),\ u \geq 1 \text{ on } E \right\}.$$

The converse inequality of (iii) follows from the observation that if $u \in \mathcal{C}_c^\infty(\mathcal{O})$, $u \geq 1$ on E, then for any constant $a > 1$ we have $au \geq 1$ on an open set containing E.

To prove (iv), set

$$s = \sup_j \gamma(E_j), \quad E = \bigcup_j E_j.$$

Then clearly

$$s \leq \gamma(E)$$

and we will prove the converse. Find $u_j \in \mathcal{U}$ such that

$$E_j \subset \{u_j \geq 1\}^\circ,$$
$$\int_\mathcal{O} \left(|\nabla u_j|^p + r^{-p}|u_j|^p\right) dx \leq \gamma(E_j) + 2^{-j}.$$

We may suppose that $s < \infty$, and thus $\{u_j\}$ is bounded in $\mathcal{U}$. Since $\mathcal{U}$ is reflexive, there is a subsequence of $\{u_j\}$ which converges weakly to a function $u \in \mathcal{U}$. Using Mazur's lemma, there is a sequence v_j of convex combinations of u_i's $(i \geq j)$ which converges strongly to u. We observe that again

$$E_j \subset \{v_j \geq 1\}^\circ.$$

Convexity arguments show that

$$\int_\mathcal{O} \left(|\nabla v_j|^p + r^{-p}|v_j|^p\right) dx \leq \sup_{i \geq j} \int_\mathcal{O} \left(|\nabla v_i|^p + r^{-p}|v_i|^p\right) dx \leq s + 2^{-j}.$$

Passing to a subsequence, we may assume that

$$\|v_{j+1} - v_j\|_\mathcal{U} \leq 2^{-j}.$$

We set

$$w_j = v_j + \sum_{i=j}^\infty |v_{i+1} - v_i|.$$

Then $w_j \in \mathcal{U}$, $w_j \geq 1$ on the open set

$$\bigcup_{i=j}^\infty \{v_i \geq 1\}^\circ \supset E,$$

and thus

$$\gamma(E) \leq \int_\mathcal{O} \left(|\nabla w_j|^p + r^{-p}|w_j|^p\right) dx.$$

We have

$$\int_{\mathcal{O}} \left(|\nabla w_j|^p + r^{-p}|w_j|^p\right) dx \leq \left(\|v_j\|_u + \sum_{i=j}^{\infty} \|v_{i+1} - v_i\|_u\right)^p$$

$$\leq \left[(s + 2^{-j})^{1/p} + 2^{-j+1}\right]^p \to s.$$

It follows that $\gamma(E) \leq s$.

As a first step of the proof of (v), we consider $E_1, E_2 \subset \mathcal{O}$. Let $u_i \in \mathcal{Y}(E_i)$, $i = 1, 2$, and

$$u = \max\{u_1, u_2\}, \quad w = \min\{u_1, u_2\}.$$

Then $u \in \mathcal{Y}(E_1 \cup E_2)$, $w \in \mathcal{Y}(E_1 \cap E_2)$, and by means of splitting the integration domains into sets $\{u_1 < u_2\}$, $\{u_1 > u_2\}$ and $\{u_1 = u_2\}$ we observe that

$$\int_{\mathcal{O}} \left(|\nabla u|^p + r^{-p}|u|^p\right) dx + \int_{\mathcal{O}} \left(|\nabla w|^p + r^{-p}|w|^p\right) dx$$
$$= \int_{\mathcal{O}} \left(|\nabla u_1|^p + r^{-p}|u_1|^p\right) dx + \int_{\mathcal{O}} \left(|\nabla u_2|^p + r^{-p}|u_2|^p\right) dx.$$

Passing to the infima over all competitors, we obtain

(2.3) $$\gamma(E_1 \cup E_2) + \gamma(E_1 \cap E_2) \leq \gamma(E_1) + \gamma(E_2).$$

By induction we obtain

$$\gamma\left(\bigcup_{i=1}^{k} E_i\right) \leq \sum_{i=1}^{k} \gamma(E_i)$$

for any finite family $E_1, E_2, \ldots, E_k \subset \mathcal{O}$. Now, for an infinite sequence we obtain from (iv) that

$$\gamma\left(\bigcup_{i=1}^{\infty} E_i\right) = \lim_{k \to \infty} \gamma\left(\sum_{i=1}^{k} E_i\right) \leq \lim_{k \to \infty} \sum_{i=1}^{k} \gamma(E_i)$$
$$= \sum_{i=1}^{\infty} \gamma(E_i). \qquad \square$$

Theorem 2.3 and the strong subadditivity property (2.3) verify that γ is a Choquet capacity [**Ch1**] and satisfies the axioms considered by Brelot [**Br3**], [**Br4**]. Choquet's capacitability theory yields the following result.

2.4 THEOREM. *If $E \subset \mathcal{O}$ is a Suslin set, then*

$$\gamma(E) = \sup\{\gamma(K) : K \subset E, K \text{ compact}\}.$$

2.1.1 Comparison of capacities; capacities of balls.

2.5 LEMMA. *Suppose that $1 < p \leq n$. There is a constant $C = C(n, p)$ such that*

$$C^{-1} \mathbf{c}_p(E; B(x, r)) \leq \gamma_{p;r}(E; \mathbf{R}^n) \leq C \, \mathbf{c}_p(E; B(x, r))$$

whenever $E \subset B(x, r/2)$.

PROOF. Let $u \in \mathcal{Y}_{p;r}(E; B(x,r))$. Then u, extended by zero outside $B(x,r)$, is also a competitor for $\gamma_{p;r}(E; \mathbf{R}^n)$ and using Poincaré's inequality (Corollary 1.57) we obtain
$$\gamma_{p;r}(E; \mathbf{R}^n) \leq \int_{\mathbf{R}^n} \left(|\nabla u|^p + r^{-p} |u|^p \right) dx \leq C \int_{B(x,r)} |\nabla u|^p \, dx.$$
It follows that
$$\gamma_{p;r}(E; \mathbf{R}^n) \leq C \, \mathbf{c}_p\bigl(E; B(x,r)\bigr).$$
Conversely, let $v \in \mathcal{Y}_{p;r}(E, \mathbf{R}^n)$. Consider $\eta \in \mathcal{C}^1_c\bigl(B(x,r)\bigr)$ such that $0 \leq \eta \leq 1$, $\eta = 1$ on $B(x,r/2)$ and $|\nabla \eta| \leq 3/r$. Then
$$\mathbf{c}_p\bigl(E; B(x,r)\bigr) \leq \int_{B(x,r)} |\nabla(\eta v)|^p \, dx \leq C \int_{B(x,r)} \left(|\nabla v|^p + r^{-p} |v|^p \right) dx,$$
and thus
$$\mathbf{c}_p\bigl(E; B(x,r)\bigr) \leq C \, \gamma_{p;r}(E; \mathbf{R}^n). \qquad \square$$

2.6 LEMMA. *Let $\mathcal{O} = \mathbf{R}^n$. Suppose $1 < p \leq n$ and $E \subset \mathbf{R}^n$ is an arbitrary set. Then*
  (i) *$\gamma_{p;r}(E) \geq \gamma_{p;R}(E)$ whenever $0 < r < R \leq \infty$, in particular $\gamma_{p;r}(E) \geq \mathbf{c}_p(E)$,*
  (ii) *If $p < n$, there is $C = C(n,p)$ such that $\gamma_{p;r}(E) \leq C \, \mathbf{c}_p(E)$ whenever $\operatorname{diam} E \leq r$,*
  (iii) *If $p < n$, then $\gamma_{p;r}(E) = 0$ if and only if $\mathbf{c}_p(E) = 0$.*

PROOF. (i) is obvious.

(ii) Let u be an admissible function in the definition of $\mathbf{c}_p(E)$ for which $\|\nabla u\|_p^p$ is close to $\mathbf{c}_p(E)$. There is a ball $B = B(x,r)$ such that $E \subset B$. Let η be a Lipschitz function such that $0 \leq \eta \leq 1$, $\operatorname{spt} \eta \subset 2B := B(x, 2r)$, $\eta = 1$ on B and $|\nabla \eta| \leq 1/r$. Set $v = \eta u$. Then

(2.4)
$$\begin{aligned} \gamma_{p;r}(E) &\leq \int_{\mathbf{R}^n} \left(|\nabla v|^p + r^{-p} |v|^p \right) dx \\ &\leq C \int_{2B} \left(|\nabla u|^p + r^{-p} |u|^p \right) dx \\ &\leq C \int_{\mathbf{R}^n} |\nabla u|^p \, dx + C \left(\int_{2B} |u|^{p^*} \, dx \right)^{p/p^*} \\ &\leq C \int_{\mathbf{R}^n} |\nabla u|^p \, dx. \end{aligned}$$

(iii) By (ii), $\mathbf{c}_p(E) = 0$ if $\gamma_{p;r}(E) = 0$. The converse implication is true for sets of diameter less that r. However, each set is a countable union of sets with small diameters and thus the assertion follows from Theorem 2.3 (v). $\qquad \square$

2.7 LEMMA. *Suppose that $1 < p \leq n$ and let $G \subset\subset \Omega \subset \Omega'$ be open sets. Then*
  (i) *$\gamma_{p;r}(E; \Omega') \leq \gamma_{p;r}(E, \Omega)$ for any set $E \subset \Omega$,*
  (ii) *If $1 \notin \mathcal{U}_{p;r}(\Omega')$ (which is the case if $p < n$, or $r < \infty$, or Ω is an n-Green domain), then there is $C = C(p, r, \Omega, \Omega', G)$ such that $\gamma_{p;r}(E; \Omega) \leq C \, \gamma_{p;r}(E, \Omega')$ for any $E \subset G$.*

PROOF. (i) is easy. For (ii), we take a function $\eta \in \mathcal{C}_c^1(\Omega)$ such that $0 \leq \eta \leq 1$ and $\eta = 1$ on G. If $u \in \mathcal{Y}_{p;r}(E, \Omega')$, then $v := u\eta \in \mathcal{Y}_{p;r}(E, \Omega)$ and

$$\int_\Omega \left(|\nabla v|^p + r^{-p}|v|^p\right) dx \leq C \int_{\Omega'} \left(|\nabla u|^p + r^{-p}|u|^p\right) dx + C \int_{\operatorname{spt}\eta} |u|^p \, dx.$$

We need only to show that

$$\int_{\operatorname{spt}\eta} |u|^p \, dx \leq C \int_{\Omega'} \left(|\nabla u|^p + r^{-p}|u|^p\right) dx.$$

This is trivial if $r < \infty$. If $p < n$, then

$$\int_{\operatorname{spt}\eta} |u|^p \, dx \leq C \left(\int_{\operatorname{spt}\eta} |u|^{p^*} \, dx\right)^{p/p^*} \leq C \int_{\Omega'} |\nabla u|^p \, dx.$$

Finally, if Ω' is an n-Green domain we can use Lemma 1.84. $\square$

2.8 THEOREM. Let $\mathcal{O} = \mathbf{R}^n$.
(i) If $1 < p < n$, $\mathbf{c}_p(B(x,\rho)) = C \rho^{n-p}$ with $C = n\boldsymbol{\alpha}(n)\left(\frac{n-p}{p-1}\right)^{p-1}$.
(ii) $\mathbf{c}_n(B(x,\rho); B(x,1)) = n\boldsymbol{\alpha}(n) \ln^{1-n}(1/\rho)$ if $\rho < 1$,
(iii) There is a constant $C(n,p)$ such that

$$C^{-1} \rho^n (r^{-p} + \rho^{-p}) \leq \gamma_{p;r}(B(x,\rho)) \leq C \rho^n (r^{-p} + \rho^{-p})$$

if $1 < p < n$, and

$$C^{-1} g(\rho/r) \leq \gamma_{n;r}(B(x,\rho)) \leq C g(\rho/r)$$

where

$$g(t) = \begin{cases} \ln^{1-n}(t), & t < 1/2 \\ t^n, & t \geq 1/2. \end{cases}$$

PROOF. (i) We may suppose that $x = 0$. Let $u \in \mathcal{C}_c^\infty(\mathbf{R}^n)$, $u \geq 1$ on $B(0,r)$. If $z \in \partial B(0,1)$, then

$$1 \leq \int_\rho^\infty |\nabla u(tz)| \, dt.$$

Integrating with respect to the spherical measure we obtain

$$n\boldsymbol{\alpha}(n) \leq \int_{\mathbf{R}^n \setminus B(0,\rho)} |y|^{1-n} |\nabla u(y)| \, dy$$

$$\leq \left(\int_{\mathbf{R}^n} |\nabla u|^p \, dy\right)^{1/p} \left(\int_{\mathbf{R}^n \setminus B(0,\rho)} |y|^{p'(1-n)}\right)^{1/p'}$$

$$= \left(\int_{\mathbf{R}^n} |\nabla u|^p \, dy\right)^{1/p} \left(n\boldsymbol{\alpha}(n) \int_\rho^\infty t^{(p'-1)(1-n)} \, dt\right)^{1/p'}$$

$$= \left(n\boldsymbol{\alpha}(n) \tfrac{p-1}{n-p} \rho^{\frac{p-n}{p-1}}\right)^{1/p'} \left(\int_{\mathbf{R}^n} |\nabla u|^p \, dy\right)^{1/p}.$$

Using Theorem 2.3 (iii) it follows that

$$\mathbf{c}_p(B(0,\rho)) \geq C \rho^{n-p}.$$

2.1 CAPACITY

Now, we will test the capacity of $B(0,\rho)$ by the function
$$v(x) = \begin{cases} (|x|/\delta)^{\frac{p-n}{p-1}} & |x| > \delta, \\ 1 & |x| \leq \delta, \end{cases}$$
with $\delta > \rho$. Computing $\int_{\mathbf{R}^n} |\nabla v|^p \, dy$ we obtain
$$\mathbf{c}_p\big(B(0,\rho)\big) \leq C\,\delta^{n-p}$$
with the right constant. Letting $\delta \downarrow \rho$ we obtain the result.

(ii) The proof is similar to that of (i), we merely notice the differences. Let $u \in \mathcal{C}_c^\infty\big(B(0,1)\big)$, $u \geq 1$ on $B(0,\rho)$. Then
$$n\boldsymbol{\alpha}(n) \leq \int_{B(0,1)\setminus B(0,\rho)} |y|^{1-n} |\nabla u(y)| \, dy$$
$$\leq \Big(\int_{B(0,1)} |\nabla u|^n \, dy\Big)^{1/n} \Big(\int_{B(0,1)\setminus B(0,\rho)} |y|^{-n}\Big)^{(n-1)/n}$$
$$= \big[n\boldsymbol{\alpha}(n) \ln(1/\rho)\big]^{(n-1)/n} \Big(\int_{B(0,1)} |\nabla u|^n \, dy\Big)^{1/n}$$

and thus
$$\mathbf{c}_n\big(B(0,\rho); B(0,1)\big) \geq C \ln^{1-n}(1/\rho).$$
Conversely,
$$\mathbf{c}_n\big(B(0,\rho); B(0,1)\big) \leq C \ln^{1-n}(1/\rho),$$
which we get by testing with
$$v(x) = \begin{cases} \frac{\ln|x|}{\ln \delta} & |x| > \delta, \\ 1 & |x| \leq \delta, \end{cases}$$
with $\delta > \rho$.

(iii) First, assume that $1 < p < n$. Let $\eta \in \mathcal{C}_c^\infty(\mathbf{R}^n)$ be a cutoff function such that $0 \leq \eta \leq 1$, $\operatorname{spt}\eta \subset B(0,2\rho)$, $\eta = 1$ on an open set containing $B(0,\rho)$ and $|\nabla \eta| < 2/\rho$. Then
$$\gamma_{p;r} B(0,\rho) \leq \int_{\mathbf{R}^n} \big(|\nabla \eta|^p + r^{-p}\eta^p\big) \, dy \leq C\rho^n(\rho^{-p} + r^{-p}).$$
Conversely, if $v \in \mathcal{C}_c^\infty(\mathbf{R}^n)$, $v \geq 1$ on $B(0,\rho)$, then
$$\int_{\mathbf{R}^n} \big(|\nabla v|^p + r^{-p}|v|^p\big) \, dy \geq r^{-p} \int_{B(0,\rho)} dy = \boldsymbol{\alpha}(n)\rho^n r^{-p}.$$
It follows that
$$\gamma_{p;r} B(0,\rho) \geq C\rho^n r^{-p}.$$
From (i) and Theorem 2.6(ii) we get the remaining inequality
$$\gamma_{p;r} B(0,\rho) \geq C\rho^{n-p}.$$

Now, let $p = n$. Then we observe that $\gamma_{n;r} B(0,\rho)$ depends only on the ratio ρ/r. Hence we may assume that $r = 1$. If $\rho < 1/2$, appealing to Lemma 2.5 and (ii) we observe that the quantities $\mathbf{C}_n\big(B(0,\rho)\big)$, $\mathbf{c}_n\big(B(0,\rho); B(0,1)\big)$, $\ln^{n-1}(1/\rho)$ and

$g(\rho)$ are equivalent. On the other hand, if $\rho \geq 1/2$, then as in the $p < n$ part we obtain
$$\rho^n \leq C\, \mathbf{C}_n(B(0,\rho))$$
and
$$\mathbf{C}_n(B(0,\rho)) \leq C\rho^n. \qquad \square$$

2.9 COROLLARY. *If $1 < p \leq n$, then*
$$\mathbf{C}_p(\{x\}) = 0$$
for any $x \in \mathbf{R}^n$.

PROOF. It is enough to notice that a singleton is contained in an arbitrarily small ball and pass to limits in the estimates obtained in Theorem 2.8. $\square$

2.10 LEMMA. *Suppose that $p > n - 1$. Let $u \in \mathcal{C}^1(B(x_0, r_0))$. Then for every $r \in (0, R)$ we have*
$$(2.5) \qquad \operatorname{osc}_{\partial B(x_0,r)} u \leq C \left[r^{p+1-n} \int_{\partial B(x_0,r)} |\nabla u|^p(x)\, dH^{n-1}(x) \right]^{1/p},$$
with $C = C(n,p)$.

PROOF. We may assume that $x_0 = 0$ and $r = 1$. In the general case, the dependence on r may be obtained by an easy scaling argument.

We will now distinguish n-dimensional and $n - 1$-dimensional balls by subscripts. We denote by S^+ the upper halfsphere $\{x \in \partial B_n(0,1), x_n > 0\}$. We use the mapping φ defined by
$$\varphi(tz) = (z_1 \sin t, \ldots, z_{n-1} \sin t, \cos t), \quad z \in \partial B_{n-1}(0,1),\ t \in (0, \pi/2)$$
and observe that φ is a globally bilipschitz diffeomorphism of $U := B_{n-1}(0, \pi/2)$ onto S^+. Denote $v = u \circ \varphi$. Using Lemma 1.50, Hölder's inequality and change of variable estimates for bilipschitz mappings, for any $y_0 \in U$ we obtain
$$|u(\varphi(y_0)) - \bar{v}_U| \leq C \int_U \frac{|\nabla v(y)|}{|y_0 - y|^{n-1}}\, dy$$
$$\leq C \left[\int_U |\nabla v|^p\, dy \right]^{1/p}$$
$$\leq C \left[\int_{S^+} |\nabla u(x)|^p\, dH^{n-1}(x) \right]^{1/p}.$$
It follows that
$$|u(x) - u(x')| \leq C \left[\int_{\partial B(x)} |\nabla u(x)|^p\, dH^{n-1}(x) \right]^{1/p}$$
for any $x, x' \in \overline{S}^+$. Now, it remains to notice that the estimate is rotation-invariant and any two points of $\partial B(0,1)$ share a closed halfsphere. $\square$

2.11 LEMMA. *Let $\mathcal{O} = \mathbf{R}^n$. Let $E \subset B(x, r)$, $r < 1$. Then*
$$\min\{\gamma_{n;r}(E), \ln^{1-n}(1/r)\} \leq C\, \mathbf{C}_n(E),$$
where $C = C(n)$.

PROOF. We may suppose that $r < 1/2$. Also, by Theorem 2.3 (i) and (iv), we may assume that E is compact. Thus, by Theorem 2.3 (iii), the capacity of E may be defined by means of smooth competitors. Find smooth $u \in \mathcal{Y}_{n;1}(E)$ such that

$$\int_{\mathbf{R}^n} \left(|\nabla u|^n + |u|^n\right) dy$$

is close to $\mathbf{C}_n(E)$. We distinguish three cases.

(i) Suppose that there is $\rho \in (r, 2r)$ such that

$$\inf_{\partial B(x,\rho)} u > 1/3.$$

Set

$$v = \begin{cases} 1 & \text{on } B(x,\rho), \\ \min(1, 3u) & \text{outside } B(x,\rho). \end{cases}$$

Then

$$\mathbf{C}_n\bigl(B(x,r)\bigr) \leq \int_{\mathbf{R}^n} \left(|\nabla v|^n + |v|^n\right) dy \leq 3^n \int_{\mathbf{R}^n} \left(|\nabla u|^n + |u|^n\right) dy$$

and we use Theorem 2.8 (iii) to estimate $\mathbf{C}_n\bigl(B(x,r)\bigr)$ from below.

(ii) Suppose that there is $\rho \in (r, 2r)$ such that

$$\sup_{\partial B(x,\rho)} u < 2/3.$$

We set

$$w = \begin{cases} (3u-2)^+ & \text{on } B(x,\rho), \\ 0 & \text{outside } B(x,\rho). \end{cases}$$

Then $v \in \mathcal{Y}_{n,r}\bigl(E; B(x,2r)\bigr)$ and, appealing to Lemma 2.5, we obtain

$$\gamma_{n;r}(E) \leq C\,\mathbf{c}_n\bigl(E, B(x,2r)\bigr) \leq C \int_{B(x,2r)} \left(|\nabla w|^n + |w|^n\right) dy$$

$$\leq 3^n C \int_{\mathbf{R}^n} \left(|\nabla u|^n + |u|^n\right) dy.$$

(iii) Finally, consider the case when neither (i), nor (ii) can be used. Then, using Lemma 2.10, for every $\rho \in (r, 2r)$ we have

$$3^{-n} \leq \bigl(\mathrm{osc}_{\partial B(x,\rho)}\, u\bigr)^n \leq C\rho \int_{\partial B(x_0,\rho)} |\nabla u|^n(y)\, dH^{n-1}(y).$$

Dividing by ρ and integrating with respect to ρ we obtain

$$3^{-n} \ln 2 = 3^{-n} \int_r^{2r} \rho^{-1}\, d\rho \leq C \int_{B(x,2r)} |\nabla u|^n\, dy.$$

Since $\ln^{1-n}(1/r)$ is upper bounded on $(0, 1/2)$, the estimate holds. □

2.1.2 Polar sets.

Here we introduce p-polar sets and establish a few basic results including their relation to n-Green domains. In this subsection we assume that $\mathcal{O} = \mathbf{R}^n$.

2.12 DEFINITION. Let $1 < p \leq n$. A set $E \subset \mathbf{R}^n$ is said to be a *p-polar set* if $\mathbf{C}_p(E) = 0$.

2.13 REMARKS. In view of results of preceding sections, there are various possibilities in defining p-polar sets. Notice that a set E is p-polar if and only if it is "locally p-polar". If $p < n$, we may replace $\mathbf{C}_p$ by $\mathbf{c}_p$ in the definition of a p-polar set. A set E is n-polar if and only $\mathbf{c}_n(E \cap B(x,r); B(x,2r)) = 0$ for any ball $B(x,r) \subset \mathbf{R}^n$.

By Corollary 2.9, singletons are p-polar sets. It is immediately seen from the definition that any p-polar set has Lebesgue measure zero.

2.14 LEMMA. *Let E be a p-polar set. Then E is q-polar for any $q \in (1,p)$.*

PROOF. We may assume that $E \subset B(0,r)$. By Lemma 2.5, $\mathbf{c}_p(E; B(0,2r)) = 0$ and
$$\int_{B(0,2r)} |\nabla u|^q \, dx \le C \Big(\int_{B(0,2r)} |\nabla u|^p \, dx \Big)^{q/p}$$
for all $u \in \mathcal{Y}_{p;\infty}(E, B(0,2r))$. It follows that $\mathbf{c}_q(E) = 0$ for each $q \in (1,p)$. □

The following removability result shows that, roughly speaking, $W^{1,p}(\Omega) = W^{1,p}(\Omega')$ if and only if Ω differs from Ω' by at most a p-polar set.

2.15 THEOREM. *Let $\Omega \subset \Omega' \subset \mathbf{R}^n$ be open sets with $|\Omega' \setminus \Omega| = 0$. Then $v \llcorner \Omega \in W_0^{1,p}(\Omega)$ for each $v \in W_0^{1,p}(\Omega')$ if and only if $\Omega' \setminus \Omega$ is p-polar.*

PROOF. Suppose that $v \llcorner \Omega \in W_0^{1,p}(\Omega)$ for each $v \in W_0^{1,p}(\Omega')$. Consider a compact set $K \subset \Omega'$ and pick a function $w \in C_c^\infty(\mathbf{R}^n)$ such that spt $w \subset \Omega'$ and such that $w = 1$ on K. Then $w \in W_0^{1,p}(\Omega)$ and thus there are $w_j \in C_c^\infty(\mathbf{R}^n)$ with spt $w_j \subset \Omega$ such that $\|w_j - w\|_{1,p;\Omega} \to 0$. Since $w_j - w = 0$ a.e. outside Ω, we have
$$\mathbf{C}_p(K \setminus \Omega) \le \int_{\mathbf{R}^n} \big(|\nabla(w - w_j)|^p + |w - w_j|^p \big) \, dx \to 0.$$
We can express $\Omega' \setminus \Omega$ as $\bigcup (K_j \setminus \Omega)$ where $K_j \subset \Omega'$ are compact, which yields the result.

Now, suppose that $\Omega' \setminus \Omega$ is p-polar. Then there are $\varphi_j \in W^{1,p}(\mathbf{R}^n)$ such that $0 \le \varphi_j \le 1$, $\varphi_j = 1$ on an open set containing $\Omega' \setminus \Omega$ and
$$\int_{\mathbf{R}^n} \big(|\nabla \varphi_j|^p + |\varphi_j|^p \big) \, dx \to 0.$$
If $u \in C_0^\infty(\mathbf{R}^n)$ with spt $u \subset \Omega'$, then $u_j := u(1 - \varphi_j)$ has support in Ω and thus $u_j \in W_0^{1,p}(\Omega)$. Since $\|u - u_j\|_{1,p} \to 0$, it follows that $u \llcorner \Omega \in W_0^{1,p}(\Omega)$. Consequently, $v \llcorner \Omega \in W_0^{1,p}(\Omega)$ for any $v \in W_0^{1,p}(\Omega)$. □

2.16 THEOREM. *Let $\Omega \subset \mathbf{R}^n$ be an open set. The Ω is an n-Green domain if and only if $\mathbf{R}^n \setminus \Omega$ is not n-polar.*

PROOF. Suppose that Ω is not an n-Green domain. Then $1 \in Y_0^{1,n}$. An approximation argument using Lemma 1.83 shows that $f := \chi_\Omega$ is weakly differentiable on $\mathbf{R}^n$ and $\nabla f = 0$ a.e. in $\mathbf{R}^n$. By Corollary 1.42, f is constant on $\mathbf{R}^n$ and thus $|\mathbf{R}^n \setminus \Omega| = 0$. Now we use the "only if" part of Theorem 2.15 to show that $\mathbf{R}^n \setminus \Omega$ is n-polar. The converse implication follows from the "if" part of Theorem 2.15. □

2.1.3 Quasicontinuity.

The role played by quasicontinuity in the theory of Sobolev spaces is analogous to that played by Lusin's theorem in real analysis. In this subsection, we return to a general definition of capacity with reference domain $\mathcal{O}$. Throughout the exposition, we suppose that either $p < n$, or $r < \infty$ or $\mathcal{O}$ is an n-Green domain.

Our aim is to prove existence of p-quasicontinuous representatives of Sobolev functions and show further relations between capacity, quasicontinuity and Sobolev spaces.

2.17 DEFINITION. A function u is called *p-quasicontinuous* on E if for each $\varepsilon > 0$, there exists an open set V with $\mathbf{C}_p(V) < \varepsilon$ such that u restricted to $E \setminus V$ is finite and continuous.

In connection with the study of p-quasicontinuous functions the following terminology is common: We say that a property hold *p-quasi everywhere* (or at *p-quasi every x*, etc., abbreviated p-q.e.) if it holds except for a p-polar set.

2.18 REMARK. Using countable subadditivity and an exhaustion argument, we easily observe that a function u is p-quasicontinuous on an open set Ω if and only if it is p-quasicontinuous "locally on Ω", i.e. on all $\Omega' \subset\subset \Omega$. Now, on relatively compact subsets, all p-capacities that we consider are comparable (this follows from Lemma 2.6 and Lemma 2.7), so that we may use any of them to define p-quasicontinuity. Working with a reference domain $\mathcal{O}$ and a space $\mathcal{U}$, it is convenient to use just the $\mathcal{U}$-capacity γ.

2.19 LEMMA. *Let $u \in \mathcal{U}$. If u_j is a sequence of functions from $\mathcal{C}_c^\infty(\mathcal{O})$ such that*

$$\sum_{j=1}^\infty 2^{jp} \|u - u_j\|_\mathcal{U}^p < \infty,$$

then the pointwise limit

$$\lim_{j \to \infty} u_j = u_1 + \sum_{j=1}^\infty (u_{j+1} - u_j)$$

converges p-q.e. and defines a p-quasicontinuous representative of u.

PROOF. We refer to Remark 2.18 to justify that we can work with γ instead of $\mathbf{C}_p$. Set

$$G_j = \{x \in \mathcal{O} \colon |u_{j+1}(x) - u_j(x)| > 2^{-j}\}.$$

Then G_j are open sets and

$$2^j |u_{j+1} - u_j| \in \mathcal{Y}(G_j).$$

Hence

$$\gamma(G_j) \leq 2^{jp} \|u_{j+1} - u_j\|_\mathcal{U}^p$$

and

$$\sum_{j=1}^\infty \mathbf{C}_p(G_j) < \infty.$$

Given $\varepsilon > 0$ we find j_0 such that

$$\gamma(G) < \varepsilon \quad \text{with} \quad G = \bigcup_{j \geq j_0} G_j.$$

We have $|u_{j+1} - u_j| < 2^{-j}$ for $j \geq j_0$ on $\mathcal{O} \setminus G$. This implies that $\{u_j\}$ is a sequence of continuous functions which converges uniformly on $\mathcal{O} \setminus G$. It follows that $\lim_j u_j$ is continuous on $\mathcal{O} \setminus G$. This proves p-quasicontinuity of $\lim_j u_j$ on $\mathcal{O}$. □

2.20 THEOREM. *Let $u \in W^{1,p}_{\text{loc}}(\Omega)$. Then there is a p-quasicontinuous representative of u.*

PROOF. Since the property is local, we may assume that $\Omega = \mathbf{R}^n$ and $u \in W^{1,p}(\mathbf{R}^n)$. Since u can be approximated as required in Lemma 2.19, the assertion follows. □

Since the set $\mathcal{Y}(E)$ of all competitors in the definition of capacity is not closed, we will investigate its closure $\overline{\mathcal{Y}}(E)$. This will help us later to obtain existence of capacitary extremals.

2.21 LEMMA. *Let $E \subset \mathcal{O}$. Suppose that $u \in \mathcal{U}$ is a p-quasicontinuous function such that $u \geq 1$ p-q.e. on E. Then for any $\varepsilon > 0$ there is $v \in \mathcal{Y}(E)$ such that*

$$\|v - u\|_{\mathcal{U}} < \varepsilon.$$

PROOF. Choose an $\varepsilon > 0$ and find a corresponding positive integer k such that

$$\|u - u_k\|_{1,p} < \varepsilon \quad \text{where } u_k = \max\{u, -k\}.$$

There is an open set V such $\mathcal{O}$ with

$$\gamma(V) < (k+1)^{-p}\varepsilon$$

such that u restricted to $\mathcal{O} \setminus V$ is continuous and $u \geq 1$ on $E \setminus V$. Find $w \in \mathcal{Y}(V)$ such that

$$\int_{\mathcal{O}} (|\nabla w|^p + r^{-p}|w|^p)\, dx < (k+1)^{-p}\varepsilon$$

and set

$$v = (1+\varepsilon)u + (k+1)w.$$

Obviously v is close to u in the norm of $\mathcal{U}$. The set

$$G := V \cup (\{u > 1\} \setminus V)$$

is an open set containing A and $v \geq 1$ on G, thus implying that

$$v \in \mathcal{Y}(E).$$

2.22 LEMMA. *If $u \in \mathcal{U}(\mathcal{O})$ is p-quasicontinuous and $0 < a < \infty$, then*

$$\gamma(\{u > a\}; \mathcal{O}) \leq a^{-p} \int_{\{u>0\}} (|\nabla u|^p + r^{-p}|u|^p)\, dx.$$

PROOF. By Lemma 2.21, there are $v_j \in \mathcal{Y}(\{u > a\})$ such that

$$v_j \to a^{-1}u^+ \quad \text{in } \mathcal{U}.$$

A routine passage to limits proves the assertion. □

2.23 COROLLARY. *If u and v are p-quasicontinuous functions in $W^{1,p}(\Omega)$, and $u \leq v$ a.e in Ω, then $u \leq v$ p-q.e. in Ω.*

PROOF. It is enough to show that $\varphi := (u-v)^+ \eta = 0$ p-q.e. in Ω for any nonnegative $\eta \in \mathcal{C}_c^\infty(\Omega)$. But φ can be extended by zero to a p-quasicontinuous function in $W^{1,p}(\mathbf{R}^n)$ and the assertion follows from Lemma 2.22. $\square$

2.24 THEOREM. *Let $E \subset \mathcal{O}$. Then a function $u \in \mathcal{U}$ belongs to $\overline{\mathcal{Y}}(E)$ if and only if its p-quasicontinuous representative satisfies $u \geq 1$ q.e. on E.*

PROOF. In view of Remark 2.18, we may work with "γ-quasicontinuity". Sufficiency follows from Lemma 2.21. If $u \in \overline{\mathcal{Y}}(E)$ is p-quasicontinuous, then there exist $u_j \in \mathcal{Y}(E)$ such that
$$\|u_j - u\|_\mathcal{U}^p < 4^{-j}.$$
If v_j is a p-quasicontinuous representative of u_j, then $v_j \geq 1$ a.e. on an open set G_j containing E. By Corollary 2.23, $v_j \geq 1$ p-q.e. on G_j, in particular $v_j \geq 1$ p-q.e. on E. Set
$$E_j = \bigcup_{i=j}^\infty \{|u - v_i| > 2^{-i}\}.$$
Using Lemma 2.22 we obtain
$$\gamma(E_j) \leq \sum_{i=j}^\infty 2^{ip} \|u - v_i\|_{1,p} \to 0.$$
By Lemma 2.3 (i) there are open sets V_j containing E_j such that
$$\gamma(V_j) \to 0.$$
Now, if $x \in \mathbf{R}^n \setminus V_j$, then $|v_i(x) - u(x)| < 2^{-i}$ for all $i \geq j$ and thus $v_i \to u$ uniformly on $\mathbf{R}^n \setminus V_j$. It follows that
$$u = \lim_{j \to \infty} v_j \geq 1$$
p-q.e. on $E \setminus V_j$. $\square$

2.25 COROLLARY. *If $A \subset \mathcal{O}$, then*
$$\gamma(A) = \inf\left\{\int_\mathcal{O} (|u|^p + |\nabla u|^p) \, dx\right\},$$
where the infimum is taken over all p-quasicontinuous functions $u \in \mathcal{U}(\mathcal{O})$ such that $u(x) \geq 1$ for p-quasi all $x \in A$.

The following Lemma is a version of Theorem 2.147, which will be proved later. Let $\mathcal{O} = \mathbf{R}^n$.

2.26 LEMMA. *Let $\Omega \subset \mathbf{R}^n$ be an open set. A p-quasicontinuous function $u \in W^{1,p}(\Omega)$ belongs to $W_0^{1,p}(\Omega)$ if and only if the function*
$$v := \begin{cases} u & \text{on } \Omega, \\ 0 & \text{on } \mathbf{R}^n \setminus \Omega \end{cases}$$
is p-quasicontinuous on $\mathbf{R}^n$.

PROOF. First, assume that v is p-quasicontinuous and observe that without loss of generality, we may assume u is nonnegative. With each positive integer i we associate an open set G_i with $\mathbf{C}_p(G_i) < 2^{-i}$ such that

$$v \mathbin{\llcorner} (\mathbf{R}^n \setminus G_i) \quad \text{is continuous.} \tag{2.6}$$

Let φ_i be a competitor for $\mathbf{C}_p(G_i)$ such that $0 \leq \varphi_i \leq 1$ and

$$\int_{\mathbf{R}^n} \left(|\nabla \varphi_i|^p + |\varphi_i|^p\right) dx < 2^{-i},$$

and $\eta_i \in \mathcal{C}_c^\infty(\mathbf{R}^n)$ a cutoff function such that

$$0 \leq \eta_i \leq 1, \quad \operatorname{spt} \eta_i \subset B(0, 2^{i+1}), \quad \eta_i = 1 \text{ on } B(0, 2^i) \quad \text{and} \quad |\nabla \eta_i| \leq 2^{-i+1}.$$

Now define

$$w_i := \min\{v, i\} - \min\{v, i^{-1}\} \quad \text{and} \quad u_i = \eta_i(1 - \varphi_i) w_i.$$

Then

$$\Omega \cap \{w_i > 0\} \subset \Omega \cap B(0, 2^{i+1}) \cap \{u > 1/i\} \setminus G_i.$$

Let $\{x_i\}$ be a sequence of points of $\Omega \cap \{w_i > 0\}$ convergent to a point x. By (2.6),

$$v(x) = \lim_{i \to \infty} u(x_i) \geq 1/i.$$

It follows that $x \in \Omega$. Hence the support of u_i is compact and contained in Ω. By mollification we see that $u_i \in W_0^{1,p}(\Omega)$. We have

$$v - u_i = v - w_i + (1 - \eta_i) w_i + \eta_i \varphi_i w_i.$$

We estimate

$$|\nabla(w_i - v)|^p + |w_i - v|^p \leq \left(|\nabla v|^p + |v|^p\right) \chi_{\{0 < v < 1/i\} \cup \{v > i\}} + \min\{v, i^{-1}\},$$

$$|(1 - \eta_i) w_i|^p + |\nabla[(1 - \eta_i) w_i]|^p \leq C(|v|^p + |\nabla v|^p) \chi_{\mathbf{R}^n \setminus B(0, 2^i)},$$

$$|\eta_i \varphi_i w_i|^p + |\nabla(\eta_i \varphi_i w_i)|^p \leq C i^p \left(|\varphi_i|^p + |\nabla \varphi_i|^p\right) + C|\varphi_i|^p |\nabla v|^p.$$

Using Lebesgue's dominated convergence theorem, we conclude that

$$\|u - u_i\|_{1,p;\Omega} = \|v - u_i\|_{1,p} \to 0,$$

and so $u \in W_0^{1,p}(\Omega)$.

Conversely, suppose that $u \in W_0^{1,p}(\Omega)$. Then there is a sequence $\{u_i\}$ of functions from $\mathcal{C}_c^\infty(\mathbf{R}^n)$ with support in Ω such that

$$\|u - u_i\|_{1,p;\Omega} \to 0.$$

Let w be the p-quasicontinuous representative of $\lim u_i$ in $\mathbf{R}^n$. By Lemma 2.22,

$$\mathbf{C}_p(\{|w| \geq \varepsilon\} \setminus \Omega) \leq \varepsilon^{-p} \|w - u_i\|_{1,p;\mathbf{R}^n}^p \to 0$$

for any $\varepsilon > 0$. Hence $w = 0 = v$ p-quasi everywhere outside Ω. By Corollary 2.23, $w = u = v$ p-quasi everywhere on Ω. It follows that $v = w$ p-q.e. in $\mathbf{R}^n$ and thus v is p-quasicontinuous in $\mathbf{R}^n$. □

2.1.4 Multipliers.

We introduce the space of multipliers for the purpose of allowing a very general class for the coefficients of differential equations discussed in Chapter 3.

2.27 DEFINITION. Let X, Y be Banach spaces whose elements are measurable functions on Ω. We define $M(X, Y)$ as the space of all measurable functions v on Ω such that
$$\|uv\|_Y \leq C\|u\|_X$$
for all $u \in X$. Given v, the smallest constant C with such a property is called the multiplier norm of v and denoted by $\|v\|_{M(X,Y)}$. The functions from $M(X, Y)$ are termed *multipliers*.

We already know that powers of functions from suitable Morrey spaces provide multipliers from $Y^{1,p}$ to L^q, $q > p$, according to Adams' inequality, Theorem 1.93.

In this subsection we treat the case $q = p$ and prove a result due to Maz'ya and Shaposhnikova [**MaSh**] which will be used later in Chapter 4.

2.28 THEOREM. *Let f be a nonnegative measurable function on $\mathcal{O}$. Then $f^{1/p} \in M(\mathcal{U}, L^p(\mathcal{O}))$ if and only if*

(2.7)
$$\int_E f\,dx \leq K\gamma(E)$$

for any measurable set $E \subset \Omega$. Furthermore, in this case
$$K_0^{1/p} \leq \|f^{1/p}\|_{M(\mathcal{U}, L^p)} \leq CK_0^{1/p}$$
where K_0 is the best constant in (2.7).

PROOF. Suppose that $f^{1/p} \in M(\mathcal{U}, L^p)$ with $\|f^{1/p}\| = \overline{K}$. Let $E \subset \mathbf{R}^n$ be a measurable set and $u \in \mathcal{Y}(E)$. Then
$$\int_E f\,dx \leq \int_E fu^p\,dx \leq \overline{K}^p \|u\|_{\mathcal{U}}^p.$$
Taking the infimum over $u \in \mathcal{Y}(E)$ we obtain
$$\int_E f\,dx \leq \int_E fu^p\,dx \leq \overline{K}^p \gamma(E).$$
Conversely, let $u \in \mathcal{U}$ (a p-quasicontinuous representative) and $t > 0$. Set
$$v(x) = \min\left\{1, 2\left(\frac{|u|}{t} - \frac{1}{2}\right)^+\right\}.$$
Then $v \in \overline{\mathcal{Y}}(\{|u| > t\})$ and thus
$$\gamma\{|u| > t\} \leq \int_\Omega \left(|\nabla v|^p + r^{-p} v^p\right) dx$$
$$\leq (2/t)^p \int_{\{t/2 < |u| \leq t\}} \left(|\nabla u|^p + r^{-p}|u - t/2|^p\right) dx + r^{-p}|\{|u| > t\}|.$$

Using Fubini's theorem with $W = \{(x,t)\colon x \in \Omega,\ t/2 < |u(x)| \leq t\}$ we have

$$\int_\Omega f|u|^p\,dx = p\int_0^\infty t^{p-1}\Big(\int_{\{|u|>t\}} f\,dx\Big)\,dt$$

$$\leq pK\int_0^\infty t^{p-1}\gamma(\{|u|>t\})\,dt$$

$$\leq pK2^p \int_0^\infty t^{-1} \int_{\{t/2<|u|\leq t\}} \Big(|\nabla u|^p + r^{-p}|u|^p\Big)\,dx$$

$$+ pKr^{-p}\int_0^\infty t^{p-1}|\{u>t\}|\,dt$$

$$= pK2^p \int_{\Omega\times(0,\infty)} \chi_W t^{-1}\Big(|\nabla u|^p + r^{-p}|u|^p\Big)\,dx\,dt + pKr^{-p}\int_\Omega |u|^p\,dx$$

$$= pK2^p \int_\Omega \Big(\int_{|u|}^{2|u|} \frac{dt}{t}\Big)\Big(|\nabla u|^p + r^{-p}|u|^p\Big)\,dx + pKr^{-p}\int_\Omega |u|^p\,dx$$

$$\leq pK2^p \ln 2 \int_\Omega \Big(|\nabla u|^p + r^{-p}|u|^p\Big)\,dx + pKr^{-p}\int_\Omega |u|^p\,dx.$$

This concludes the proof. □

2.1.5 Capacity and energy minimizers.

We suppose that either $p < n$, or $r < \infty$, or $\mathcal{O}$ is an n-Green domain. We first observe that

$$(2.8) \qquad \lim_{\varepsilon \to 0+} \frac{|\xi + \varepsilon\theta|^p - |\xi|^p}{\varepsilon} = \nabla(|\xi|^p) \cdot \theta = p|\xi|^{p-2}\xi \cdot \theta$$

for all $\xi, \theta \in \mathbf{R}^n$ and, due to convexity of the function $\xi \mapsto |\xi|^p$,

$$(2.9) \qquad |\xi'|^p - |\xi|^p \geq p|\xi|^{p-2}\xi \cdot (\xi' - \xi)$$

for all $\xi, \xi' \in \mathbf{R}^n$.

The capacitary extremals defined below represent an important special case of solutions of obstacle problems. In Chapter 5 we consider obstacle problems in greater generality.

2.29 DEFINITION. Let $E \subset \mathbf{R}^n$. A function $u \in \mathcal{U}$ is called a *capacitary extremal* for $\gamma(E)$ if $u \in \overline{\mathcal{Y}}(E)$ and

$$\gamma(E) = \|u\|_\mathcal{U}^p.$$

2.30 LEMMA. *A function $u \in \overline{\mathcal{Y}}(E)$ is a capacitary extremal for $\gamma(E)$ if and only if*

$$(2.10) \qquad \int_\mathcal{O} \Big(|\nabla u|^{p-2}\nabla u \cdot [\nabla v - \nabla u] + r^{-p}|u|^{p-2}u[v-u]\Big)\,dx \geq 0$$

for any $v \in \overline{\mathcal{Y}}(E)$.

PROOF. If u is a capacitary extremal for $\gamma(E)$ and $v \in \overline{\mathcal{Y}}(E)$, then for any $\varepsilon > 0$, $v_\varepsilon := u + \varepsilon(v - u)$ also belongs to $\overline{\mathcal{Y}}(E)$. We have

$$\int_{\mathcal{O}} \varepsilon^{-1}\Big[|\nabla v_\varepsilon|^p - |\nabla u|^p + r^{-p}(|v_\varepsilon|^p - |u|^p)\Big]\,dx \geq 0.$$

Letting $\varepsilon \to 0$ with the aid of (2.8) we obtain (2.10). Conversely, if (2.10) is satisfied, using (2.9) we easily derive that u is a capacitary extremal.

2.31 THEOREM. *Let $E \subset \mathcal{O}$ be a set with $\gamma(E) < \infty$. Then there is a unique capacitary extremal u for $\gamma(E)$ and its p-quasicontinuous representative satisfies $u = 1$ p-q.e. on E.*

PROOF. If $\gamma(E) < \infty$, then $\overline{\mathcal{Y}}(E)$ is a nonempty closed convex subset of $\mathcal{U}$. Let $\{u_j\}$ be a minimizing sequence. Then $\{u_j\}$ is obviously bounded and passing to a subsequence, we may suppose that u_j is weakly convergent to a $u \in \overline{\mathcal{Y}}(E)$. Since norms in Banach spaces are weakly lower semicontinuous, we have

$$\|u\|_{\mathcal{U}}^p \leq \liminf_{j \to \infty} \|u_j\|_{\mathcal{U}}^p = \gamma(E).$$

The uniqueness of the capacitary extremal follows from the strict convexity of the function $\xi \mapsto |\xi|^p$. By Theorem 2.24, $u \geq 1$ p-q.e. on E. The inequality $u \leq 1$ is easy, since the truncation $u \mapsto \min\{u, 1\}$ does not increase the norm in $\mathcal{U}$. □

2.32 DEFINITION. Let μ be a Radon measure on $\mathcal{O}$ and $\mathcal{U} = \mathcal{U}_{p;r}(\mathcal{O})$. The $p;r$-energy of μ on $\mathcal{O}$ is defined as

$$\mathcal{E}(\mu) = \mathcal{E}_{p;r}(\mu;\mathcal{O}) = \sup\Big\{\Big(\int_{\mathcal{O}} v\,d\mu\Big)^{p'} : v \in \mathcal{C}_c^\infty(\mathcal{O}), \|v\|_{\mathcal{U}} \leq 1\Big\}.$$

This means that $\mathcal{E}(\mu)$ is the p'-th power of the norm of μ in $\mathcal{U}^*$ provided that μ belongs to this dual space; otherwise $\mathcal{E}(\mu) = \infty$.

Suppose that $\mu \in \mathcal{U}^*$. Then the functional

$$v \mapsto \int_{\mathcal{O}} v\,d\mu, \quad v \in \mathcal{C}_c^\infty(\mathcal{O}),$$

can be extended in a unique way to a continuous functional on $\mathcal{U}$ and this extension will be denoted as

$$\langle \mu, v \rangle.$$

We say that u is the $\mathcal{U}$-potential of μ if u minimizes

$$\frac{1}{p}\|v\|_{\mathcal{U}}^p - \langle \mu, v \rangle$$

in $\mathcal{U}$.

The $\mathcal{U}$-potential of μ exists and is unique for any $\mu \in \mathcal{U}^*$; we omit the proof which is precisely that of Theorem 2.31.

2.33 LEMMA. *Let $\mu \in \mathcal{U}^*$ be a Radon measure. A function $u \in \mathcal{U}$ is the $\mathcal{U}$-potential of μ if and only if*

(2.11) $$\int_{\mathcal{O}} \Big(|\nabla u|^{p-2}\nabla u \cdot \nabla v + r^{-p}|u|^{p-2}uv\Big)\,dx = \langle \mu, v \rangle, \quad \text{for any } v \in \mathcal{U}.$$

PROOF. The proof follows the same lines as that of Lemma 2.30.

2.34 THEOREM. *Let μ be a measure of finite $p;r$-energy on $\mathcal{O}$ and let u be its $\mathcal{U}$-potential. Then*
$$\mathcal{E}(\mu) = \langle \mu, u \rangle = \|u\|_{\mathcal{U}}^p.$$

PROOF. By Lemma 2.33 the function u satisfies
$$\int_{\mathcal{O}} \left(|\nabla u|^{p-2} \nabla u \cdot \nabla v + r^{-p} |u|^{p-2} uv \right) dx = \langle \mu, v \rangle$$

for all $v \in \mathcal{U}$. Setting $v = u$ we obtain

(2.12)
$$\|u\|_{\mathcal{U}}^p = \int_{\mathcal{O}} \left(|\nabla u|^p + r^{-p} |u|^p \right) dx = \langle \mu, u \rangle.$$

Let $v \in \mathcal{U}$, $\|v\|_{\mathcal{U}} \leq \|u\|_{\mathcal{U}}$. Then
$$\langle \mu, v \rangle = \left(\langle \mu, v \rangle - \frac{1}{p} \|v\|_{\mathcal{U}}^p \right) + \frac{1}{p} \|v\|_{\mathcal{U}}^p$$
$$\leq \left(\langle \mu, u \rangle - \frac{1}{p} \|u\|_{\mathcal{U}}^p \right) + \frac{1}{p} \|u\|_{\mathcal{U}}^p = \langle \mu, u \rangle.$$

It follows that
$$\langle \mu, w \rangle \leq \left\langle \mu, \frac{u}{\|u\|_{\mathcal{U}}} \right\rangle$$

for all $w \in \mathcal{U}$ with $\|w\|_{\mathcal{U}} \leq 1$, and we conclude that
$$\mathcal{E}(\mu) = \left\langle \mu, \frac{u}{\|u\|_{\mathcal{U}}} \right\rangle^{p'}.$$

Using (2.12) we obtain
$$\left[\mathcal{E}(\mu) \right]^{1/p'} = \frac{\langle \mu, u \rangle}{\|u\|_{\mathcal{U}}} = \|u\|_{\mathcal{U}}^{p-1},$$

so that
$$\mathcal{E}(\mu) = \|u\|_{\mathcal{U}}^p. \qquad \square$$

2.35 DEFINITION. Let $E \subset \mathcal{O}$ be an arbitrary set and $\mu \in \mathcal{U}^*$ a Radon measure on $\mathcal{O}$. We say that μ is a *capacitary distribution* for $\gamma(E)$ if the $\mathcal{U}$-potential of μ is the capacitary extremal for $\gamma(E)$.

2.36 THEOREM. *Let $E \subset \mathcal{O}$ be a set with $\gamma(E) < \infty$ and u the capacitary extremal for $\gamma(E)$. Then there is a unique capacitary distribution μ for $\gamma(E)$. Moreover,*
$$\gamma(E) = \langle \mu, u \rangle = \mathcal{E}(\mu).$$

PROOF. By Lemma 2.30, the capacitary extremal u for $\gamma(E)$ satisfies

(2.13) $$\int_{\mathcal{O}} \bigl(|\nabla u|^{p-2}\nabla u \cdot [\nabla v - \nabla u] + r^{-p}|u|^{p-2}u[v-u]\bigr)\,dx \geq 0 \text{ for all } v \in \overline{\mathcal{Y}}(E).$$

By Lemma 2.33, μ is the capacitary distribution μ for $\gamma(E)$ if and only if

(2.14) $$\int_{\mathcal{O}} \bigl(|\nabla u|^{p-2}\nabla u \cdot \nabla w + r^{-p}|u|^{p-2}uw\bigr)\,dx = \langle \mu, w\rangle \text{ for all } w \in \mathcal{U}.$$

If $w \in \mathcal{C}_c^{\infty}$ is nonnegative, then $u + w \in \overline{\mathcal{Y}}(E)$ and by (2.13),

$$\int_{\mathcal{O}} \bigl(|\nabla u|^{p-2}\nabla u \cdot \nabla w + r^{-p}|u|^{p-2}uw\bigr)\,dx \geq 0.$$

In other words

$$w \mapsto \int_{\mathcal{O}} \bigl(|\nabla u|^{p-2}\nabla u \cdot \nabla w + r^{-p}|u|^{p-2}uw\bigr)\,dx$$

is a nonnegative linear functional on $\mathcal{C}_c^{\infty}$ and by the Riesz representation theorem and a routine approximation argument, there is a unique nonnegative Radon measure μ such that (2.14) holds. By Theorem 2.34, substitution $w = u$ in (2.14) shows that

$$\gamma(E) = \int_{\mathcal{O}} \bigl(|\nabla u|^p + r^{-p}|u|^p\bigr)\,dx = \langle \mu, u\rangle = \mathcal{E}(\mu). \qquad \square$$

2.37 THEOREM. *Let $\mu \in \mathcal{U}^*$ be a Radon measure and $E \subset \mathcal{O}$ be μ-measurable. Then*

(2.15) $$[\mu(E)]^p \leq [\mathcal{E}(\mu)]^{p-1}\gamma(E).$$

PROOF. If E is compact, then for any $v \in \mathcal{C}_c^{\infty}(\mathcal{O})$ with $v \geq 1$ on E we have

$$\mu(E) = \int_{\mathcal{O}} v\,d\mu = \langle \mu, v\rangle \leq \|v\|_{\mathcal{U}} \|\mu\|_{\mathcal{U}^*}.$$

Taking infimum over all v, by Theorem 2.3 (iii) we obtain

$$\mu(E) \leq \bigl[\gamma(E)\bigr]^{1/p}\bigl[\mathcal{E}(\mu)\bigr]^{1/p'}$$

which is equivalent to (2.15). If E is open, we consider an increasing sequence $\{E_j\}$ of compact sets such that $E = \bigcup E_j$ and get (2.15) by passing to the limit according to Theorem 2.3 (iv). Now, by Theorem 2.3 (i), (2.15) holds for an arbitrary μ-measurable set E.

2.38 COROLLARY. *Let $\mu \in \mathcal{U}^*$ be a Radon measure. Then $\mu(E) = 0$ for any p-polar set $E \subset \mathcal{O}$ (μ is "absolutely continuous with respect to γ").*

2.39 THEOREM. *Let $\mu \in \mathcal{U}^*$ be a Radon measure and $u \in \mathcal{U}$. Then each p-quasicontinuous representative of u belongs to $L^1(\mathcal{O}, \mu)$ and satisfies*

$$\langle \mu, u\rangle = \int_{\mathcal{O}} u\,d\mu.$$

PROOF. Let $u \in \mathcal{U}$. Let v_j be a sequence of functions from $\mathcal{C}_c^\infty(\mathcal{O})$ which converges to u so rapidly that

$$\|v_{j+1} - v_j\|_\mathcal{U} < 4^{-j}, \qquad j = 1, 2, \ldots.$$

Then

$$\int_\mathcal{O} \sum_{j=1}^\infty |v_{j+1} - v_j| \, d\mu = \sum_{j=1}^\infty \int_\mathcal{O} |v_{j+1} - v_j| \, d\mu < \infty.$$

It follows that the sequence $\{v_j\}$ converges μ-almost everywhere and in $L^1(\mathcal{O}, \mu)$ to a function v. By Lemma 2.19, v is a p-quasicontinuous representative of u. We have

$$\int_\mathcal{O} v \, d\mu = \lim_{j \to \infty} \int_\mathcal{O} v_j \, d\mu = \lim_{j \to \infty} \langle \mu, v_j \rangle = \langle \mu, v \rangle.$$

For any other p-quasicontinuous representative of u we have

$$\int_\mathcal{O} u \, d\mu = \langle \mu, u \rangle$$

as well, since by Corollary 2.23 $u = v$ p-q.e., and by Corollary 2.38, μ is absolutely continuous with respect to γ. □

2.40 NOTATION. In the rest of this subsection, the capacitary extremal for $\gamma(E)$ will be denoted by u_E and the corresponding capacitary distribution by μ_E.

2.41 COROLLARY. *Let $E \subset \mathcal{O}$ be a set with $\gamma(E) < \infty$. Then*

$$\gamma(E) = \int_\mathcal{O} u_E \, d\mu_E.$$

PROOF. This follows from Theorem 2.39 and Theorem 2.36. □

2.42 COROLLARY.
Let $E \subset \mathcal{O}$ and $u \in \overline{\mathcal{Y}}(E)$. Then

$$\int_\mathcal{O} u_E \, d\mu_E \leq \int_\mathcal{O} u \, d\mu_E.$$

PROOF. Using test function $u - u_E$, by combination of Lemma 2.33 and Lemma 2.30 we obtain

$$\langle \mu_E, u_E \rangle \leq \langle \mu_E, u \rangle.$$

In view of Lemma 2.39, this is equivalent to the assertion. □

2.43 LEMMA. *Let $U \subset \mathcal{O}$ be an open set, $E \subset \mathcal{O}$ and $u \in W^{1,p}_{\mathrm{loc}}(\Omega)$. Suppose that $u = 1$ p-q.e. on $E \cap U$. Then*

(2.16) $$\int_U u \, d\mu_E = \int_U u_E \, d\mu_E.$$

In particular,

(2.17) $$\gamma(E) = \int_\mathcal{O} u_E \, d\mu_E = \mu_E(\mathcal{O}).$$

PROOF. Let η be a smooth function such that $0 \le \eta \le 1$ and spt $\eta \subset U$. Then by Theorem 2.24, $u_E + \eta(u - u_E)$ and $u_E - \eta(u - u_E)$ are in $\overline{\mathcal{Y}}(E)$ and thus by Corollary 2.42
$$\int_{\mathcal{O}} \eta(u - u_E)\, d\mu_E = 0.$$
Letting $\eta \uparrow \chi_U$ we obtain (2.16). The consequence (2.17) follows by using Corollary 2.41. □

2.44 COROLLARY. *Let $E \subset \mathcal{O}$ be a compact set. Then the capacitary distribution μ for $\gamma(E)$ is supported by E and*

(2.18) $$\gamma(E) = \mu(E) = \mu(\mathcal{O}).$$

2.45 THEOREM. *Let $U \subset \mathcal{O}$ be an open set and $E \subset \mathcal{O}$ be a set with $\gamma(E) < \infty$. Then*
$$\mu_E(U) \le \gamma(E \cap U).$$

PROOF. Let $\nu = \mu_E \llcorner U$ and u be the $\mathcal{U}$-potential of ν. First, we will prove that $u \le u_E$. Proceeding by contradiction we assume that this is not the case. Denote $v_1 = \min\{u_E, u\}$, $v_2 = \max\{u_E, u\}$. Then v_1 differs from u and v_2 differs from u_E. Since u and u_E are unique minimizers of the associated integrals, we have

(2.19) $$\tfrac{1}{p}\|u\|_{\mathcal{U}}^p - \int_\Omega u\, d\nu < \tfrac{1}{p}\|v_1\|_{\mathcal{U}}^p - \int_\Omega v_1\, d\nu$$

and

(2.20) $$\tfrac{1}{p}\|u_E\|_{\mathcal{U}}^p - \int_\Omega u_E\, d\mu_E < \tfrac{1}{p}\|v_2\|_{\mathcal{U}}^p - \int_\Omega v_2\, d\mu_E.$$

Adding these inequalities and observing that
$$u - v_1 = v_2 - u_E \ge 0$$
we obtain
$$\int_\Omega (v_2 - u_E)\, d\mu_E < \int_\Omega (u - v_1)\, d\nu = \int_U (v_2 - u_E)\, d\mu_E \le \int_\Omega (v_2 - u_E)\, d\mu_E,$$
which is the contradiction.

As a second step, we denote $w := u_{E \cap U}$ and using Lemma 2.43 and the first step we estimate

(2.21)
$$\begin{aligned}
\|u\|_{\mathcal{U}}^p &= \|u\|_{\mathcal{U}}^p - p\int_\Omega u\, d\nu + p\int_\Omega u\, d\nu \\
&\le \|w\|_{\mathcal{U}}^p - p\int_\Omega w\, d\nu + p\int_\Omega u_E\, d\nu \\
&= \|w\|_{\mathcal{U}}^p - p\int_U w\, d\mu_E + p\int_U u_E\, d\mu_E \\
&= \|w\|_{\mathcal{U}}^p.
\end{aligned}$$

Finally, we will estimate $\mu_E(U)$. By Lemma 2.43 we have
$$\mu_E(U) = \int_\Omega w\, d\nu.$$

We test (2.11) for ν by w and with the help of (2.21) we obtain

$$\mu_E(U) = \int_\Omega w\,d\nu = \int_\Omega (|\nabla u|^{p-2}\nabla u \cdot \nabla w + r^{-p}|u|^{p-2}uw)\,dx$$

$$\leq \left(\int_\Omega (|\nabla u|^p + r^{-p}|u|^p)\,dx\right)^{1/p'} \left(\int_\Omega (|\nabla w|^p + r^{-p}|w|^p)\,dx\right)^{1/p}$$

$$\leq \int_\Omega (|\nabla w|^p + r^{-p}|w|^p)\,dx = \gamma_{p;r}(E \cap U).\qquad \square$$

2.46 THEOREM. *Let $E \subset \mathcal{O}$ be a compact set. Then*

$$\gamma(E) = \sup\{[\nu(E)]^p \colon \nu \geq 0 \text{ is a Radon measure on } \mathcal{O},\ \mathcal{E}(\nu) \leq 1\}.$$

PROOF. We may assume that $\gamma(E) > 0$. Let μ be the capacitary distribution for $\gamma(E)$ and u the corresponding capacitary extremal. Denote $\nu_0 = \left[\mathcal{E}(\mu)\right]^{-1/p'}\mu$. Then $\mathcal{E}(\nu_0) \leq 1$ and by Corollary 2.44 and Theorems 2.39, 2.34 and 2.36 we have

$$\nu_0(E) = \nu_0(\mathcal{O}) \geq \int_\mathcal{O} u\,d\nu_0$$

$$= \mathcal{E}(\mu)^{-1/p'}\int_\mathcal{O} u\,d\mu = \mathcal{E}(\mu)^{1-1/p'} = \gamma(E)^{1/p}.$$

Hence

$$\gamma(E) \leq [\nu_0(E)]^p \leq \sup\{[\nu(E)]^p \colon \mathcal{E}(\nu) \leq 1\}.$$

The converse inequality follows from Theorem 2.37. $\qquad \square$

2.1.6 Thinness.

In measure theory, a set is sparse or "dispersed" at a point if the metric density of the set at this point is zero. In the framework of potential theory the analogous concept is a set being thin at a point. In this subsection we assume that $1 < p \leq n$ and $\mathcal{O} = \mathbf{R}^n$.

2.47 DEFINITION. A set $A \subset R^n$ is said to be *p-thin* $(1 < p \leq n)$ at $x \in \mathbf{R}^n$ if

$$(2.22) \qquad \int_0^1 \left[\frac{\mathbf{C}_p(A \cap B(x,r))}{r^{n-p}}\right]^{1/(p-1)} \frac{dr}{r} < \infty.$$

For any set $A \subset \mathbf{R}^n$ we denote by $b_p A$ the set of all points of $\mathbf{R}^n$ at which A is not thin (the *p-base* of A).

2.48 THEOREM. *If $E \subset \mathbf{R}^n$ is p-thin at $z \in \mathbf{R}^n \setminus E$, then there is an open set U containing E such that U is p-thin at z.*

PROOF. We may assume that $E \subset B(z,1)$. We write $r_j = 2^{-j+1}$ and $B_j = B(z, r_j)$. For every $j = 1, 2, \ldots$ we find an open set $U_j \supset \overline{B}_j \cap E$ such that

$$\mathbf{C}_p(\overline{B}_j \cap U_j) \leq \mathbf{C}_p(\overline{B}_j \cap E) + r_j^{n-1}.$$

Set

$$U = (U_1 \setminus \overline{B}_2) \cup (U_1 \cap U_2 \setminus \overline{B}_3) \cup (U_1 \cap U_2 \cap U_3 \setminus \overline{B}_4) \cup \ldots.$$

Then U is an open set containing E and
$$U \cap \overline{B}_j \subset U_j \cap \overline{B}_j$$
for each $j = 1, 2, \ldots$. It follows that
$$\mathbf{C}_p(B(z,r) \cap U) \leq \mathbf{C}_p(\overline{B}_j \cap U) \leq \mathbf{C}_p(\overline{B}_j \cap U_j)$$
$$\leq \mathbf{C}_p(\overline{B}_j \cap E) + r_j^{n-1} \leq \mathbf{C}_p(B(z,2r) \cap E) + (2r)^{n-1}$$
for each $r \in (r_{j+1}, r_j]$, $j = 1, 2, \ldots$, and thus
$$\int_0^1 \left(\frac{\mathbf{C}_p(B(z,r) \cap U)}{r^{n-p}} \right)^{1/(p-1)} \frac{dr}{r}$$
$$\leq \int_0^1 \left(\frac{\mathbf{C}_p(B(z,2r) \cap E) + (2r)^{n-1}}{r^{n-p}} \right)^{1/(p-1)} \frac{dr}{r} < \infty.$$
We conclude that E is p-thin at z. □

2.49 THEOREM. *Let $A \subset \mathbf{R}^n$ and $x \in \mathbf{R}^n$. Then A is p-thin at x if and only if*

(2.23) $$\int_0^1 \left[\frac{\gamma_{p;r}(A \cap B(x,r))}{r^{n-p}} \right]^{1/(p-1)} \frac{dr}{r} < \infty.$$

PROOF. One part is trivial, because
$$\mathbf{C}_p(E) \leq \gamma_{p;r}(E)$$
for any $E \subset \mathbf{R}^n$ and $r \in (0,1)$. Also the converse follows easily from Lemma 2.6 if $p < n$. Suppose that $p = n$ and A is n-thin at z. By Lemma 2.11,
$$\min\{\gamma_{n;r}(A \cap B(x,r)), \ln^{1-n}(1/r)\} \leq C\, \mathbf{C}_n(A \cap B(x,r))$$
for all $r \in (0,1)$. If this minimum is $\gamma_{n;r}(A \cap B(x,r))$ for all r small enough, then clearly
$$\int_0^1 [\gamma_{n;r}(A \cap B(x,r))]^{1/(n-1)} \frac{dr}{r} < \infty.$$
Otherwise, there is a decreasing sequence $r_j \to 0$ such that
$$\ln^{1-n}(1/r_j) \leq C\, \mathbf{C}_n(A \cap B(x,r_j)), \quad j = 1, 2, \ldots$$
and
$$\ln r_{j-1} < \frac{1}{2} \ln r_j, \quad j = 2, 3, \ldots.$$
With $r_0 := 1$, we have
$$\int_0^1 [\mathbf{C}_n(A \cap B(x,r))]^{1/(n-1)} \frac{dr}{r} \geq \sum_{j=1}^\infty \int_{r_j}^{r_{j-1}} [\mathbf{C}_n(A \cap B(x,r_j))]^{1/(n-1)} \frac{dr}{r}$$
$$\geq \sum_{j=1}^\infty \int_{r_j}^{r_{j-1}} [\ln^{1-n}(1/r_j)]^{1/(n-1)} \frac{dr}{r} = \sum_{j=1}^\infty \ln^{-1}(1/r_j) \int_{r_j}^{r_{j-1}} \frac{dr}{r}$$
$$= \sum_{j=1}^\infty \frac{\ln r_{j-1} - \ln r_j}{\ln(1/r_j)} = \sum_{j=1}^\infty \left(1 - \frac{\ln r_{j-1}}{\ln r_j} \right) = \infty.$$

This contradicts that A is n-thin at x. □

2.50 REMARK. Observe also that (2.22) is equivalent to the analogous integral with $\mathbf{C}_p$ replaced by $\mathbf{c}_p$ if $p < n$.

2.51 COROLLARY. *Let $M \subset \mathbf{R}^n$ be a measurable set. If $\mathbf{R}^n \setminus M$ is p-thin at $x \in \mathbf{R}^n$, then x is a Lebesgue density point for M.*

PROOF. It follows from the elementary inequality

$$\frac{|E \cap B(x,r)|^{(n-p)/n}}{|B(x,r)|} \leq C \frac{\gamma_{p;r}(E \cap B(x,r))}{r^{n-p}}$$

and Theorem 2.49. □

2.1.7 Capacity and Hausdorff measure.

In this subsection we set $\mathcal{O} = \mathbf{R}^n$. It follows immediately from Theorem 2.8 and definitions that

$$\mathbf{c}_p(E) \leq CH^{n-p}(E)$$

whenever $E \subset \mathbf{R}^n$ and $1 < p \leq n$. The next result gives a bit more information.

2.52 THEOREM. *If $1 < p \leq n$, $E \subset B$ and $H^{n-p}(E) < \infty$, then $\mathbf{C}_p(E) = 0$.*

PROOF. Since the 0-dimensional Hausdorff measure is just the cardinality, from Corollary 2.9 it follows that $\mathbf{C}_n(E) = 0$ for each set E with $H^0(E) < \infty$. Hence we may assume that $p < n$ and in view of Lemma 2.6 (iii), we may replace $\mathbf{C}_p$ with $\mathbf{c}_p$.

First, we will show that for each open set $V \supset E$, there is an open set $W \subset \mathbf{R}^n$ and a function $v \in Y^{1,p}(\mathbf{R}^n)$ such that

$$\chi_W \leq v \leq \chi_V \quad \text{and} \quad \int_{\mathbf{R}^n} |\nabla v|^p \, dx < C(H^{n-p}(E) + 1)$$

for some constant C. For this, we construct a sequence of balls $B(a_i, r_i)$ with $B(a_i, 2r_i) \subset V$ such that

$$E \subset W := \bigcup_{i=1}^{\infty} B(a_i, r_i) \quad \text{and} \quad \sum_{i=1}^{\infty} r_i^{n-p} \leq C(H^{n-p}(E) + 1).$$

For each positive integer i, let $f_i \in Y^{1,p}(\mathbf{R}^n)$ be such that

$$\chi_{B(a_i, r_i)} \leq f_i \leq \chi_{B(a_i, 2r_i)} \quad \text{and} \quad |\nabla f_i| \leq 1/r_i.$$

The desired function is defined by

$$v = \sup_{i \geq 1} f_i = \lim_{j \to \infty} \sup_{1 \leq i \leq j} f_i.$$

Indeed, using Fatou's lemma we obtain

$$\int_{\mathbf{R}^n} |\nabla v|^p \, dx \leq \liminf_{j \to \infty} \int_{\mathbf{R}^n} \sum_{i=1}^{j} |\nabla f_i|^p \, dx \leq C(H^{n-p}(E) + 1).$$

Starting with $V_1 = \mathbf{R}^n$, we inductively define open sets $V_i \supset E$ and functions $v_i \in Y^{1,p}(\mathbf{R}^n)$ such that

$$\chi_{V_{i+1}} \leq v_i \leq \chi_{V_i} \quad \text{and} \quad \int_{\mathbf{R}^n} |\nabla v_i|^p \, dx < C$$

for each positive integer i. Since

$$|\{\nabla v_i \neq 0\} \setminus (V_i \setminus V_{i+1})| = 0$$

and the sets $V_i \setminus V_{i+1}$ are disjoint, we conclude that

$$\int_{\mathbf{R}^n} \left(\sum_{i=1}^\infty i^{-1} |\nabla v_i|\right)^p dx = \int_{\mathbf{R}^n} \sum_{i=1}^\infty i^{-p} |\nabla v_i|^p \, dx < C\delta$$

where

$$\delta = \sum_{i=1}^\infty i^{-p} < \infty.$$

Therefore,

$$v := \sum_{i=1}^\infty i^{-1} v_i \in Y^{1,p}(\mathbf{R}^n) \quad \text{with} \quad \int_{\mathbf{R}^n} |\nabla v|^p \, dx < C\delta.$$

Since

$$v(x) \geq \sum_{i=1}^j i^{-1}$$

for $x \in V_j$, we obtain

$$\mathbf{c}_p(E) \leq \left(\sum_{i=1}^j i^{-1}\right)^{-p} C\delta$$

where

$$\left(\sum_{i=1}^j i^{-1}\right)^{-p} \to 0 \quad \text{as} \quad j \to \infty. \qquad \square$$

A final comparison between capacity and Hausdorff measure is given by the next result. The following notation will be used in this development: for $v \in L^1(\mathbf{R}^n)$, we define

(2.24) $$\overline{v}(a,r) := \fint_{B(a,r)} v(x) \, dx.$$

2.53 THEOREM. *Suppose $E \subset \mathbf{R}^n$ and $1 < p \leq n$. If $\mathbf{C}_p(E) = 0$, then $H^s(E) = 0$ for all $s > n - p$.*

2. POTENTIAL THEORY

PROOF. By Lemma 2.14, we may take $q < n$ such that $n - q < s$, thus reducing the problem to case $p < n$. In view of Lemma 2.5 (iii), we may consider $\mathbf{c}_p$ instead of $\mathbf{C}_p$.

Since $\mathbf{c}_p(E) = 0$, there exist $v_i \in Y^{1,p}(\mathbf{R}^n)$ such that

$$E \subset \{v_i \geq 1\}^\circ \quad \text{and} \quad \sum_{i=1}^\infty \|\nabla v_i\|_p^p < \infty.$$

With $v := \sum_{i=1}^\infty v_i$, we have by Sobolev's inequality,

$$\|v\|_{p^*} \leq \sum_{i=1}^\infty \|v_i\|_{p^*} \leq C \sum_{i=1}^\infty \|\nabla v_i\|_p < \infty.$$

Also, since

$$\|\nabla v\|_p \leq \sum_{i=1}^\infty \|\nabla v_i\|_p < \infty,$$

it follows that $v \in Y^{1,p}(\mathbf{R}^n)$.

Note that $E \subset \{v \geq k\}^\circ$ for each positive integer k. Thus, for each $a \in E$ there exists $r > 0$ such that $B(a,r) \subset \{v \geq k\}^\circ$ and therefore,

$$\fint_{B(a,r)} v \, dx \geq k.$$

Consequently,

$$\fint_{B(a,r)} v \, dx \to \infty \quad \text{as} \quad r \to 0.$$

For $s > n - p$ and $a \in E$, we claim

$$\limsup_{r \to 0} \frac{1}{r^s} \int_{B(a,r)} |\nabla v|^p \, dx = \infty.$$

If this were not true, there would exist $M < \infty$ such that

$$\frac{1}{r^s} \int_{B(a,r)} |\nabla v|^p \, dx \leq M$$

for all $0 < r \leq 1$. Then, for all such r,

$$\fint_{B(a,r)} |v - \bar{v}(a,r)|^p \, dx \leq C r^p \fint_{B(a,r)} |\nabla v|^p \, dx \leq C r^\theta$$

where $\theta = s - (n - p) > 0$. Thus,

$$\left| \bar{v}(a, r/2) - \bar{v}(a, r) \right| \leq \frac{1}{|B(a, r/2)|} \left| \int_{B(a, r/2)} [v - \bar{v}(a,r)] \, dx \right|$$

$$\leq 2^n \fint_{B(a,r)} |v - \bar{v}(a,r)| \, dx$$

$$\leq 2^n \left(\fint_{B(a,r)} |v - \bar{v}(a,r)|^p \, dx \right)^{1/p}$$

$$\leq C r^{\frac{\theta}{p}}.$$

Hence, for $k > j$,

$$|\bar{v}(a, 1/2^k) - \bar{v}(a, 1/2^j)| \leq \sum_{i=j+1}^{k} |\bar{v}(a, 1/2^i) - \bar{v}(a, 1/2^{i-1})|$$

$$\leq C \sum_{i=j+1}^{k} \left(\frac{1}{2^{i-1}}\right)^{\frac{\theta}{p}}.$$

This implies that the sequence $\{\bar{v}(a, 1/2^k)\}$ converges to a finite number, a contradiction. Therefore, we have

$$E \subset \left\{a : \limsup_{r \to 0} r^{-s} \int_{B(a,r)} |\nabla v|^p \, dx = \infty\right\}$$

$$\subset \left\{a : \limsup_{r \to 0} r^{-s} \int_{B(a,r)} |\nabla v|^p \, dx > 0\right\} := E_s.$$

Now refer to Corollary 1.19 to conclude that $H^s(E_s) = 0$. □

2.1.8 Lebesgue points for Sobolev functions.

The Lebesgue point concept makes it possible to define a locally integrable function at almost all points in a way that is independent of any representative in the equivalence class determined by the function. Another concept that achieves the same end without requiring the function to be locally integrable is that of approximate continuity. A measurable function u is said to be *approximately continuous* at x_0 if there is a measurable set E whose density at x_0 is one and for which $u \llcorner E$ is continuous at x_0. Within the class of locally integrable functions, the former concept is stronger. In this subsection it will be shown that p-quasicontinuous representatives of Sobolev functions possess Lebesgue points everywhere except for a set of capacity zero.

2.54 THEOREM.
Let $u \in W^{1,p}(\mathbf{R}^n)$, $1 < p \leq n$, be nonnegative. For $\varepsilon > 0$ let

$$A := \{x : Mu(x) > \varepsilon\}.$$

Then

$$\mathbf{C}_p(A) \leq \frac{C}{\varepsilon^p} \int_{\mathbf{R}^n} (|\nabla u|^p + |u|^p) \, dx.$$

PROOF. Assume first $\varepsilon = 1$. Note that if $x \in A$, then $\bar{u}(x,r) > 1$ for some $r > 0$. For such an r,

$$|B(x,r)| \leq \int_{B(x,r)} u \, dx \leq |B(x,r)|^{1/p'} \|u\|_p,$$

so that r is bounded above by some constant C. Consequently, we may appeal to the Besicovitch covering lemma (Lemma 1.16) to conclude that there exists an integer $N > 1$ and sequence $\{B_j\}$ of disjoint balls such that

(2.25) $$A \subset \bigcup_{j=1}^{\infty} B_j, \quad \sum_j \chi_{B_j} \leq N,$$

and
$$\fint_{B_j} u\, dx > 1 \quad \text{for each} \quad j = 1, 2, \ldots.$$

We refer to both Theorem 1.63 and Poincaré's inequality to obtain $v_j \in W^{1,p}(\mathbf{R}^n)$ such that
$$v_j = (\overline{u}_{B_j} - u)^+ \quad \text{on} \quad B_j$$
and
$$\int_{\mathbf{R}^n} \left(|\nabla v_j|^p + |v_j|^p\right) dx \leq C \int_{B_j} \left(|\nabla u|^p + |u|^p\right) dx$$

where $C = C(n,p)$. Note that
$$u + v_j \geq \overline{u}_{B_j} \geq 1 \quad \text{on} \quad B_j.$$

Therefore, defining
$$w_j = \sup\{v_1, \ldots, v_j\},$$
we see that
$$u + w_j > 1 \text{ on } A \cap (B_1 \cup \cdots \cup B_j)$$
and
$$\int_{\mathbf{R}^n} \left(|\nabla(u+w_j)|^p + |u+w_j|^p\right) dx$$
$$\leq C \int_{\mathbf{R}^n} \left(|\nabla u|^p + |u|^p\right) dx + C \sum_{i=1}^{j} \int_{\mathbf{R}^n} \left(|\nabla v_i|^p + |v_i|^p\right) dx$$
$$\leq C \int_{\mathbf{R}^n} \left(|\nabla u|^p + |u|^p\right) dx + C \sum_{i=1}^{j} \int_{B_i} \left(|\nabla u|^p + |u|^p\right) dx$$
$$\leq C \int_{\mathbf{R}^n} \left(|\nabla u|^p + |u|^p\right) dx + CN \int_{\mathbf{R}^n} \left(|\nabla u|^p + |u|^p\right) dx.$$

Since A is open,
$$\mathbf{c}_p\big(A \cap (B_1 \cup \cdots \cup B_j)\big) \leq C \int_{\mathbf{R}^n} \left(|\nabla u|^p + |u|^p\right) dx$$

and we pass to the union using Lemma 2.3 (iv). In case ε is a positive number not equal to 1, set $w := \varepsilon^{-1} u$ so that
$$A := \{x : Mu(x) > \varepsilon\} = \{x : Mw(x) > 1\}.$$

Thus,
$$\mathbf{C}_p(A) \leq C \int_{\mathbf{R}^n} \left(|\nabla w|^p + |w|^p\right) dx \leq \frac{C}{\varepsilon^p} \int_{\mathbf{R}^n} \left(|\nabla u|^p + |u|^p\right) dx. \qquad \square$$

2.55 THEOREM. *Suppose $u \in W^{1,p}_{\text{loc}}(\mathbf{R}^n)$, $1 < p \leq n$, and q is a finite exponent with $1 \leq q \leq p^*$. Then any p-quasicontinuous representative of u satisfies*

(2.26) $$\lim_{r \to 0} \fint_{B(x,r)} |u(y) - u(x)|^q \, dy = 0$$

at p-quasi every point of $\mathbf{R}^n$.

REMARK. Observe that if x is a point for which (2.26) holds, then
$$u(x) = \lim_{r \to 0} \fint_{B(x,r)} u(y)\, dy.$$

PROOF. Since our result is of a local nature, we may assume that $\operatorname{spt} u$ is compact.

Let
$$A := \Big\{ x \in \mathbf{R}^n : \limsup_{r \to 0} \frac{1}{r^{n-p}} \int_{B(x,r)} |\nabla u|^p\, dx > 0 \Big\}.$$

Corollary 1.19 asserts that $H^{n-p}(A) = 0$ and therefore $\mathbf{C}_p(A) = 0$. Furthermore, it follows from Poincaré's inequality (Corollary 1.64) that

(2.27) $$\lim_{r \to 0} \fint_{B(x,r)} |u(y) - \overline{u}(x,r)|^q\, dy = 0$$

for each $x \notin A$.

For each positive integer i, choose $u_i \in C_c^\infty(\mathbf{R}^n)$ so that
$$\|u - u_i\|_{1,p}^p \leq 2^{-(p+1)i}$$

and let
$$N_i := \{x : M(u - u_i)(x) > 2^{-i}\}.$$

It follows from Theorem 2.54 that
$$2^{-pi}\mathbf{C}_p(N_i) \leq C \int_{\mathbf{R}^n} \big(|\nabla u - \nabla u_i|^p + |u - u_i|^p\big)\, dx \leq C\, 2^{-(p+1)i},$$

thus implying that $\mathbf{C}_p(N_i) \leq C\, 2^{-i}$. Since
$$|\overline{u}(x,r) - u_i(x)| \leq \fint_{B(x,r)} |u(y) - \overline{u}(x,r)|\, dy + \fint_{B(x,r)} |u(y) - u_i(y)|\, dy$$
$$+ \fint_{B(x,r)} |u_i(y) - u_i(x)|\, dy,$$

it follows from (2.27) and the definition of N_i that

(2.28) $$\limsup_{r \to 0} |\overline{u}(x,r) - u_i(x)| \leq 2^{-i}$$

for all $x \notin A \cup N_i$. With
$$E_k := A \cup \Big(\bigcup_{j=k}^\infty N_j \Big)$$

for each positive integer k, we have
$$\mathbf{C}_p(E_k) \leq \mathbf{C}_p(A) + \sum_{j=k}^\infty \mathbf{C}_p(N_j) \leq C \sum_{j=k}^\infty \frac{1}{2^j}.$$

Thus, $\lim_{i \to \infty} u_i$ exists and is uniform on $\mathbf{R}^n \setminus E_k$. We define

(2.29) $$v(x) := \lim_{i \to \infty} u_i(x), \qquad x \notin E := \bigcap_{k=1}^\infty E_k.$$

Note that $\mathbf{C}_p(E) = 0$ and by Lemma 2.19, v is a p-quasicontinuous representative of u. By Theorem 2.23, $v = u$ outside a p-polar set Z. Moreover, for $x \notin E \cup Z$,

$$\lim_{r \to 0} |u(x) - \overline{u}(x,r)| = \lim_{r \to 0} |v(x) - \overline{u}(x,r)| = 0$$

because $x \notin E_k$ for some k and then

$$\limsup_{r \to 0} |v(x) - \overline{u}(x,r)| \leq |v(x) - u_i(x)| + \limsup_{r \to 0} |u_i(x) - \overline{u}(x,r)|$$

whenever $i \geq k$. In view of (2.29) and (2.28), this sum can be made arbitrarily small for sufficiently large i. Finally, for $x \notin A \cup E \cup Z$ we have with the help of (2.27),

$$\lim_{r \to 0} \fint_{B(x,r)} |u(y) - u(x)|^q \, dy \leq 2^{q-1} \lim_{r \to 0} \fint_{B(x,r)} |u(y) - \overline{u}(x,r)|^q \, dy$$
$$+ 2^{q-1} \lim_{r \to 0} \fint_{B(x,r)} |\overline{u}(x,r) - v(x)|^q \, dy$$
$$= 0. \qquad \square$$

2.2 Laplace equation

The Laplace equation is the classical prototype of a second order elliptic equation. Its special symmetry allows one to quickly establish a position in the development which requires greater effort with more general equations. The main goal of this section is to show a typical method of proving the classical Wiener criterion. Imitation of this method is not the right way to proceed in the theory of quasilinear elliptic equations.

2.56 DEFINITION. By the *Laplace equation* we understand the equation

$$\Delta u = 0, \quad \text{where } \Delta u = \sum_i \frac{\partial^2 u}{\partial x_i^2}.$$

Let $\Omega \subset \mathbf{R}^n$ be an open set.
  A function $u \in L^1_{\text{loc}}(\Omega)$ is called a
  — *distributional solution* of the Laplace equation if

$$\int_\Omega u \, \Delta \varphi \, dx = 0$$

  for each $\varphi \in \mathcal{C}_c^\infty(\Omega)$,
  — *distributional supersolution* of the Laplace equation if

(2.30) $$\int_\Omega u \, \Delta \varphi \, dx \leq 0$$

  for each nonnegative $\varphi \in \mathcal{C}_c^\infty(\Omega)$,
  — *distributional subsolution* of the Laplace equation if (2.30) holds for any nonpositive $\varphi \in \mathcal{C}_c^\infty(\Omega)$.

A function $u \in W^{1,2}_{\mathrm{loc}}(\Omega)$ is called a
— *weak solution* of the Laplace equation if
$$\int_\Omega \nabla u \cdot \nabla \varphi \, dx = 0$$
for each $\varphi \in W^{1,2}_c(\Omega)$,
— *weak supersolution* of the Laplace equation if

(2.31)
$$\int_\Omega \nabla u \cdot \nabla \varphi \, dx \geq 0$$
for each nonnegative $\varphi \in W^{1,2}_c(\Omega)$,
— *weak subsolution* of the Laplace equation if (2.31) holds for any nonpositive $\varphi \in W^{1,2}_c(\Omega)$.

We denote
$$\Omega_r = \{x \in \Omega \colon B(x,r) \subset\subset \Omega\} = \{x \in \Omega \colon \mathrm{dist}\,(x, \partial\Omega) > r\}.$$
For a locally integrable function u on Ω, we denote by $\overline{u}_r$ the function
$$\overline{u}_r(x) = \fint_{B(x,r)} u \, dy, \qquad x \in \Omega_r.$$
We say that a extended-real valued pointwise defined function $u \in L^1_{\mathrm{loc}}(\Omega)$ on Ω is
— *harmonic*, if u is continuous and
$$u(x) = \overline{u}_r(x)$$
for each ball $B(x,r) \subset\subset \Omega$,
— *superharmonic* if u is lower semicontinuous and

(2.32)
$$u(x) \geq \overline{u}_r(x)$$
for each ball $B(x,r) \subset\subset \Omega$,
— *subharmonic* if u is upper semicontinuous and
$$u(x) \leq \overline{u}_r(x)$$
for each ball $B(x,r) \subset\subset \Omega$.

2.57 Observations.

1. Any weak supersolution u is a distributional supersolution. The converse implication holds if $u \in W^{1,2}_{\mathrm{loc}}(\Omega)$.

2. A function u is a distributional (weak, respectively) subsolution of the Laplace equation, if and only if $-u$ is a distributional (weak, respectively) supersolution.

3. A function u is is a distributional (weak, respectively) solution of the Laplace equation if and only if u is both distributional (weak, respectively) subsolution and distributional (weak, respectively) supersolution.

4. A function u is a weak solution of the Laplace equation if and only if u is a local weak minimizer of the *Dirichlet integral*
$$v \mapsto \int_\Omega |\nabla v|^2 \, dx;$$

this means that
$$\int_{\text{spt}\,\varphi} |\nabla u|^2\, dx \le \int_{\text{spt}\,\varphi} |\nabla [u+\varphi]|^2\, dx$$
for each $\varphi \in W^{1,2}_c(\Omega)$.

5. A function u is subharmonic on Ω if and only if $-u$ is superharmonic.

6. A function u is harmonic on Ω if and only if u is both superharmonic and subharmonic.

7. The values of a superharmonic function are in $(-\infty, \infty]$; the value $u(x) = -\infty$ is excluded by local integrability and (2.32).

8. Linear polynomials (in particular constants) are harmonic.

9. The family of all superharmonic functions on Ω is a convex cone.

10. If u and v are superharmonic functions, then $\min\{u,v\}$ is superharmonic.

2.58 Theorem. *Let u be a distributional supersolution of the Laplace equation on Ω. Consider a nonnegative function $\omega \in L^\infty(\mathbf{R})$ which vanishes outside $[0,1]$. Then for each $x \in \Omega$,*
$$r \mapsto \phi_r * u, \qquad \text{where} \qquad \phi_r(y) = r^{-n}\omega(|y|/r),$$
is nonincreasing on $(0, \operatorname{dist}(x, \partial\Omega))$. In particular, $r \mapsto \bar{u}_r$ is nonincreasing on $(0, \operatorname{dist}(x, \partial\Omega))$.

PROOF. Suppose first that ω is a function in $\mathcal{C}^\infty_c(\mathbf{R})$ with support in $(0,1)$ and write
$$\omega_r(t) = r^{-n}\omega(t/r).$$
Choose $x \in \Omega$ and radii $0 < \rho < R < \delta$ such that $B(x,\delta) \subset\subset \Omega$. Set
$$f(s) = \int_0^s t^{n-1}\left[\omega_\rho(t) - \omega_R(t)\right] dt, \qquad s > 0,$$
$$\varphi(y) = \int_{|y-x|}^\infty s^{1-n} f(s)\, ds, \qquad y \in \mathbf{R}^n.$$

By an elementary change of variables we obtain that
$$\int_0^s t^{n-1} \omega_\rho(t)\, dt - \int_0^s t^{n-1} \omega_R(t)\, dt = \int_0^{s/\rho} \tau^{n-1}\omega(\tau)\, d\tau - \int_0^{s/R} \tau^{n-1}\omega(\tau)\, d\tau$$
$$= \int_{s/R}^{s/\rho} \tau^{n-1}\omega(\tau)\, d\tau$$
for any $s > 0$. Hence
$$f(s) \ge 0 \text{ on } (0, \infty)$$
and
$$f(s) = 0 \text{ on } (R, \infty) \text{ and on a right neighborhood of } 0.$$
We compute that $\varphi \ge 0$, φ is infinitely differentiable on $\mathbf{R}^n$, $\operatorname{spt} \varphi \subset \Omega$ and
$$-\Delta \varphi(y) = \omega_\rho(|y-x|) - \omega_R(|y-x|) = \phi_\rho(x-y) - \phi_R(x-y).$$
Now,
$$0 \le -\int_\Omega u\, \Delta\varphi\, dy = [u * \phi_\rho](x) - [u * \phi_R](x).$$

This proves the assertion under our temporary assumption about smoothness and support of ω; the general case follows by a standard approximation argument. □

2.59 Theorem. *Let u be a distributional supersolution of the Laplace equation on Ω. Then*
$$x \mapsto \lim_{r \to 0+} \overline{u}_r(x)$$
defines a superharmonic representative of u.

Proof. Let
$$v(x) = \lim_{r \to 0+} \overline{u}_r(x).$$
By Theorem 2.58,
$$r \mapsto \overline{u}_r(x)$$
is nonincreasing and thus, in particular, the limit exists. By Theorem 1.24, $v = u$ a.e. and thus v is a representative of u. Since the functions $\overline{u}_r$ are continuous and $\overline{u}_r \uparrow v$, v is lower semicontinuous. We have verified that v is superharmonic. □

2.60 Corollary. *Let u be a distributional supersolution of the Laplace equation on Ω. Then u has an unique superharmonic representative.*

Proof. The existence follows from Theorem 2.59. If u is superharmonic, then the functions $\overline{u}_r$ do not depend on a choice of representative,
$$\lim_{r \to 0+} \overline{u}_r \leq u,$$
and, by lower semicontinuity,
$$u \leq \lim_{r \to 0+} \overline{u}_r.$$
□

2.61 Corollary. *Let u be a distributional solution of the Laplace equation on Ω. Then u has a harmonic representative.* □

2.62 Corollary. *Let u, v be superharmonic functions on Ω. If $\Delta u = \Delta v$ in the sense of distributions, then there exists a harmonic function h on Ω such that $v = u + h$.*

Proof. By Corollary 2.61, there is a harmonic function h such that $h = v - u$ a.e. Since both the functions v and $u + h$ are superharmonic, by Corollary 2.60 the equality holds everywhere on Ω. □

2.63 Lemma. *Let u be a C^2-smooth superharmonic function on Ω. Then $\Delta u \leq 0$ in Ω.*

Proof. Assume that the assertion is false. Then, since u is smooth, there exists a ball $B = B(x, \delta) \subset \Omega$ and $\varepsilon > 0$ such that $\Delta u(y) > 2n\varepsilon$ on B. Let $v(y) = \varepsilon |y - x|^2$. Then $\Delta[v - u] = \Delta v - \Delta u \leq 0$ and thus by Theorem 2.59, $v - u$ is superharmonic. It follows that $v = u + (v - u)$ is superharmonic, but an easy experiment with $B(x, \frac{1}{2}\delta)$ shows that we arrived at a contradiction. □

The following lemma enables us to approximate superharmonic functions by smooth superharmonic functions, or by smooth functions with nonpositive Laplacian.

2.64 LEMMA. *There is a $\mathcal{C}_c^\infty$-smooth mollifier ϕ such that if u is a superharmonic function on Ω and $\varepsilon > 0$, then $u * \phi_\varepsilon$ is superharmonic on Ω_ε and*

$$u * \phi_\varepsilon \leq u$$

on Ω_ε.

PROOF. Let $\psi \in \mathcal{C}_c^\infty(\mathbf{R})$ be a nonnegative function with a support in $(0,1)$, such that

$$\int_0^1 \psi \, dt = 1.$$

Set

$$\phi(y) = \int_{|y|}^\infty \frac{\psi(t)}{\boldsymbol{\alpha}(n) t^n} \, dt.$$

Then $\phi \in \mathcal{C}_c^\infty(\mathbf{R}^n)$ (the only singularity could be at the origin but this is excluded as ϕ is constant at a neighborhood of 0). Let $B(x, \delta) \subset\subset \Omega$. Using Fubini's theorem with $W = \{[y, t] : |y - x| < \delta t\}$ we obtain

$$[u * \phi_\delta](x) = \delta^{-n} \int_{B(x,\delta)} u(y) \, \phi([x-y]/\delta) \, dy$$

$$= \delta^{-n} \int_{B(x,\delta)} u(y) \left(\int_{|x-y|/\delta}^1 \frac{\psi(t)}{\boldsymbol{\alpha}(n) t^n} \, dt \right) dy$$

$$= \int_{B(x,\delta) \times (0,1)} \frac{u(y) \psi(t)}{\boldsymbol{\alpha}(n) \, \delta^n t^n} \chi_W(y, t) \, dy \, dt$$

$$= \int_0^1 \left(\int_{B(x,\delta t)} \frac{u(y)}{\boldsymbol{\alpha}(n) \, \delta^n t^n} \psi(t) \, dy \right) dt$$

$$= \int_0^1 \overline{u}_{\delta t}(x) \, \psi(t) \, dt$$

$$\leq \int_0^1 u(x) \, \psi(t) \, dt = u(x).$$

It remains to show that $\phi_\delta * u$ is superharmonic on Ω_δ. Let $r > 0$ and

$$\eta(y) := \frac{\chi_{B(0,r)}}{|B(0,r)|}.$$

If $B(x, r) \subset\subset \Omega_\delta$, then

$$\fint_{B(x,r)} \phi_\delta * u \, dy = [\eta * (\phi_\delta * u)](x) = [\phi_\delta * (\eta * u)](x) = [\phi_\delta * \overline{u}_r)](x)$$

$$\leq [\phi_\delta * u](x). \qquad \square$$

2.65 THEOREM. *Let u be a superharmonic function on Ω. Then u is a distributional supersolution of the Laplace equation.*

PROOF. Choose a nonnegative test function $\varphi \in \mathcal{C}_c^\infty(\Omega)$. There is a $\delta > 0$ such that the support of φ is contained in $\Omega_{2\delta}$. Since u is lower bounded on Ω_δ, we henceforth may assume that u is nonnegative. We consider a sequence $\varepsilon_j \downarrow 0$ with $\varepsilon_1 < \delta$ and set $u_j = \phi_{\varepsilon_j} * u$. Then by Lemma 2.64 and Lemma 2.63, u_j are smooth superharmonic functions (and thus also distributional supersolutions of the Laplace equation) on Ω_δ and $0 \leq u_j \leq u$ on Ω_δ. Using the Lebesgue dominated convergence theorem we obtain

$$\int_\Omega u \Delta \varphi \, dy = \lim_{j \to \infty} \int_\Omega u_j \Delta \varphi \, dy \leq 0. \qquad \square$$

2.66 COROLLARY. *Let u be a harmonic function on Ω. Then $u \in \mathcal{C}^\infty(\Omega)$ and $\Delta u = 0$.*

PROOF. Choose a mollifier ϕ as in Lemma 2.64. Then for any $\varepsilon > 0$,

$$u = \phi_\varepsilon * u \quad \text{on } \Omega_\varepsilon.$$

It follows that u is infinitely differentiable in Ω. By Lemma 2.63, $\Delta u = 0$ in Ω. $\square$

2.67 THEOREM. *Let $\{u_j\}$ be a sequence of superharmonic functions on Ω and $u := \sup u_j$. Suppose that $u < \infty$ on a dense subset of Ω, the sequence $\{u_j\}$ is uniformly locally lower bounded and*

$$u = \lim_{j \to \infty} u_j.$$

Then u is a superharmonic function. If moreover the u_j are harmonic on an open set $\Omega' \subset \Omega$, then u is harmonic on Ω'.

PROOF. Obviously u is lower semicontinuous. Let $B(x, r) \subset\subset \Omega$. By Fatou's lemma,

$$\fint_{B(x,r)} u \, dy = \liminf_{j \to \infty} \fint_{B(x,r)} u_j \, dy \leq \lim_{j \to \infty} u_j(x) = u(x).$$

If $B(x, r) \subset\subset \Omega$, we find another ball $B(y, R) \subset\subset \Omega$ such that $B(x, r) \subset B(y, R)$ and $u(y) < \infty$. Then

$$(2.33) \qquad \int_{B(x,r)} u \, dy \leq \int_{B(y,R)} u \, dy \leq |B(y, R)| \, u(y).$$

Since u is locally lower bounded, (2.33) is sufficient to verify local integrability. Now, suppose that u_j are harmonic on Ω'. Then for any ball $B(x, r) \subset\subset \Omega'$ we have

$$u(x) = \lim_{j \to \infty} u_j(x) = \lim_{j \to \infty} \fint_{B(x,r)} u_j \, dy \leq \fint_{B(x,r)} u \, dy \leq u(x).$$

Since $\overline{u}_r$ is continuous on Ω'_r for any $r > 0$ and $\Omega' = \bigcup_{r>0} \Omega'_r$, u is continuous on Ω. Hence u is harmonic on Ω'. $\square$

2.68 THEOREM. *Let u be a superharmonic function on Ω. Then u is quasi-continuous, in particular, u is finite q.e. (we mean 2-quasicontinuous, 2-q.e.).*

PROOF. Let $v \geq 1$ be a smooth superharmonic function on Ω and $\eta \in \mathcal{C}_c^\infty(\Omega)$. Testing v with $v^{-1}\eta^2$ we obtain

$$\int_\Omega v^{-2}|\nabla v|^2 \eta^2 \, dx \leq 2 \int_\Omega v^{-1} \eta \nabla v \cdot \nabla \eta \, dx$$
$$\leq 2 \Big(\int_\Omega v^{-2}|\nabla v|^2 \eta^2 \, dx\Big)^{1/2} \Big(\int_\Omega |\nabla \eta|^2 \, dx\Big)^{1/2},$$

and thus

(2.34) $$\int_\Omega |\nabla(\ln v)|^2 \eta^2 \, dx \leq 4 \int_\Omega |\nabla \eta|^2 \, dx.$$

Let $\Omega' \subset\subset \Omega'' \subset\subset \Omega$, $m = \inf_{\Omega''} u - 1$ and $\{u_j\}$ a sequence of smooth superharmonic functions on Ω' obtained by mollification of u (Theorem 2.64), so that $u_j \to u$ and $m + 1 \leq u_j \leq u$. If $\eta \in \mathcal{C}_c^\infty(\Omega')$, then the estimate (2.34) applied to $u_j - m$ shows that the sequence $\{(u_j - m)\eta\}$ is bounded in $W^{1,2}(\Omega')$. We can select a subsequence which converges weakly to $(u - m)\eta$ and using convex combinations (Mazur's lemma), we find a sequence v_j of smooth superharmonic functions such that $v_j \leq u$ and $\eta \ln(v_j - m)$ converge strongly to $\eta \ln(u - m)$. We may assume the convergence to be so rapid that

$$\sum_{j=1}^\infty 4^j \|\eta \ln(v_j - m) - \eta \ln(u - m)\|_{1,2}^2 < \infty.$$

By Lemma 2.19, $\eta \ln(u - m)$ is quasicontinuous. Hence, u is quasicontinuous on $\{\eta = 1\}$. Since quasicontinuity is a local property, the proof is complete. □

2.2.1 Green potentials.

In this subsection we discuss Green potentials as preparation for an analysis of classical potential theory concepts that follow in the next subsection. We fix a reference domain $\mathcal{O}$; if $n = 2$, we suppose that $\mathcal{O}$ is a 2-Green domain. We write

$$\mathcal{U} = Y_0^{1,2}(\mathcal{O}).$$

If $\mu \in \mathcal{U}^*$ is a nonnegative Radon measure on $\mathcal{O}$, then by Lemma 2.33, the $\mathcal{U}$-potential u of μ is a weak supersolution of the Laplace equation. By Corollary 2.60, u has a unique superharmonic representative which will be denoted by $P\mu$. By Theorem 2.68, $P\mu$ is one of the quasicontinuous representatives of u.

2.69 DEFINITIONS. Let $B(z, r) \subset\subset \mathcal{O}$. We denote by $\lambda_{z,r}$ the measure with the density $\chi_{B(z,r)}/|B(z,r)|$ and write

$$g_{z,r}(x) = P\lambda_{z,r}.$$

Further, we define

$$g_z(x) = \liminf_{r \to 0+} g_{z,r}(x), \qquad x, z \in \Omega.$$

The function $(x, z) \mapsto g_z(x)$ on $\mathcal{O} \times \mathcal{O}$ is called the *Green function* on $\mathcal{O}$ and denoted by G, so that

$$G(x, z) = g_z(x).$$

If μ is a nonnegative Radon measure on $\mathcal{O}$, the *Green potential* of μ is defined by
$$G\mu(x) = \int_{\mathcal{O}} G(x,y)\, d\mu(y).$$
Special cases of Green potentials are Newtonian potentials for $\Omega = \mathbf{R}^n$. If $n=2$, the entire space is not a Green domain and then logarithmic potentials are used instead. Newtonian, or logarithmic potentials (respectively) are convolutions with the fundamental harmonic function Γ_0. The *fundamental harmonic function with pole at z* is defined by
$$\Gamma_z(x) = \Gamma(|z-x|),$$
where for $r > 0$,

(2.35) $$\Gamma(r) = \begin{cases} \frac{1}{n(n-2)\alpha(n)} r^{2-n}, & n > 2, \\ -\frac{1}{2\pi} \log r, & n = 2. \end{cases}$$

2.70 LEMMA. *Let $\mu, \nu \in \mathcal{U}^*$ are nonnegative Radon measures on $\mathcal{O}$. Then*

(2.36) $$\int_{\mathcal{O}} P\mu\, d\nu = \int_{\mathcal{O}} P\nu\, d\mu.$$

PROOF. We test $u := P\mu$ by $v := P\nu$ and conversely and obtain
$$\int_{\mathcal{O}} u\, d\nu = \int_{\mathcal{O}} \nabla v \cdot \nabla u\, dx = \int_{\mathcal{O}} v\, d\mu. \qquad \square$$

2.71 THEOREM. *For any $z, y \in \mathcal{O}$,*
$$G(z,y) = \lim_{R \to 0+} g_{y,R}(z) = \sup_{0 < R < \text{dist}(y, \partial \mathcal{O})} g_{y,R}(z) = G(y,z).$$

PROOF. Consider balls $B(z,r)$, $B(y,R) \subset\subset \mathcal{O}$. By Lemma 2.70 we have

(2.37) $$\fint_{B(y,R)} g_{z,r}\, dx = \fint_{B(z,r)} g_{y,R}\, dx.$$

We apply Fatou's lemma to (2.37) and obtain

(2.38) $$\fint_{B(y,R)} g_z\, dx \le \liminf_{r \to 0+} \fint_{B(y,R)} g_{z,r}\, dx = \lim_{r \to 0+} \fint_{B(z,r)} g_{y,R}\, dx = g_{y,R}(z).$$

Since $g_{z,r}$ is superharmonic, by (2.37)
$$\fint_{B(z,r)} g_{y,R}\, dx = \fint_{B(y,R)} g_{z,r}\, dx \le g_{z,r}(y)$$
and passing to limit for $r \to 0+$ we obtain

(2.39) $$g_{y,R}(z) = \lim_{r \to 0+} \fint_{B(z,r)} g_{y,R}\, dx \le \liminf_{r \to 0+} g_{z,r}(y) = g_z(y).$$

By (2.38) and (2.39),

(2.40) $$\fint_{B(y,R)} g_z\, dx \le g_{y,R}(z) \le g_z(y),$$

and both lower and upper estimates of $g_{y,R}(z)$ converge to $g_z(y)$ as $R \to 0^+$. Taking into account (2.39) again, we have

$$g_z(y) = \sup_{0<R<\mathrm{dist}\,(y,\partial\mathcal{O})} g_{y,R}(z) = \lim_{R \to 0+} g_{y,R}(z) = g_y(z). \qquad \square$$

2.72 LEMMA. *In the sense of distributions on* $\mathbf{R}^n$,

$$-\Delta \Gamma_z = \delta_z.$$

PROOF. An easy calculation yields

$$\nabla \Gamma_z(x) = -\frac{1}{n\boldsymbol{\alpha}(n)} \frac{x-z}{|x-z|^n}$$

and $-\Delta \Gamma_z = 0$ outside z. If $\varphi \in \mathcal{C}_c^\infty(\mathbf{R}^n)$ then, by integration by parts,

$$-\int_{\mathbf{R}^n} \Gamma_z \Delta\varphi\, dx = -\lim_{r \to 0+} \int_{\mathbf{R}^n \setminus B(z,r)} \Gamma_z \Delta\varphi\, dx$$

$$= \lim_{r \to 0+} \int_{\partial B(z,r)} \bigl(\Gamma_z(y)\nabla\varphi(y) - \varphi(y)\nabla\Gamma_z(y)\bigr) \cdot \frac{y-z}{|y-z|}\, dH^{n-1}(y)$$

$$- \lim_{r \to 0+} \int_{\mathbf{R}^n \setminus B(z,r)} \varphi \Delta \Gamma_z\, dx$$

$$= \lim_{r \to 0+} \int_{\partial B(z,r)} \frac{\varphi(y)}{n\boldsymbol{\alpha}(n) r^{n-1}}\, dH^{n-1}(y) = \varphi(z). \qquad \square$$

2.73 THEOREM. *Let* $z \in \mathcal{O}$. *Then*
(i) g_z *is superharmonic on* $\mathcal{O}$ *and harmonic on* $\mathcal{O} \setminus \{z\}$,
(ii) $-\Delta g_z = \delta_z$ *in the sense of distributions*,
(iii) *there is a harmonic function* h *on* $\mathcal{O}$ *such that*

$$g_z = h + \Gamma_z.$$

PROOF. (i) By Theorem 2.71,

$$g_z = \lim_{r \to 0+} g_{z,r} = \sup_{0<r<\mathrm{dist}\,(z,\partial\mathcal{O})} g_{z,r}$$

on $\mathcal{O}$. The functions $g_{z,r}$ are superharmonic on $\mathcal{O}$ and harmonic on $\mathcal{O} \setminus B(z,r)$. By Theorem 2.67, g_z is superharmonic on $\mathcal{O}$ and harmonic on $\mathcal{O} \setminus \{z\}$.

(ii) For any test function $\varphi \in \mathcal{C}_c^\infty \mathcal{O}$,

(2.41) $$-\int_\mathcal{O} g_{z,r} \Delta\varphi\, dx = \fint_{B(z,r)} \varphi\, dx.$$

Since

$$|g_{z,r} \Delta\varphi| \leq g_z |\Delta\varphi|,$$

we may pass to limit in (2.41) according to Lebesgue's dominated convergence theorem. It follows

$$-\int_\mathcal{O} g_z \Delta\varphi\, dx = \varphi(z).$$

(iii) We have
$$-\Delta g_z = \delta_z = -\Gamma_z$$
in the sense of distributions on $\mathcal{O}$. By Corollary 2.62, there is a harmonic function h such that
$$g_z = h + \Gamma_z \text{ on } \mathcal{O}.$$
□

2.74 THEOREM. *Let $\mu \in \mathcal{U}^*$ be a Radon measure. Then $G\mu = P\mu$.*

PROOF. Consider a ball $B(z,r) \subset\subset \Omega$. Then by Lemma 2.70 and Theorem 2.71,
$$\fint_{B(z,r)} G\mu(x)\,dx = \int_{\mathcal{O}\times\mathcal{O}} G(x,y)\,d\mu(y)\,d\lambda_{z,r}(y) = \int_{\mathcal{O}} g_{z,r}\,d\mu$$
$$= \int_{\mathcal{O}} P\lambda_{z,r}\,d\mu = \int_{\mathcal{O}} P\mu\,d\lambda_{z,r}$$
$$= \fint_{B(z,r)} P\mu\,dx.$$
Letting $r \to 0+$ we obtain the assertion. □

2.75 EXAMPLE. *Let $\mathcal{O} = \mathbf{R}^n$, $n > 2$. Then*
$$G(x,y) = \Gamma_y(x). \tag{2.42}$$

PROOF. It follows from Theorem 2.73 that there is a harmonic function h on $\mathbf{R}^n$ such that $G_y = \Gamma_y + h$. Obviously, $h \in \mathcal{U}$. Hence h is the $\mathcal{U}$-potential of the zero function and thus $h = 0$. □

2.76 EXAMPLE. *Let $\mathcal{O}$ be the unit ball in $\mathbf{R}^2$. Then*
$$G(x,y) = \frac{1}{2\pi}\left[\frac{1}{2}\ln\bigl(|x|^2|y|^2 - 2x\cdot y + 1\bigr) - \ln|x-y|\right]. \tag{2.43}$$

PROOF. We denote by $\tilde{G}(x,y)$ the candidate defined in (2.43). We have
$$\frac{1}{2}\ln\bigl(1 - 2x\cdot y + |x|^2|y|^2\bigr) = \ln\left|y - \frac{x}{|x|^2}\right| + \ln|x|,$$
which shows that
$$\tilde{G}(x,y) = 0 \qquad \text{when } |x| = 1.$$
Symmetrically,
$$\tilde{G}(x,y) - \Gamma_y(x) = \begin{cases} \frac{1}{2\pi}\left(\ln\left|x - \frac{y}{|y|^2}\right| + \ln|y|\right), & y \neq 0, \\ 0, & y = 0, \end{cases}$$
which proves that
$$\tilde{G}(x,y) = \Gamma_y(x) + \tilde{h}_y(x),$$
where $\tilde{h}_y(x)$ is harmonic. Now, by Theorem 2.73, there is a harmonic function h_y such that
$$G(x,y) = \Gamma_y(x) + h_y(x).$$
It is easy to see that $\tilde{h}_y - h_y \in \mathcal{U}$, and thus $\tilde{h}_y = h_y$. It follows that $\tilde{G} = G$. □

2.2.2 Classical thinness.

In the following discussion of thinness in the classical context, we will simply write $\gamma := \mathbf{c}_2(\cdot, \mathcal{O})$ and generally we will omit the prefix "2" in expressions such as 2-capacity, 2-thin, etc.. Given a set $E \subset \mathcal{O}$, we will denote by u_E the capacitary extremal for $\gamma(E)$ and by μ_E the capacitary distribution for $\gamma(E)$. When referring to the capacitary extremal, we tacitly assume that we are considering its superharmonic representative, which is quasicontinuous by Theorem 2.68.

2.77 LEMMA. *Let $E \subset \mathcal{O}$ and $u \in \overline{\mathcal{Y}}(E)$. If u is superharmonic, then $u_E \leq u$ on $\mathcal{O}$.*

PROOF. We test u_E and u by $v := (u_E - u)^+ \in \mathcal{U}$. The function $u_E - v$ belongs to $\overline{\mathcal{Y}}(E)$ and v is nonnegative. We obtain

$$\int_{\mathcal{O}} \nabla u_E \cdot \nabla v \, dx \leq 0,$$

$$\int_{\mathcal{O}} \nabla u \cdot \nabla v \, dx \geq 0,$$

$$\int_{\mathcal{O}} |\nabla v|^2 \, dx = \int_{\mathcal{O}} (\nabla u_E - \nabla u) \cdot \nabla v \, dx \leq 0,$$

and thus $u_E \leq u$ a.e. Since both u and u_E are superharmonic, in fact $u_E \leq u$ everywhere. $\square$

2.78 LEMMA. *A set E is thin at a point $z \in \mathcal{O}$ if and only if*

(2.44) $$\lim_{r \to 0+} u_{E \cap B(z,r)}(z) = 0.$$

PROOF. We start with consideration which are common for the both parts of the proof. For $0 < r < \mathrm{dist}\,(z, \partial \mathcal{O})$, we write

(2.45)
$$E_r = E \cap B(z,r),$$
$$\mu_r = \mu_{E \cap B(z,r)} \quad \text{(the capacitary distribution for } \gamma(E_r)\text{)},$$
$$u_r = u_{E \cap B(z,r)} = P\mu_r = G\mu_r \quad \text{(the capacitary extremal for } \gamma(E_r)\text{)},$$
$$v_r = \Gamma_0 * \mu_r.$$

According to Theorem 2.62 and Theorem 2.71, there is a harmonic function h on $\mathcal{O}$ such that

(2.46) $$G(z, x) = \Gamma(|x - z|) + h(x) \quad \text{on } \mathcal{O}.$$

Let $0 < r < R < \mathrm{dist}\,(z, \partial \mathcal{O})$. It is easily computed that

(2.47) $$\Gamma(r) = \Gamma(R) + \frac{1}{n\boldsymbol{\alpha}(n)} \int_r^R t^{1-n} \, dt.$$

By Fubini's theorem, (2.46) and (2.47), we have

$$u_R(z) = \int_O G(z-x)\, d\mu_R(x)$$

$$= \int_O \Big[h(x) + \Gamma(|z-x|)\Big]\, d\mu_R(x)$$

(2.48)

$$= \int_{\overline{B}(z,R)} \Big[h(x) + \Gamma(R) + \frac{1}{n\boldsymbol{\alpha}(n)} \int_{|x-z|}^R t^{1-n}\, dt\Big]\, d\mu_R(x)$$

$$= \int_{\overline{B}(z,R)} \big[h(x) + \Gamma(R)\big]\, d\mu_R(x) + \frac{1}{n\boldsymbol{\alpha}(n)} \int_0^R t^{1-n} \mu_R[B(z,t)]\, dt.$$

By Lemma 2.43,

(2.49) $$\mu_R[B(z,r)] = \int_{B(z,r)} u_R\, d\mu_R = \int_{B(z,r)} u_r\, d\mu_R.$$

We use (2.49) and Lemma 2.70 to obtain

(2.50)
$$\gamma(E_r) = \int_O u_r\, d\mu_r = \int_O u_R\, d\mu_r = \int_O u_r\, d\mu_R$$

$$= \mu_R[B(z,r)] + \int_{\overline{B}(z,R)\setminus B(z,r)} u_r\, d\mu_R.$$

This shows that
$$\mu_R(B(z,r)) \leq \gamma(E_r) = \gamma[B(z,r) \cap E]$$

and therefore

(2.51) $$\int_0^R r^{1-n} \mu_R(B(z,r))\, dr \leq \int_0^R r^{1-n} \gamma[B(z,r) \cap E]\, dr.$$

Now we are ready to prove the implications. Suppose that E is thin at z. Let R be so small that $h \leq h(z) + 1$ on $\overline{B}(z,R)$. Then by (2.48) and (2.51),

(2.52)
$$u_R(z) \leq \Big(h(z) + 1 + \Gamma(R)\Big)\mu_R[\overline{B}(z,R)] + \int_0^R r^{1-n} \mu_R[B(z,r)]\, dr$$

$$\leq \Big(h(z) + 1 + \Gamma(R)\Big)\gamma[B(z,2R) \cap E] + \int_0^R r^{1-n} \gamma[B(z,r) \cap E]\, dr.$$

Since the thinness of E at z implies that
$$R \mapsto [\Gamma(R) + h(z) + 1]\, \gamma[B(z,2R) \cap E]$$
cannot be bounded away from 0, and, by Lemma 2.77, $R \mapsto u_R(z)$ is nondecreasing, it follows that the right-hand side of (2.52) can be made arbitrarily small.

Now assume (2.44). Let $R > 0$ be so small that
$$u_R(z) < 2^{-n-1}$$
and

(2.53) $$\Gamma(R) + h(x) \geq 0 \qquad \text{for all } x \in B(z,R).$$

Let $0 < r \leq R$. Let $\nu = \mu_R \mathop{\llcorner} [\overline{B}(z,R) \setminus B(z,r)]$. Then $G\nu$ is harmonic on $B(z,r)$ and superharmonic on $\mathcal{O}$. If $x \in B(z, r/2)$, then

(2.54)
$$G\nu(x) = \fint_{B(x,r/2)} G\nu\, dy \leq 2^n \fint_{B(z,r)} G\nu\, dy \leq 2^n G\nu(z)$$
$$\leq 2^n G\mu_R(z) = 2^n u_R(z) \leq 1/2.$$

Referring to (2.50) and Lemma 2.43, we have

(2.55)
$$\gamma(E_{r/2}) = \mu_R[B(z, r/2)] + \int_{\overline{B}(z,R) \setminus B(z,r/2)} u_{r/2}\, d\mu_R$$
$$= \mu_R[B(z, r/2)] + \int_{B(z,r) \setminus B(z,r/2)} u_{r/2}\, d\mu_R$$
$$+ \int_{\overline{B}(z,R) \setminus B(z,r)} u_{r/2}\, d\mu_R$$
$$\leq \mu_R[B(z,r)] + \int_{\overline{B}(z,R) \setminus B(z,r)} u_{r/2}\, d\mu_R.$$

By Lemma 2.70,

(2.56)
$$\int_{\overline{B}(z,R) \setminus B(z,r)} u_{r/2}\, d\mu_R = \int_{\mathcal{O}} u_{r/2}\, d\nu = \int_{\mathcal{O}} G\nu\, d\mu_{r/2}.$$

From (2.54) we infer that $\frac{3}{2} u_{r/2} - G\nu \in \overline{\mathcal{Y}}(E_{r/2})$, and thus by Corollary 2.42

$$\int_{\mathcal{O}} u_{r/2}\, d\mu_{r/2} \leq \int_{\mathcal{O}} \left[\frac{3}{2} u_{r/2} - G\nu \right] d\mu_{r/2}.$$

It follows

(2.57)
$$\int_{\mathcal{O}} G\nu\, d\mu_{r/2} \leq \frac{1}{2} \int_{\mathcal{O}} u_{r/2}\, d\mu_{r/2} = \frac{1}{2} \gamma(E_{r/2}).$$

By (2.55), (2.56) and (2.57) we have

(2.58)
$$\frac{1}{2} \gamma(E_{r/2}) \leq \mu_R[B(z,r)].$$

From (2.48) and (2.53) we obtain

(2.59)
$$u_R(z) = \int_{\overline{B}(z,R)} [h(x) + \Gamma(R)]\, d\mu_R(x) + \frac{1}{n\boldsymbol{\alpha}(n)} \int_0^R t^{1-n} \mu_R[B(z,t)]\, dt$$
$$\geq \frac{1}{n\boldsymbol{\alpha}(n)} \int_0^R t^{1-n} \mu_R[B(z,t)]\, dt.$$

By (2.58) and (2.59),

$$\int_0^R \gamma(E_{r/2}) r^{1-n}\, dr \leq \int_0^R t^{1-n} \mu_R[B(z,t)]\, dt \leq C u_R(z) < \infty,$$

so that E is thin at z. $\square$

2.79 Theorem. *A set E is thin at $z \in \mathcal{O}$ if and only if there is a nonnegative superharmonic function u on $\mathcal{O}$ such that*

$$u(z) < \liminf_{\substack{x \to z \\ x \in E}} u(x).$$

Proof. If E is thin at z, then by Lemma 2.48 there is an open set U containing E such that U is thin at z. According to Lemma 2.78, there exists $r > 0$ such that

$$u := u_{U \cap B(z,r)}(z) < 1$$

whereas $u \geq 1$ on $U \cap B(z,r) \supset E \cap B(z,r)$.

Conversely, suppose that there is a nonnegative superharmonic function u on $\mathcal{O}$ such that

$$u(z) < \liminf_{\substack{x \to z \\ x \in E}} u(x).$$

We may assume that

$$\liminf_{\substack{x \to z \\ x \in E}} u(x) > u(z) + 1.$$

Let $\mu = -\Delta u$ and write

$$\mu_j = \mu \mathbin{\llcorner} B(z, 2^{-j})$$
$$w_j = G\mu_j.$$

By Corollary 2.62, there exist harmonic functions h on $\mathcal{O}$ and h_j on $B(z, 2^{-j})$ such that

(2.60) $$u = G\mu + h \quad \text{on } \mathcal{O},$$
$$u = G\mu_j + h_j = w_j + h_j \text{ on } B(z, 2^{-j}).$$

Using Lebesgue's dominated convergence theorem we obtain

$$\lim_{j \to \infty} w_j(z) = \lim_{j \to \infty} \int_{B(z, 2^{-j})} G(z, x) \, d\mu(x) = 0.$$

Given $\varepsilon > 0$, we fix j such that $w_j(z) < \varepsilon$. Then by (2.60),

$$\liminf_{\substack{x \to z \\ x \in E}} w_j(x) > 1 + w_j(z) \geq 1$$

and thus there is $r > 0$ such that

$$w_j(x) > 1 \quad \text{on } B(z,r) \cap E.$$

Let u_r be the capacitary extremal for $\gamma[E \cap B(z,r)]$. Since $w_j \in \overline{\mathcal{Y}}(E_r)$, by Lemma 2.77 $u_r \leq w_j$, in particular

$$u_r(z) \leq w_j(z) < \varepsilon.$$

By Theorem 2.78, E is thin at z. □

2.2.3 Dirichlet problem and the Wiener criterion.

In this subsection we establish the Wiener criterion for harmonic functions by using the classical methods of potential theory. Later, in Chapter 4, we will give another proof of boundary regularity which holds for a wide class of equations in divergence form and uses no techniques of potential theory. Throughout the subsection we suppose that $\mathcal{O}$ is as in Definition 2.1 and $\Omega \subset\subset \mathcal{O}$.

2.80 LEMMA (Strong minimum principle). *Let u be a nonnegative superharmonic function on Ω and $x \in \Omega$. If $u(x) = 0$, then u vanishes on the connected component of Ω containing x.*

PROOF. Let $B := B(z, r) \subset\subset \Omega$ and $u(z) = 0$. Then $u \geq 0$ on B and
$$\fint_B u \leq 0.$$
It follows that $u = 0$ a.e. in B and by Corollary 2.60, $u = 0$ everywhere in B. Then the assertion immediately follows. □

2.81 LEMMA. *Let u be a superharmonic function on Ω. Suppose that*
$$\liminf_{\substack{x \to y \\ x \in \Omega}} u(x) \geq 0$$
for each $y \in \partial\Omega$. Then u is nonnegative.

PROOF. The function
$$\overline{u} := \begin{cases} u & \text{on } \Omega, \\ 0 & \text{on } \partial\Omega \end{cases}$$
is lower semicontinuous, and thus it attains its minimum at a point $z \in \overline{\Omega}$. Suppose that $u(z) < 0$. Then $z \in \Omega$ and by Lemma 2.80, $u = u(z)$ on the connected component of Ω containing z which contradicts the behavior on $\partial\Omega$. □

2.82 REMARK. Another minimum principle is almost obvious: If $u \in W^{1,2}(\Omega)$ is a distributional supersolution of the Laplace equation and $u^- \in W_0^{1,2}(\Omega)$, then u is nonnegative. This is observed by testing with u^-.

2.83 DEFINITION. Consider a function $f : \partial\Omega \to \mathbf{R}$. We say the u is an *upper function* to f if u is either superharmonic or $+\infty$ on each connected component of Ω and
$$\liminf_{\substack{x \to y \\ x \in \Omega}} u(x) \geq f(y)$$
for all $y \in \partial\Omega$. Similarly, we say the u is an *lower function* to f if u is either subharmonic or $-\infty$ on each connected component of Ω and
$$\limsup_{\substack{x \to y \\ x \in \Omega}} u(x) \leq f(y)$$
for all $y \in \partial\Omega$.

We denote by $\overline{H}_f$ the pointwise infimum of all upper functions to f and by $\underline{H}_f$ the pointwise supremum of all lower functions to f.

Let us notice that $\underline{H}_f \leq \overline{H}_f$ for each f. Indeed, if u is an upper function to f and v is a lower function to f, the $u - v \geq 0$ according to Lemma 2.81.

We say that f is *resolutive* if $\overline{H}_f$ is harmonic and $\overline{H}_f = \underline{H}_f$. In this case the common value is denoted by H_f and called the PWB (Perron-Wiener-Brelot) solution of the Dirichlet problem with boundary function f.

2.84 LEMMA. *Let $f \in C_c^\infty(\mathcal{O})$. Then f is resolutive and $f - H_f \in W_0^{1,2}(\Omega)$.*

PROOF. Let w be the minimizer of
$$\int_\Omega (|\nabla u|^2 + 2u\Delta f)\, dx$$
in $W_0^{1,2}(\Omega)$. Then
$$\Delta w = \Delta f$$
in the sense of distributions on Ω. By Corollary 2.61, there is a harmonic representative h of $f - w$ on Ω. Also $|h| \leq M := \|f\|_\infty$ by Remark 2.82. We shall show that $h = H_f$. Choose a ball $B := B(z,r)$ such that $2B = B(z,2r) \subset\subset \Omega$ and $\varepsilon > 0$. We extend h be setting $h = f$ outside Ω. By Lemma 2.26, h is quasicontinuous on $\mathbf{R}^n$ and thus there is open set $G \subset \mathbf{R}^n$ such that $\gamma(G) < \varepsilon^2$ and $h \llcorner \mathcal{O} \setminus G$ is continuous. Let v be the capacitary extremal for $\gamma(G \setminus \overline{2B})$ and $u = h + 2Mv$. Then u is superharmonic in Ω. Since
$$\liminf_{\substack{x \to y \\ x \in \Omega \cap G}} u(x) \geq M \geq f(y)$$
and
$$\liminf_{\substack{x \to y \\ x \in \Omega \setminus G}} u(x) \geq \liminf_{\substack{x \to y \\ x \in \Omega \setminus G}} h(x) \geq h(y) = f(y),$$
u is an upper function to f. Since v is harmonic on $2B$, using Hölder's and Sobolev's inequality (if $n > 2$) we infer that
$$v(z) = \fint_B v\, dx \leq C\|v\|_{2^*} \leq C\|\nabla v\|_2 \leq C(\gamma(G))^{1/2} \leq C\varepsilon,$$
where $C = C(n, \Omega, B)$. If $n = 2$, we use Lemma 1.84 with the same effect. It follows
$$\overline{H}_f \leq h + CM\varepsilon.$$
Letting $\varepsilon \to 0$ and using similar estimates from below we conclude that
$$h \leq \underline{H}_f \leq \overline{H}_f \leq h$$
which means that $h = H_f$. □

2.85 THEOREM. *Every $f \in \mathcal{C}(\partial\Omega)$ is resolutive.*

PROOF. In view of Tietze's extension theorem, we may suppose that f is given on $\mathbf{R}^n$ and $f \in \mathcal{C}_c(\mathbf{R}^n)$. By means of mollification we find a sequence $\{f_j\}$ of functions from $\mathcal{C}_c^\infty(\mathbf{R}^n)$ such that $|f_j - f| < 2^{-j}$ on $\mathbf{R}^n$. Then any upper function to f_j is an upper function to $f - 2^{-j}$ and any lower function to f_j is a lower function to $f + 2^{-j}$. By Lemma 2.84, f_j are resolutive. Hence
$$\overline{H}_f - 2^{-j} \leq \overline{H}_{f_j} = \underline{H}_{f_j} \leq \underline{H}_f + 2^{-j}.$$
This shows that $\overline{H}_f = \underline{H}_f$ and
$$H_{f_j} - 2^{-j+2} \uparrow \overline{H}_f.$$
By Theorem 2.67, $\overline{H}_f$ is harmonic. We have verified that f is resolutive. □

2.86 DEFINITION. A point $z \in \partial\Omega$ is called *regular* for Ω if

(2.61)
$$\lim_{\substack{x \to z \\ x \in \Omega}} H_f(x) = f(z)$$

for any continuous f defined on $\partial\Omega$.

2.87 DEFINITION. A function $u \in W^{1,2}(\Omega)$ is called a *barrier* at $z \in \partial\Omega$ if
(i) u is harmonic in Ω,
(ii) for every $y \in \partial\Omega$ except $y = z$,
$$\liminf_{\substack{x \to y \\ x \in \Omega}} u(x) > 0,$$

(iii)
$$\lim_{\substack{x \to y \\ x \in \Omega}} u(x) = 0.$$

2.88 LEMMA. *A point $z \in \partial\Omega$ is regular if and only if there exists a barrier u at z.*

PROOF. Suppose z is regular. Let $f(x) := |x - z|^2$ and note that f is smooth and subharmonic. Then H_f is easily seen to be a barrier at z. Indeed, property (ii) follows from the fact that f is a lower function to itself and (iii) is a consequence of the regularity of z.

Now suppose that a barrier u exists. From the definition of barrier we infer that there is a smooth function v on $\mathbf{R}^n$ such that $v(z) = 0$, $v(x) > 0$ for all $x \neq z$ and $v \leq u$ on Ω. Let $f \in \mathcal{C}(\partial\Omega)$. Then for each $\varepsilon > 0$ there exists $r > 0$ such that
$$|f(x) - f(z)| < \frac{\varepsilon}{2}$$
when ever $x \in B(z,r) \cap \partial\Omega$. Moreover, there exist $M, m > 0$ such that $|f(x)| \leq M$ for all $x \in \partial\Omega$ and $v \geq m$ on $\partial\Omega \setminus B(z,r)$. It follows that
$$f \leq f(z) + \varepsilon + \frac{2M}{m} v \quad \text{on } \partial\Omega.$$
Then $f(z) + \varepsilon + \frac{2M}{m} u$ is an upper function to f. Since $u(x) \to 0$ as $x \to z$,
$$\limsup_{\substack{x \to z \\ x \in \Omega}} H_f(x) \leq f(z) + \varepsilon.$$

Similarly we estimate the lower limit of H_f at z. Hence, z is regular. □

2.89 THEOREM (Wiener criterion). *A point $z \in \partial\Omega$ is regular for Ω if and only if $\mathbf{R}^n \setminus \Omega$ is not thin at z.*

PROOF. Suppose that z is regular and $\mathbf{R}^n \setminus \Omega$ is thin at z. By Theorem 2.79, there is a nonnegative superharmonic function u on $\mathcal{O}$ and a positive constant a such that
$$u(z) < a < \liminf_{\substack{x \to z \\ x \in \mathcal{O} \setminus \Omega}} u(x).$$
We find a ball $B(z,r) \subset\subset \mathcal{O}$ such that $u \geq a$ on $B(z,r) \setminus (\Omega \cup \{z\})$, and a function $f \in \mathcal{C}_c^\infty(\mathcal{O})$ such that
$$0 \leq f \leq a, \quad \operatorname{spt} f \subset B(z,r) \quad \text{and} \quad f(z) = a.$$

Then for any $\varepsilon > 0$, $u + \varepsilon G(\cdot, z)$ is an upper function to f and thus $H_f \leq u$ in Ω. Since z is regular,
$$a = \lim_{\substack{x \to z \\ x \in \Omega}} h(x) \leq \liminf_{\substack{x \to z \\ x \in \Omega}} u(x).$$
Hence
$$a \leq \liminf_{x \to z} u(x) \leq \lim_{r \to 0+} \overline{u}_r(z) \leq u(z),$$
which is a contradiction. Now, suppose that $\mathbf{R}^n \setminus \Omega$ is not thin at z. Let u_r be the capacitary extremal for $\gamma[B(z,r) \setminus \Omega]$ and $E_r = \{u_r = 1\}$. By Theorem 2.31, quasi every point of $B(z,r) \setminus \Omega$ belongs to E_r and thus E_r is not thin at z. By Theorem 2.79,

(2.62) $$\liminf_{x \to z} u_r(x) \geq u_r(z) = \liminf_{\substack{x \to z \\ x \in E_r}} u_r(x) = 1$$

and thus

(2.63) $$\lim_{\substack{x \to z \\ x \in \Omega}} (1 - u_r(x)) = 0.$$

Since $1 - u_r$ is nonnegative and harmonic on $\mathcal{O} \setminus \overline{B}(z,r)$, by Lemma 2.80 we have $u_r < 1$ thereon and
$$\liminf_{\substack{x \to y \\ x \in \Omega}} (1 - u_r(x)) = 1 - u_r(y) > 0$$
for each $y \in \partial \Omega \setminus \overline{B}(z,r)$. Consider a sequence $r_j \downarrow 0$ and set
$$h = \sum_{j=1}^{\infty} 2^{-j}(1 - u_{r_j}).$$

Then h is harmonic by Theorem 2.67. By (2.62) and (2.63), h is a barrier at z. Hence z is regular. $\square$

2.3 Regularity of minimizers

One of the major results established to resolve Hilbert's twentieth problem is due, independently, to De Giorgi and Nash [**DeG**], [**Na**] which states that weak solutions of the linear elliptic equation
$$\sum_{i=1}^{n} D_i \left[\sum_{j=1}^{n} a^{ij}(x) D_j u \right] = 0, \quad a^{ij} \in L^{\infty}$$
are Hölder continuous. This result is the keystone of our book. To demonstrate the crucial part it plays in the solution of Hilbert's twentieth problem, we will give a brief (and simplified) development of the regularity of solutions to certain variational problems. This will include De Giorgi's proof of the Hölder continuity of solutions of (2.87). We will also include Moser's proof of this result, [**Mos1**]. Although these proofs will be subsumed by those that apply to equations of greater generality in the next chapter, we choose to include them to clearly exhibit their fundamental components, for these components will be used extensively throughout the remainder of this book.

2.3.1 Abstract minimization.

2.90 DEFINITION. We say that functional J on a Banach space X is coercive if
$$\lim_{\|u\|\to\infty} \frac{Ju}{\|u\|} = \infty.$$

2.91 THEOREM. *Let X be a reflexive Banach space, $H \subset X$ be a nonempty closed convex set and J be a coercive strictly convex continuous functional on X. Then there is a unique minimizer of J on H.*

PROOF. Let $a = \inf J$. Let $\{a_j\}$, $a_j \neq a$, be a sequence of real numbers such that $a_j \downarrow a$. Let $E_j = \{u \in H \colon J(u) \leq a_j\}$. Then $\{E_j\}$ is a decreasing sequence of closed convex sets. Due to coercivity of J, the sets E_j are bounded, and from reflexivity of X we deduce that the E_j are weakly compact. It implies that the intersection of E_j's must be nonempty. This proves the existence, because any point in $\bigcap E_j$ is a minimizer. Uniqueness of minimizers is an immediate consequence of strict convexity. $\square$

2.92 DEFINITION. Let X be a Banach space, J be a (nonlinear) functional on X. A linear functional $f \in X^*$ is said to be a Gâteaux derivative of J at a point $u \in X$, denoted by $\partial J(u)$, if
$$\partial J(u) = \lim_{t\to 0} \frac{J(u+tv) - J(u)}{t}$$
for all $v \in X$.

2.93 THEOREM. *Let X be a Banach space, $H \subset X$ be a nonempty closed convex set and J be a nonnegative Gâteaux differentiable convex continuous functional on X. Then u is a minimizer of J on H if and only if $u \in H$ and*
$$\partial J(u)(v - u) \geq 0$$
for each $v \in H$.

PROOF. The only if part is a trivial consequence of definitions. Suppose that $u \in H$ and that
$$\partial J(u)(v - u) \geq 0$$
for each $v \in H$. If $v \in H$, consider the real function
$$\varphi(t) = J(u + tv)$$
defined of the interval $\{t \colon u + tv \in H\}$. As J is convex and Gâteaux differentiable, φ is also convex and differentiable and thus
$$J(v) - J(u) \geq \varphi(1) - \varphi(0) \geq \varphi'(0) = \partial J(u)(v - u) \geq 0.$$
This shows that u is a minimizer of J.

2.3.2 Minimizers and weak solutions.

To demonstrate the role played by the De Giorgi-Nash result in the regularity problem of minimizers to certain variational problems, in this subsection we begin by showing the existence of weak minimizers. We suppose that $\Omega \subset \mathbf{R}^n$ is an open set and $1 < p \leq n$.

2.94 DEFINITION. In this definition, we introduce a situation general enough to cover all cases needed in the sequel. Let $\mathbf{F} = \mathbf{F}(x, \zeta, \xi)$ be a nonnegative function on $\Omega \times \mathbf{R} \times \mathbf{R}^n$ such that
$$\mathbf{F}(x, u, \nabla u)$$
is measurable for each $u \in W^{1,p}_{\text{loc}}(\Omega)$. We say that a function $w \in W^{1,p}_{\text{loc}}(\Omega)$ is a *local minimizer* of the functional associated with $\mathbf{F}$ if
$$\int_{\text{spt } v} \mathbf{F}(x, w, \nabla w)\, dx \leq \int_{\text{spt } v} \mathbf{F}(x, w+v, \nabla[w+v])\, dx$$
for all $v \in W^{1,p}_c(\Omega)$. Let $\mathbf{A}$, $\mathbf{B}$ be functions on $\Omega \times \mathbf{R} \times \mathbf{R}^n$ such that
$$\mathbf{A}(x, u, \nabla u) \cdot \nabla v + \mathbf{B}(x, u, \nabla u) v \in L^1_{\text{loc}}$$
for any $u \in W^{1,p}_{\text{loc}}(\Omega)$ and $v \in W^{1,p}_c(\Omega)$ and consider the equation
$$(2.64) \qquad -\operatorname{div} \mathbf{A}(x, u, \nabla u) + \mathbf{B}(x, u, \nabla u) = 0.$$
A function $u \in W^{1,p}_{\text{loc}}(\Omega)$ is called a *weak solution* of (2.64) if
$$\int_\Omega \Big(\mathbf{A}(x, u, \nabla u) \cdot \nabla\varphi + \mathbf{B}(x, u, \nabla u)\varphi\Big)\, dx = 0$$
for each $\varphi \in W^{1,p}_c(\Omega)$. Similarly we define weak *subsolutions* and *supersolutions*. Namely, we test by nonnegative functions $\varphi \in W^{1,p}_c(\Omega)$ and require
$$(2.65) \qquad \int_\Omega \mathbf{A}(x, u, \nabla u) \cdot \nabla\varphi + \mathbf{B}(x, u, \nabla u)\,\varphi\, dx \begin{cases} \geq 0 & \text{for supersolutions,} \\ \leq 0 & \text{for subsolutions} \end{cases}$$
for each test function.

Notice that a function u is is a weak solution of (2.64) if and only if u is both weak subsolution and weak supersolution.

We shall call particular attention to functionals associated with $\mathbf{F} : \Omega \times \mathbf{R}^n \to \mathbf{R}$, where
(2.66) $\mathbf{F}(\cdot, \xi)$ is measurable for any $\xi \in \mathbf{R}^n$,
(2.67) $\mathbf{F}(x, \cdot)$ is strictly convex and differentiable for all $x \in \Omega$,
(2.68) there are constants $\alpha > 0$, β such that
$$\alpha |\xi|^p \leq \mathbf{F}(x, \xi) \leq \beta |\xi|^p$$
for all $x \in \Omega$ and $\xi \in \mathbf{R}^n$.
We will write
$$\mathbf{J}[u, \Omega] = \int_\Omega \mathbf{F}(x, u)\, dx.$$
Also, we will consider an equation
$$(2.69) \qquad -\operatorname{div} \mathbf{A}(x, \nabla u) = 0,$$
where
(2.70) $\mathbf{A}(\cdot, \xi)$ is measurable for all $\xi \in \mathbf{R}^n$,
(2.71) $\mathbf{A}(x, \cdot)$ is continuous for all $x \in \Omega$,
(2.72) $|\mathbf{A}(x, \xi)| \leq a|\xi|^{p-1}$ for all $x \in \Omega$ and $\xi \in \mathbf{R}^n$.
(2.73) $\mathbf{A}(x, \xi) \cdot \xi \geq c|\xi|^p$ for all $x \in \Omega$ and $\xi \in \mathbf{R}^n$,
(2.74) $[\mathbf{A}(x, \xi') - \mathbf{A}(x, \xi)] \cdot (\xi' - \xi) > 0$ for all $x \in \Omega$ and $\xi \neq \xi' \in \mathbf{R}^n$.

2.95 LEMMA. *Let $\mathbf{F}$ satisfy (2.66)–(2.68) and*

$$J(g) = \int_\Omega \mathbf{F}(x, g(x))\, dx, \quad g \in L^p(\Omega; \mathbf{R}^n).$$

Then
  (i) *J is continuous on $L^p(\Omega; \mathbf{R}^n)$,*
  (ii) *J is strictly convex and coercive on $L^p(\Omega; \mathbf{R}^n)$,*
  (iii) *J is Gâteaux differentiable on $L^p(\Omega; \mathbf{R}^n)$ and*

(2.75) $$\partial J(g)(h) = \int_\Omega \mathbf{A}(x, g) \cdot h\, dx, \quad h \in L^p(\Omega; \mathbf{R}^n),$$

where $\mathbf{A} = (\mathbf{A}_1, \ldots, \mathbf{A}_n)$,

$$\mathbf{A}_i(x, \xi) = \frac{\partial}{\partial \xi_i} \mathbf{F}(x, \xi).$$

PROOF. Suppose that $y_j \to y$ in $L^p(\Omega, \mathbf{R}^n)$. Due to the growth

$$0 \leq \mathbf{F}(x, \xi) \leq C\, |\xi|^p$$

we have

$$|\mathbf{F}(x, g_j) - \mathbf{F}(x, g)| \leq C(|g|^p + |g_j - g|^p).$$

Thus there are sets E_j such that

$$|\mathbf{F}(x, g_j) - \mathbf{F}(x, g)| \leq C\, |g|^p \quad \text{on } E_j$$

and

$$|\mathbf{F}(x, g_j)| \leq |\mathbf{F}(x, g_j) - \mathbf{F}(x, g)|\, \chi_{E_j} + C\, |g_j - g|^p.$$

Now,

$$\int_\Omega |\mathbf{F}(x, g_j) - \mathbf{F}(x, g)|\, \chi_{E_j}\, dx \to 0$$

by the Lebesgue dominated convergence theorem and

$$\int_\Omega |g_j - g|^p\, dx \to 0$$

as $g_j \to g$ in $L^p(\Omega; \mathbf{R}^n)$. This proves the continuity property (i). The strict convexity and coercivity (ii) are evident. It remains to prove (iii). First, notice that as

$$\mathbf{A}_i(x, \xi) = \lim_{k \to \infty} k\big[\mathbf{F}(x, \xi + e_i/k) - \mathbf{F}(x, \xi)\big],$$

the mapping $\mathbf{A}_i(\cdot, \xi)$ is measurable for any $\xi \in \mathbf{R}^n$. Further, any convex differentiable function is continuously differentiable, so that $\mathbf{A}_i(x, \cdot)$ is continuous for any $x \in \Omega$. We express the convexity assumption in the form

(2.76) $$\mathbf{F}(x, \xi + \theta) - \mathbf{F}(x, \xi) \geq \mathbf{A}(x, \xi) \cdot \theta, \quad \xi, \theta \in \mathbf{R}^n,\ x \in \Omega.$$

We want to estimate $|\mathbf{A}(x, \xi)|$; for this we assume that $\mathbf{A}(x, \xi) \neq 0$. Setting $\theta = |\xi|\, \mathbf{A}(x, \xi)/|\mathbf{A}(x, \xi)|$ in (2.76) we obtain

$$|\xi||\mathbf{A}(x, \xi)| = \frac{|\xi|\, \mathbf{A}(x, \xi) \cdot \mathbf{A}(x, \xi)}{|\mathbf{A}(x, \xi)|} = \mathbf{A}(x, \xi) \cdot \theta$$
$$\leq \mathbf{F}(x, \xi + \theta) - \mathbf{F}(x, \xi) \leq \mathbf{F}(x, \xi + \theta) \leq \beta|\xi + \theta|^p \leq \beta(|\xi| + |\theta|)^p$$
$$= 2^p \beta |\xi|^p,$$

so that

(2.77) $$|A(x,\xi)| \leq 2^p \beta |\xi|^{p-1}.$$

It follows that

$$\int_\Omega A(x,g) \cdot h\, dx \leq C \int_\Omega |g|^{p-1}|h|\, dx \leq C\|g\|_p^{p-1}\|h\|_p$$

and thus

(2.78) $$h \mapsto \int_\Omega A(x,g) \cdot h\, dx$$

is a continuous linear functional on $L^p(\Omega; \mathbf{R}^n)$. Since

$$\frac{J(g+th) - J(g)}{t} = \int_\Omega \frac{\mathbf{F}(x, g+th) - \mathbf{F}(x,g)}{t}\, dx$$
$$= \int_\Omega \left(\frac{1}{t}\int_0^t A(x, g+sh) \cdot h\, ds\right) dx$$

and

$$|A(x, g+sh) \cdot h| \leq C(|g| + |h|)^{p-1}|h|, \qquad 0 < s < 1$$

using Lebesgue dominated convergence theorem we verify that (2.78) defines a Gâteaux derivative of J at g. $\square$

2.96 THEOREM. *Suppose $|\Omega| < \infty$. Let $\mathbf{F}$ be a function on $\Omega \times \mathbf{R}^n$ satisfying assumptions (2.66)–(2.68),*

$$\mathbf{J}(u) := \int_\Omega \mathbf{F}(x, \nabla u)\, dx, \qquad u \in W^{1,p}(\Omega)$$

and let

$$A(x, \xi) = [A_1(x, \xi), \ldots, A_n(x, \xi)]$$

be defined by

$$A_i(x, \xi) = \frac{\partial}{\partial \xi_i}\mathbf{F}(x, \xi).$$

Given $u_0 \in W^{1,p}(\Omega)$, let

$$\mathcal{H} = \{u \in W^{1,p}(\Omega): u - u_0 \in W_0^{1,p}(\Omega)\}.$$

Then

(2.79) $$\int_\Omega \mathbf{F}(x, \nabla u)\, dx$$

has an unique minimizer w in $\mathcal{H}$. A function u minimizes $\mathbf{J}$ in $\mathcal{H}$ if and only if it solves the problem

(2.80) $$\begin{cases} u - u_0 \in W_0^{1,p}(\Omega) \\ \int_\Omega A(x, \nabla u) \cdot \nabla \varphi\, dx = 0 \text{ for all } \varphi \in W_0^{1,p}(\Omega). \end{cases}$$

PROOF. Let
$$H = \{\nabla u \colon u \in \mathcal{H}\} \subset L^p(\Omega; \mathbf{R}^n),$$
$$J(g) = \int_\Omega \mathbf{F}(x, g)\, dx, \qquad g \in L^p(\Omega; \mathbf{R}^n).$$
Then H is a closed convex subset of $L^p(\Omega; \mathbf{R}^n)$ and the mapping $u \mapsto \nabla u \colon \mathcal{H} \to H$ is one-to-one. By Lemma 2.95, J is strictly convex, continuous and a coercive functional on $L^p(\Omega; \mathbf{R}^n)$; moreover, its Gâteaux derivative is given by (2.75). Using Theorems 2.91 and 2.93 we obtain the assertion. □

2.97 REMARK. Note that the trick of passing to $L^p(\Omega; \mathbf{R}^n)$ is useful. Indeed, **J** is neither strictly convex, not coercive on $W^{1,p}(\Omega)$; the problems occur on lines of the form $\{u + c : c \in \mathbf{R}\}$. There are also other ways to avoid this difficulty, for example by a factorization of $W^{1,p}(\Omega)$ by locally constant functions or by considering coercivity and strict convexity relative to closed convex subsets of $W^{1,p}(\Omega)$.

2.98 THEOREM. *Let* **F** *satisfy* (2.66)–(2.68) *and* $\mathbf{A} = (\mathbf{A}_1, \ldots, \mathbf{A}_n)$, *where*
$$A_i(x, \xi) = \frac{\partial}{\partial \xi_i} \mathbf{F}(x, \xi).$$
Let $a = 2^p \beta$ *and* $c = \alpha$. *Then* **A** *satisfies* (2.70)–(2.74).

PROOF. The properties (2.70), (2.71) and (2.72) have been verified in the course of the proof of Lemma 2.95. From (2.76) we obtain
$$\mathbf{F}(x, 0) - \mathbf{F}(x, \xi) \geq \mathbf{A}(x, \xi) \cdot (-\xi).$$
We infer that
$$\mathbf{A}(x, \xi) \cdot \xi \geq \mathbf{F}(x, \xi) \geq \alpha |\xi|^p,$$
which verifies (2.73). If $\xi' \neq \xi$, then the strict convexity of **F** gives
$$\mathbf{A}(x, \xi) \cdot (\xi' - \xi) < \mathbf{F}(x, \xi') - \mathbf{F}(x, \xi),$$
$$\mathbf{A}(x, \xi') \cdot (\xi - \xi') < \mathbf{F}(x, \xi) - \mathbf{F}(x, \xi'),$$
and by addition we obtain (2.74). □

2.99 REMARK. Note that the question of regularity of local minimizers reduces to that of interior regularity of minimizers on subdomains. Indeed, w is a local minimizer of **J** on Ω if and only $w \, \llcorner \, \Omega'$ is a minimizer of $\mathbf{J}[\cdot, \Omega']$ in
$$\{u \in W^{1,p}(\Omega') \colon u - w \in W_0^{1,p}(\Omega')\}$$
for any $\Omega' \subset\subset \Omega$. The idea is connected with the following theorem.

2.100 THEOREM. *Let* **F**, **A** *be as in Theorem* 2.96. *A function* $u \in W^{1,p}_{\mathrm{loc}}(\Omega)$ *is a local minimizer of* **J** *if and only if* u *is a weak solution of the Euler-Lagrange equation*
$$\operatorname{div} \mathbf{A}(x, \nabla u) = 0$$
in Ω.

PROOF. It is an immediate consequence of Theorem 2.96 and definitions. □

2.101 EXAMPLE. Let a_{ij}, $i,j = 1, \ldots, n$, be bounded measurable functions on Ω. We assume that the matrix $\mathbf{a} = (a_{ij})_{i,j=1}^n$ is symmetric ($a_{ij} = a_{ji}$) and satisfies at each $x \in \Omega$ the inequalities

$$\lambda|\xi|^2 \leq \sum_{i,j} a_{ij}(x)\xi_i\xi_j \leq \Lambda|\xi|^2, \quad \xi \in \mathbf{R}^n \tag{2.81}$$

where $0 < \lambda \leq \Lambda < \infty$ are independent of x. We write

$$\mathbf{F}(x,\xi) = \tfrac{1}{p}\left(\mathbf{a}\,\xi \cdot \xi\right)^{\frac{p}{2}}$$

where $\mathbf{a}\,\xi \cdot \xi'$ means

$$\sum_{i,j=1}^n a_{ij}\,\xi_j\,\xi'_i.$$

Let

$$\mathbf{J}(u) := \int_\Omega \mathbf{F}(x, \nabla u)\,dx. \tag{2.82}$$

Then $\mathbf{F}$ satisfies (2.66)–(2.68) and if we associate $\mathbf{J}$ and $\mathbf{A}$ with $\mathbf{F}$ as in Theorem 2.96, then

$$A(x,\xi) = \left(\mathbf{a}\,\xi \cdot \xi\right)^{\frac{p}{2}-1}\mathbf{a}\,\xi. \tag{2.83}$$

PROOF. Obviously, the measurability and differentiability assumptions are satisfied and a routine calculation gives formula (2.83). For fixed x, it is known from linear algebra that

$$\xi \mapsto \left(\mathbf{a}(x)\xi \cdot \xi\right)^{q/2}$$

is convex for $q = 1$; the exponent $q = p > 1$ makes the convexity strict. From the assumption (2.81) we obtain (2.68) with $\alpha = \tfrac{1}{p}\lambda^{p/2}$, $\beta = \tfrac{1}{p}\Lambda^{p/2}$. □

2.102 DEFINITION. Let $\mathbf{F}(\xi) = \tfrac{1}{p}|\xi|^p$, which is the case of identity matrix in Example 2.101. This functional is called the *p-Dirichlet integral*. Differentiation leads to

$$A(\xi) = |\xi|^{p-2}\xi.$$

We define the *p-Laplacian* Δ_p as the differential operator

$$\Delta_p = \operatorname{div}(|\nabla u|^{p-2})\nabla u.$$

The equation

$$-\Delta_p = 0$$

is called the *p-Laplace equation*.

2.3.3 Higher regularity.

Here we show the weak minimizer that was found in the previous subsection is actually an element of $W^{2,2}_{\text{loc}}(\Omega)$ and that its partial derivatives, which are elements of $W^{1,2}_{\text{loc}}(\Omega)$, satisfy a linear equation. For the sake of brevity, we do not prove this result in full generality, since the method indicates how more general results could be obtained.

2.103 THEOREM. *Let $\boldsymbol{A}_1,\ldots,\boldsymbol{A}_n$ be continuously differentiable functions on $\mathbf{R}^n$ and*
$$\boldsymbol{A} = [\boldsymbol{A}_1,\ldots,\boldsymbol{A}_n].$$
Suppose that there are constants $a, c > 0$ such that
   (i) $|\frac{\partial}{\partial \xi_j}\boldsymbol{A}(\xi)| \leq a$ *for all* $\xi \in \mathbf{R}^n$,
   (ii) $\sum_{i,j=1}^{n} \frac{\partial}{\partial \xi_j}\boldsymbol{A}_i(\xi)\theta_i\theta_j \geq c|\theta|^2$ *for all* $\xi, \theta \in \mathbf{R}^n$.

If $u \in W^{1,2}(\Omega)$ is a weak solution of

(2.84) $$\operatorname{div} \boldsymbol{A}(\nabla u) = 0,$$

and $k \in \{1,\ldots,n\}$, then $u \in W^{2,2}_{\mathrm{loc}}(\Omega)$ and $w := D_k u$ is a weak solution of the equation

(2.85) $$\sum_{i=1}^{n} D_i\left[\sum_{j=1}^{n} a^{ij}(x) D_j w\right] = 0$$

with

(2.86) $$a^{ij}(x) = \frac{\partial}{\partial \xi_j}\boldsymbol{A}_i(\nabla u(x)).$$

PROOF. For each test function $\varphi \in W^{1,2}_c(\Omega)$,
$$\int_\Omega \boldsymbol{A}(\nabla u) \cdot \nabla \varphi \, dx = 0$$

and
$$\int_\Omega \left[\boldsymbol{A}(\nabla u(x + he_k)) \cdot \nabla \varphi - \boldsymbol{A}(\nabla u(x)) \cdot \nabla \varphi\right] dx = 0$$

for all $|h|$ sufficiently small. Now,

$$\boldsymbol{A}(\nabla u(x + he_k)) - \boldsymbol{A}(\nabla u(x)) = \int_0^1 \frac{d}{dt}\boldsymbol{A}\bigl(t\nabla u(x + he_k) + (1-t)\nabla u(x)\bigr) dt$$
$$= \int_0^1 \sum_{j=1}^n \frac{\partial}{\partial \xi_j}\boldsymbol{A}\bigl(t\nabla u(x + he_k) + (1-t)\nabla u(x)\bigr) D_j[(u(x) + he_k) - u(x)] dt.$$

This implies
$$\sum_{i,j=1}^{n} \int_\Omega A^{ij}(x) D_j[(u(x) + he_k) - u(x)] D_i \varphi \, dt = 0$$

where
$$A^{ij}(x) := \int_0^1 \frac{\partial}{\partial \xi_j}\boldsymbol{A}_i\bigl(t\nabla u(x + he_k) + (1-t)\nabla u(x)\bigr) dt.$$

Choose open sets $U \subset\subset V \subset\subset \Omega$ and let η be a smooth cutoff function satisfying
$$\chi_U \leq \eta \leq \chi_V.$$

Choose as a test function
$$\varphi := \frac{u(x + he_k) - u(x)}{h} \eta^2.$$

Then the equation
$$\sum_{i,j=1}^n \int_\Omega A^{ij}(x) \frac{D_j[(u(x)+he_k)-u(x)]}{h} D_i\varphi \, dt = 0$$
implies
$$\sum_{i,j=1}^n \int_\Omega A^{ij}(x) \frac{D_j[(u(x)+he_k)-u(x)]}{h} \frac{D_i[(u(x)+he_k)-u(x)]}{h} \eta^2 dx$$
$$= -2 \sum_{i,j=1}^n \int_\Omega A^{ij}(x) \frac{D_j[(u(x)+he_k)-u(x)]}{h} \frac{[(u(x)+he_k)-u(x)]}{h} \eta D_i\eta dx.$$

From the ellipticity condition,
$$\sum_{i,j=1}^n \int_\Omega A^{ij}(x) \frac{D_j[(u(x)+he_k)-u(x)]}{h} \frac{D_i[(u(x)+he_k)-u(x)]}{h} \eta^2 dx$$
$$\geq c \int_\Omega \left|\frac{\nabla[u(x+he_k)-u(x)]}{h}\right|^2 \eta^2 dx,$$

and by Young's inequality
$$\sum_{i,j=1}^n \int_\Omega A^{ij}(x) \frac{D_j[(u(x)+he_k)-u(x)]}{h} \frac{[(u(x)+he_k)-u(x)]}{h} \eta D_i\eta dx$$
$$\leq \varepsilon \int_\Omega \left|\frac{\nabla[u(x+he_k)-u(x)]}{h}\right|^2 \eta^2 dx$$
$$+ C(\varepsilon) \int_\Omega \left|\frac{u(x+he_k)-u(x)}{h}\right|^2 |\nabla\eta|^2 dx.$$

By Theorem 1.49, the last term remains bounded for all small $|h|$, and therefore combining the estimates above leads to
$$\int_\Omega \left|\frac{\nabla u(x+he_k)-\nabla u(x)}{h}\right|^2 \eta^2 dx \leq C$$
for all sufficiently small $|h|$. Appealing again to Theorem 1.49, we have $\nabla u \in W^{1,2}(U)$, or equivalently, $u \in W^{2,2}(U)$. Thus, $u \in W^{2,2}_{\text{loc}}(\Omega)$.

Now, choose $\psi \in C_c^\infty(\Omega)$ and test the equation (2.84) by $\varphi = D_k\psi$. Then
$$\sum_{i=1}^n \int_\Omega \boldsymbol{A}_i(\nabla u) D_i D_k\psi \, dx = 0$$
or, by integration by parts,
$$\sum_{i,j=1}^n \int_\Omega \frac{\partial}{\partial \xi_j} \boldsymbol{A}_i(\nabla u) D_k D_j u \, D_i\psi \, dx = 0,$$
which shows that $w = D_j u$ is a weak solution of (2.85). □

2.104 REMARK. Let **J** be a variational integral as in Lemma 2.95 and suppose that the derivatives $A_i = \frac{\partial}{\partial \xi_i}\mathbf{F}$ satisfy the conditions of Theorem 2.103. If u is a local minimizer of **J** and $w = D_k u$, then according to Theorem 2.103, w is a weak solution of the equation

$$\sum_{i=1}^n D_i \Big[\sum_{j=1}^n \frac{\partial^2}{\partial \xi_i \partial \xi_j} \mathbf{F}(\nabla u) D_j w\Big] = 0$$

which takes the form (2.85), a linear equation in divergence form with nonsmooth coefficients. Since $u \in W_{\text{loc}}^{2,2}(\Omega)$, the coefficients given by (2.86) are bounded and measurable.

From the De Giorgi-Nash result that will be proved in the next subsection, it follows that w is locally Hölder continuous, which implies that the coefficients in (2.85) are Hölder continuous. Appealing to Schauder theory, cf. [**GT**], we conclude that $u \in C^{2,\alpha}$. This provides the first step in a boot-strap argument that u possesses the same regularity as assumed by **F**.

2.3.4 The De Giorgi method.

In this subsection we give a brief treatment of De Giorgi's remarkable proof of the result that weak solutions of the equation

(2.87) $$\sum_{i=1}^n D_i \Big[\sum_{j=1}^n a^{ij}(x) D_j u\Big] = 0,$$

are locally Hölder continuous. We will suppose that

(2.88) $$\|a^{ij}\|_\infty \leq a$$
$$\sum_{i,j=1}^n a^{ij}\theta_i \theta_j \geq c|\theta|^2, \quad \theta \in \mathbf{R}^n.$$

The proof is based on careful estimates of quantities associated with level sets. We fix a point x_0 and write $B(r) := B(x_0, r)$. For any level h and radius r, we denote

$$A(h,r) := B(r) \cap \{x : u(x) > h\}.$$

In this subsection, α is reserved for the positive solution of the equation

(2.89) $$2\alpha^2 - n(\alpha + 1) = 0.$$

2.105 LEMMA. *If $u \in W_{\text{loc}}^{1,2}(\Omega)$ is a weak solution of (2.87), then there exists $C = C(n, c, a)$ such that*

(2.90) $$\int_{B(z,r)} |\nabla(u-k)^+|^2 \, dx \leq \frac{C}{(R-r)^2} \int_{B(z,R)} |(u-k)^+|^2 \, dx$$

for $k \in \mathbf{R}$, $0 < r < R < \infty$ and $B(z, R) \subset \Omega$.

PROOF. Let η be a smooth function satisfying $\chi_{B(r)} \leq \eta \leq \chi_{B(R)}$ with $|\nabla\eta| \leq C(R-r)^{-1}$ and define a test function $\varphi := (u-k)^+ \eta^2$. Then

$$\sum_{i,j=1}^n \int_\Omega a^{ij} D_j u D_i (u-k)^+ \eta^2\, dx = -2\sum_{i,j=1}^n \int_\Omega a^{ij} \eta (u-k)^+ D_j u D_i \eta\, dx.$$

Applying Young's inequality, we obtain

$$c\int_\Omega \left[|\nabla(u-k)^+|^2 \eta^2\, dx \leq Ca \int_\Omega \eta(u-k)^+ |\nabla u||\nabla\eta|\, dx\right.$$
$$\leq Ca \int_\Omega \left[\varepsilon^2 |\nabla(u-k)^+|^2 \eta^2 + \varepsilon^{-2}((u-k)^+)^2 |\nabla\eta|^2\right] dx.$$

Choosing ε appropriately yields the desired result. $\square$

2.106 THEOREM. *If u satisfies (2.90), then*

$$\sup_{B(R/2)} u \leq k_0 + C \left(\frac{1}{R^n} \int_{A(k_0,R)} |u-k_0|^2\, dx\right)^{1/2} \left(\frac{|A(k_0,R)|}{R^n}\right)^{\frac{1}{2\alpha}},$$

where α is defined by (2.89) and

$$C = C(n,c,a).$$

PROOF. Let $\bar{r} = \frac{r+R}{2}$. If $\eta \in \mathcal{C}_c^\infty(\mathbf{R}^n)$ is chosen so that $\chi_{B(\bar{r})} \leq \eta \leq \chi_{B(R)}$ and $|\nabla\eta| \leq C(R-r)^{-1}$, then

$$\int_{B(\bar{r})} |\nabla(\eta(u-k)^+)|^2\, dx \leq \int_{B(\bar{r})} |\nabla(u-k)^+|^2 \eta^2\, dx$$
$$+ \int_{B(\bar{r})} |(u-k)^+|^2 |\nabla\eta|^2\, dx$$
$$\leq \frac{C}{(R-r)^2} \int_{B(R)} |(u-k)^+|^2\, dx.$$

Thus, by Sobolev's inequality we have

$$\left(\int_{B(\bar{r})} |(u-k)^+ \eta|^2\, dx\right) \leq C \left(\int_{B(\bar{r})} |\nabla(\eta(u-k)^+)|^{2n/(n+2)}\, dx\right)^{(n+2)/n}$$

while Hölder's inequality yields

$$\left(\int_{B(\bar{r})} |\nabla(\eta(u-k)^+)|^{2n/(n+2)}\, dx\right)^{(n+2)/n}$$
$$\leq |A(k,r)|^{2/n} \int_{B(\bar{r})} |\nabla(\eta(u-k)^+)|^2\, dx.$$

Hence,

$$\int_{A(k,r)} |u-k|^2\, dx \leq \frac{C}{(R-r)^2} |A(k,R)|^{2/n} \int_{A(k,R)} |u-k|^2\, dx.$$

Using the notation

$$a(h,r) := |A(h,r)|,$$

(2.91)
$$u(h,r) := \int_{A(h,r)} |u-h|^2 \, dx,$$

the last inequality takes the form

$$u(h,r) \le u(k,r) \le \frac{C}{(R-r)^2} u(k,R) \, a(k,R)^{2/n}$$

where $h > k$. The inequality

$$a(h,r) \le \frac{1}{(h-k)^2} u(k,R)$$

follows immediately from definitions, and therefore

$$u(h,r)^\alpha a(h,r) \le \frac{C^\alpha}{(R-r)^{2\alpha}} \frac{1}{(h-k)^2} u(k,R)^{\alpha+1} a(k,R)^{2\alpha/n}.$$

Setting

$$\Phi(h,r) := u(h,r)^\alpha a(h,r),$$

we have

$$\Phi(h,r) \le \frac{C^\alpha}{(R-r)^{2\alpha}(h-k)^2} \Phi(k,R)^{1+1/\alpha}.$$

Now, set

$$r_j := (1 + 2^{-j})R/2,$$
$$k_j := k_0 + d - 2^{-j} d,$$
$$\Phi_j := \Phi(k_j, r_j),$$

where

(2.92)
$$d^2 := \frac{2^{2(1+\alpha)^2} C^\alpha}{(R/2)^{2\alpha}} \Phi(k_0, R)^{1/\alpha}.$$

Then

$$\Phi_j \le \frac{C^\alpha}{[2^{-j-1} R]^{2\alpha} [2^{-j} d]^2} \Phi_{j-1}^{1+1/\alpha}$$
$$= 2^{\frac{\beta}{\alpha}(j-1-\alpha)} \Phi(k_0, R)^{-1/\alpha} \Phi_{j-1}^{1+1/\alpha} \quad \text{with } \beta = 2\alpha(\alpha+1).$$

By induction, conclude that

$$\Phi_j \le 2^{-\beta j} \Phi(k_0, R).$$

Letting $j \to \infty$ we obtain

$$u(k_0 + d, R/2)^\alpha a(k_0 + d, R/2) = \Phi(k_0 + d, R/2) = 0.$$

It follows that either $u(k_0 + d, R/2) = 0$ or $a(k_0 + d, R/2) = 0$. Thus,

$$\sup_{B(R/2)} |u| \le K_0 + d$$

and from (2.92) the result follows. $\square$

2.3 REGULARITY OF MINIMIZERS

2.107 LEMMA. *Suppose u satisfies (2.90),*

$$M(R) := \sup_{B(R)} u \quad \text{and} \quad m(R) := \inf_{B(R)} u.$$

If

$$|A(k_0, R)| \leq \tfrac{1}{2}|B(R)| \quad \text{where} \quad k_0 := \tfrac{1}{2}[M(2R) + m(2R)],$$

then for each $\varepsilon > 0$ there exists $\delta = \delta(\varepsilon, n, a, c) > 0$ such that

$$|A(k_\delta, R)| \leq \varepsilon|B(2R)|, \quad \text{where} \quad k_\delta = M(2R) - \delta\left[M(2R) - m(2r)\right].$$

PROOF. For $h > k > k_0$ let

$$v := \begin{cases} h - k, & u \geq h, \\ u - k, & k < u < h, \\ 0, & u \leq k. \end{cases}$$

Then

$$|B(r) \cap \{v = 0\}| = |\{u < k\}| \geq |\{u < k_0\}| \geq \tfrac{1}{2}|B(R)|.$$

Thus,

$$h - k - \bar{v}_{B(r)} \geq \tfrac{1}{2}(h - k)$$

and by Theorem 1.64,

$$(h-k)|A(h,r)|^{1-1/n} \leq 2\left(\int_{B(R)} |v - \bar{v}_{B(r)}|^{n/(n-1)} dx\right)^{(n-1)/n}$$

$$\leq C\int_{B(R)} |\nabla v|\, dx = C\int_{A(k,R)\setminus A(h,R)} |\nabla u|\, dx.$$

Now use Hölder's inequality and (2.90) to conclude

$$(h-k)^2 |A(h,R)|^{\frac{2n-2}{n}} \leq C|A(k,R) \setminus A(h,R)| \int_{A(k,R)\setminus A(h,R)} |\nabla u|^2\, dx$$

$$\leq C|A(k,R) \setminus A(h,R)| \int_{A(k,R)} |\nabla u|^2\, dx$$

$$\leq C|A(k,R) \setminus A(h,R)|\, R^{n-2}[M(2R) - k]^2.$$

With $M := M(2R)$ and $k_i := M - 2^{-i}(M - k_0)$, note that

$$k_i - k_{i-1} = 2^{-i}(M - k_0),$$
$$M - k_{i-1} = 2^{-i+1}(M - k_0).$$

Therefore,

$$|A(k_l, R)|^{\frac{2n-2}{n}} \leq |A(k_i, R)|^{\frac{2n-2}{n}} \leq 4CR^{n-2}[|A(k_{i-1}, R)| - |A(k_i, R)|]$$

for $i = 1, 2, \ldots, l$. Summing on i, we obtain

$$l|A(k_l, R)|^{\frac{2n-2}{n}} \leq 4CR^{n-2}[|A(k_0, R)| - |A(k_l, R)|]$$

$$\leq 4C|B(R)|^{\frac{2n-2}{n}}$$

and the result easily follows. □

2.108 THEOREM. *If u is a solution of (2.87), then u is Hölder continuous.*

PROOF. We apply Theorem 2.106 to u and $-u$ to show that u is locally bounded. With $k_0 := \frac{1}{2}[M(2R) - m(2R)]$ we may assume that

$$|B(r) \cap \{u \geq k_0\}| > \tfrac{1}{2}|B(R)|$$

for otherwise we could consider $-u$. Choose $\varepsilon > 0$ and apply Theorem 2.106 with k_0 replaced by k_δ which is as in Lemma 2.107. We obtain

$$M(R/2) \leq k_\delta + C\Big(R^{-n} \int_{A(k_\delta, R)} |u - k_\delta|^2\Big)^{1/2} \Big(\frac{|A(k_\delta, R)|}{R^n}\Big)^{\frac{1}{2\alpha}}$$

$$\leq k_\delta + C[M(2R) - k_\delta]\Big(\frac{|A(k_\delta, R)|}{R^n}\Big)^{\frac{\alpha+1}{2\alpha}}.$$

In view of the previous result, we can choose δ so small that

$$C\Big(\frac{|A(k_\delta, R)|}{R^n}\Big)^{\frac{\alpha+1}{2\alpha}} < \tfrac{1}{2}.$$

Then

$$M(R/2) < k_\delta + \tfrac{1}{2}\big[M(2R) - k_\delta\big] = M(2R) - \tfrac{\delta}{2}\big[M(2R) - m(2R)\big]$$
$$= m(2R) + (1 - \tfrac{\delta}{2})\big[M(2R) - m(2R)\big].$$

Thus, the oscillation of u satisfies

$$\operatorname{osc}_{B(R/2)} u \leq M(R/2) - m(2R) \leq (1 - \tfrac{\delta}{2})\operatorname{osc}_{B(2R)} u.$$

Iterating this inequality establishes Hölder continuity. □

2.3.5 Moser's iteration technique.

In this section we give another proof due to Moser [**Mos1**] that solutions of (2.87) are Hölder continuous. More generally, we will consider the equation (2.69), where $\boldsymbol{A}$ satisfies (2.70–2.74) with some a and $c > 0$ and $1 < p \leq n$. This technique will be used extensively in later chapters. Note that the De Giorgi's proof can be also generalized to $p \neq 2$. The first new idea here is to use powers of u to serve as a test function. The most important step of the analysis is using John-Nirenberg's estimate (Theorem 1.66) to prove the weak Harnack inequality (Theorem 2.113).

2.109 LEMMA. *Let u be a weak subsolution of (2.69) on Ω. Then u^+ is a weak subsolution of (2.69) on Ω as well.*

PROOF. Notice that $u^+ \in W^{1,p}_{\text{loc}}(\Omega)$. Let $\varphi \in \mathcal{C}^\infty_c(\Omega)$ be nonnegative. We denote $v_k = \min\{ku^+, 1\}$ and employ $v_k \varphi$ as a test function. We obtain

$$\int_\Omega v_k \boldsymbol{A}(x, \nabla u) \cdot \nabla \varphi \, dx \leq -k \int_{\Omega \cap \{0 < u < 1/k\}} \boldsymbol{A}(x, \nabla u) \cdot \nabla u \, \varphi \, dx \leq 0.$$

Letting $k \to \infty$ we obtain

$$\int_\Omega \boldsymbol{A}(x, \nabla u^+) \cdot \nabla \varphi \, dx = \int_{\Omega \cap \{u > 0\}} \boldsymbol{A}(x, \nabla u) \cdot \nabla \varphi \, dx \leq 0. \qquad \square$$

2.110 LEMMA. *Let u be a weak subsolution of (2.69) on Ω and $\beta > 0$. Let $\eta \in \mathcal{C}_c^\infty(\Omega)$, $\eta \geq 0$. Then*

(2.93) $$\int_{\Omega \cap \{u>0\}} u^{\beta-1} |\nabla u|^p \eta^p \, dx \leq C \beta^{-p} \int_{\Omega \cap \{u>0\}} u^{p+\beta-1} |\nabla \eta|^p \, dx,$$

with $C = C(n, p, a, c)$.

PROOF. In view of Lemma 2.109 we may assume that $u^+ = u$. Set $v_k = \min\{u^\beta, ku\}$, write $\Omega_k = \Omega \cap \{v_k = u^\beta\}$ and employ $v_k \eta^p$ as a test function. Using the structure and Young's inequality, we have

(2.94)
$$\beta \int_{\Omega_k} u^{\beta-1} |\nabla u|^p \eta^p \, dx + k \int_{\Omega \setminus \Omega_k} |\nabla u|^p \eta^p \, dx \leq C \int_\Omega \mathbf{A}(x, \nabla u) \cdot \nabla v_k \eta^p \, dx$$
$$\leq C \int_\Omega v_k \eta^{p-1} \mathbf{A}(x, \nabla u) \cdot \nabla \eta \, dx$$
$$\leq C \int_{\Omega_k} u^\beta \eta^{p-1} |\nabla u|^{p-1} |\nabla \eta| \, dx + Ck \int_{\Omega \setminus \Omega_k} u \eta^{p-1} |\nabla u|^{p-1} |\nabla \eta| \, dx$$
$$\leq \tfrac{\beta}{2} \int_{\Omega_k} u^{\beta-1} |\nabla u|^p \eta^p \, dx + \tfrac{k}{2} \int_{\Omega \setminus \Omega_k} |\nabla u|^p \eta^p \, dx$$
$$+ C\beta^{1-p} \int_{\Omega_k} u^{p+\beta-1} |\nabla \eta|^p \, dx + Ck \int_{\Omega \setminus \Omega_k} u^p |\nabla \eta|^p \, dx.$$

Notice that $u^{\beta-1} |\nabla u|^p$ in integrable on $\operatorname{spt} \eta \setminus \Omega_k$ as $u^{\beta-1} \leq k$. It follows that we may cancel in (2.94) and obtain

(2.95)
$$\beta \int_{\Omega_k} u^{\beta-1} |\nabla u|^p \eta^p \, dx + k \int_{\Omega \setminus \Omega_k} |\nabla u|^p \eta^p \, dx$$
$$\leq C\beta^{1-p} \int_{\Omega_k} u^{p+\beta-1} |\nabla \eta|^p \, dx + Ck \int_{\Omega \setminus \Omega_k} u^p |\nabla \eta|^p \, dx.$$

We may assume that $u^{p+\beta-1} |\nabla \eta|^p$ is integrable. Then, since $ku^p \leq u^{p+\beta-1}$ on $\Omega \setminus \Omega_k$, we may use Lebesgue's dominated convergence theorem to pass to the limit on the right-hand side (and Fatou's lemma on the left-hand side) and obtain (2.93). □

2.111 LEMMA. *Let u be a weak subsolution of (2.69) on $B(2R)$ and $q > p-1$. Then*
$$\sup_{B(R)} u \leq C \left(\fint_{B(2R)} (u^+)^q \, dx \right)^{1/q},$$
where $C = C(n, p, q, a, c)$.

PROOF. In view of Lemma 2.109 we may assume that $u^+ = u$. Let $0 < r' \leq r$ and use a cutoff function $\eta \in \mathcal{C}_c^\infty(\mathbf{R}^n)$ such that $\chi_{B(r')} \leq \eta \leq \chi_{B(r)}$, and with $|\nabla \eta| < C/(r-r')$. Denote $v = u^+$. Choose $\alpha \geq q$. Lemma 2.110 with $\beta = \alpha - p + 1$ yields

(2.96) $$\int_{B(r)} v^{\alpha-p} |\nabla v|^p \eta^p \, dx \leq C \int_{B(r)} v^\alpha |\nabla \eta|^p \, dx.$$

Choose $\sigma > 1$ such that $(n-p)\sigma \leq n$. Sobolev's inequality (Theorem 1.57) implies

(2.97)
$$\begin{aligned}
r^{n-p-\frac{n}{\sigma}} &\left(\int_{B(r)} [v^{\alpha/p}\eta]^{\sigma p}\right)^{1/\sigma} \\
&\leq C \int_{B(r)} \left(|\nabla v^{\alpha/p}|^p \eta^p + v^\alpha |\nabla \eta|^p\right) dx \\
&\leq C \int_{B(r)} \left(v^{\alpha-p}|\nabla v|^p \eta^p + v^\alpha |\nabla \eta|^p\right) dx.
\end{aligned}$$

By (2.96), (2.97) and the properties of η we have

(2.98) $$r^{n-p-\frac{n}{\sigma}} \left(\int_{B(r')} v^{\sigma\alpha} dx\right)^{1/\sigma} \leq C(r-r')^{-p} \int_{B(r)} v^\alpha dx.$$

Set
$$r_j = R + 2^{-j}R, \quad j = 0, 1, 2, \ldots$$
and substitute $\alpha = q\sigma^j$. Then
$$\left(\fint_{B(r_{j+1})} v^{\sigma^{j+1}q} dx\right)^{1/(\sigma^{j+1}q)} \leq (C 2^j)^{\sigma^{-j}p/q} \left(\fint_{B(r_j)} v^{\sigma^j q} dx\right)^{1/(\sigma^j q)}.$$

This inequality can be iterated ($j = 0, \ldots, k$) to obtain
$$\left(\fint_{B(r_{k+1})} v^{\sigma^{k+1}q} dx\right)^{1/(\sigma^{k+1}q)} \leq C^{\mu_k p/q} 2^{\nu_k p/q} \left(\int_{B(2R)} v^q dx\right)^{1/q}$$

where
$$\mu_k = 1 + \frac{1}{\sigma} + \cdots + \frac{1}{\sigma^k} \quad \text{and} \quad \nu_k = \frac{1}{\sigma} + \frac{2}{\sigma^2} + \cdots + \frac{k}{\sigma^k}.$$

Letting $k \to \infty$ we obtain
$$\sup_{B(R)} v \leq C \left(\fint_{B(2R)} v^q dx\right)^{1/q}$$

as required. $\square$

2.112 LEMMA. *Let u be a nonnegative weak supersolution of (2.69) on Ω and $\beta < 0$. Let $\eta \in C_c^\infty(\Omega)$, $\eta \geq 0$. Then*
$$\int_{\Omega \cap \{u>0\}} u^{\beta-1}|\nabla u|^p \eta^p \, dx \leq C|\beta|^{-p} \int_{\Omega \cap \{u>0\}} u^{p+\beta-1}|\nabla \eta|^p \, dx,$$
where $C = C(n, p, a, c)$.

PROOF. We write $u_k = u + \frac{1}{k}$ and use $u_k^\beta \eta^p$ as a test function. As in the proof of Lemma 2.110 we have
$$|\beta| \int_\Omega u_k^{\beta-1}|\nabla u|^p \eta^p \, dx \leq C|\beta|^{1-p} \int_\Omega u_k^{p+\beta-1}|\nabla \eta|^p \, dx.$$

Letting $k \to \infty$ we obtain the assertion. $\square$

2.113 LEMMA (Weak Harnack inequality). *Let u be a nonnegative weak supersolution of* (2.69) *on* $B(4R)$, $q > 0$ *and* $(n-p)q < n(p-1)$. *Then*

$$\left(\fint_{B(R)} u^q \, dx\right)^{1/q} \leq C \inf_{B(R)} u,$$

with $C = C(n,p,q,a,c)$.

PROOF. We may assume that $u > 0$, otherwise we replace u by $u + \varepsilon$ and let $\varepsilon \downarrow 0$ after obtaining the estimate. Choose $\sigma > 1$ such that $(n-p)\sigma \leq n$. Use Lemma 2.112 in a way similar to the way Lemma 2.110 has been used in the proof of Lemma 2.111. Observe that by Lemma 2.112, (2.96) holds with $v = u^{-1}$ and $\alpha = 1 - p - \beta$ whenever $\beta \leq \beta_0 < 0$. Hence as in the proof of Theorem 2.111,

$$(2.99) \qquad \sup_{B(R)} u^{-1} \leq C \left(\fint_{B(2R)} u^{-\lambda}\right)^{1/\lambda}$$

holds with any $\lambda > 0$. Now, let l be an positive integer and set

$$R_j = R - 2^{j-1}R, \quad j = 0, -1, -2, \ldots, -l$$

and substitute $v = u$, $\alpha = q\sigma^j$ in (2.98). Then

$$\left(\fint_{B(R_{j+1})} u^{\sigma^{j+1}q} \, dx\right)^{1/(\sigma^{j+1}q)} \leq \left(C2^{-j}\right)^{\sigma^{-j}p/q} \left(\fint_{B(R_j)} u^{\sigma^j q} \, dx\right)^{1/(\sigma^j q)}.$$

This inequality can be iterated to obtain

$$(2.100) \qquad \left(\fint_{B(R/2)} u^q \, dx\right)^{1/q} \leq C^{\mu'_l p/q} 2^{\nu'_l p/q} \left(\fint_{B(R_{-l})} u^\lambda \, dx\right)^{1/\lambda}$$

where

$$\lambda = q\sigma^{-l}, \quad \mu'_l = \sigma + \cdots + \sigma^l \text{ and } \nu'_l = \sigma + 2\sigma^2 + \cdots + l\sigma^l.$$

We now come to the most important step in the analysis. Consider a ball $B(y, 2\delta) \subset B(2R)$. Using Lemma 2.112 with $\beta = 1 - p$ we obtain

$$\int_{B(y,2\delta)} u^{-p} |\nabla u|^p \eta^p \, dx \leq C \int_{B(y,2\delta)} |\nabla \eta|^p \, dx$$

for any nonnegative $\eta \in C_c^\infty(B(y, 2\delta))$. With a suitable choice of η and $w = \ln u$ it follows

$$\left(\int_{B(y,\delta)} |\nabla w| \, dx\right)^p \leq C\delta^{n(p-1)} \int_{B(y,\delta)} |\nabla[\ln u]|^p \, dx$$

$$\leq C\delta^{n(p-1)} \int_{B(y,2\delta)} u^{-p} |\nabla u|^p \eta^p \, dx \leq C\delta^{n(p-1)} \int_{B(y,2\delta)} |\nabla \eta|^p \, dx$$

$$\leq C\delta^{n(p-1)} \delta^{n-p},$$

so that

$$\int_{B(y,\delta)} |\nabla w| \, dx \leq C\delta^{n-1}.$$

Hence by Theorem 1.66 there is a constant $\lambda_0 > 0$ such that, for any λ with $0 < \lambda < \lambda_0$ and

$$w_0 = \fint_{B(2R)} w \, dx,$$

we have
$$\int_{B(2R)} e^{\lambda|w-w_0|}\,dx \leq C.$$
Therefore,
$$\left(\int_{B(2R)} u^\lambda\,dx\right)\left(\int_{B(2R)} u^{-\lambda}\,dx\right) = \left(\int_{B(2R)} e^{\lambda w}\,dx\right)\left(\int_{B(2R)} e^{-\lambda w}\,dx\right)$$
$$= \int_{B(2R)\times B(2R)} e^{\lambda(w(x)-w_0)} e^{\lambda(w_0 - w(y))}\,dx\,dy \leq C.$$

This, along with (2.99) and (2.100) yields the assertion. □

2.114 THEOREM. *If u is a solution of (2.69), then u is Hölder continuous.*

PROOF. By Lemma 2.111 and Lemma 2.113, for any ball $B(R)$ with $B(4R) \subset \Omega$ we have
$$\sup_{B(R)}\left(u - \inf_{B(4R)} u\right) \leq C \inf_{B(R)}\left(u - \inf_{B(4R)} u\right),$$
so that
$$\operatorname{osc}_{B(R)} u \leq (1-\delta)\operatorname{osc}_{B(4R)} u$$
with some $\delta > 0$. Iterating this inequality establishes Hölder continuity. □

2.3.6 Removable singularities.

Removable singularities are relevant to our development to the extent of providing some motivation for the success of the Wiener criterion. It is well known from classical potential theory that the point at the tip of a sharp inward-pointing spine of a region is not a regular boundary point. One intuitive explanation for this is that the spine is so sharp that the solution almost "doesn't see" the spine; that is, the spine is almost considered as a removable set by the solution. The following results show that sets of capacity zero are removable and thus, it is not too surprising that capacity appears as a critical component in the Wiener condition.

2.115 LEMMA. *Let $\Omega' \subset \Omega$ be an open set and $u \in W^{1,p}_{\mathrm{loc}}(\Omega')$ be a weak subsolution of (2.69) in Ω' that is bounded below. Suppose that $\mathbf{C}_s(\Omega \setminus \Omega') = 0$, $p \leq s \leq n$. If, for some $\varepsilon > 0$,*
$$u \in L^{\frac{s(p-1+\varepsilon)}{s-p}}(\Omega),$$
then u is a weak subsolution in all of Ω. In particular, if $\mathbf{C}_p(\Omega \setminus \Omega') = 0$, then $\Omega \setminus \Omega'$ is removable for any bounded subsolution of (2.69).

PROOF. We may assume that $u \geq 1$ and $\varepsilon < 1$. Denote $E = \Omega \setminus \Omega'$. Let $\{\varphi_j\}$ be a sequence of functions from $W^{1,p}(\mathbf{R}^n)$ such that $0 \leq \varphi_j \leq 1$, $\{\varphi_j = 1\}^\circ \supset E$ and $\|\varphi_j\|_{1,p} \to 0$. Let $\eta \in \mathcal{C}^\infty_c(\Omega)$ be nonnegative. Each
$$\eta_j := (1-\varphi_j)\eta$$
belongs to $W^{1,p}_c(\Omega')$. Let $\beta > 0$. We will proceed similarly as in Lemma 2.110. Let
$$v_k = \min\{ku^\varepsilon, u^\beta\}, \qquad k = 1, 2, \ldots.$$

Using $v_k \eta_j^p$ as a test function we obtain

$$\beta \int_{\{v_k = u^\beta\}} u^{\beta-1} |\nabla u|^p \eta_j^p \, dx + k\varepsilon \int_{\{v_k = ku^\varepsilon\}} u^{\varepsilon-1} |\nabla u|^p \eta_j^p \, dx$$

(2.101)
$$\leq C \int_\Omega \mathbf{A}(x, \nabla u) \cdot \nabla v_k \, \eta_j^p \, dx \leq C \int_\Omega v_k \, \eta_j^{p-1} \mathbf{A}(x, \nabla u) \cdot \nabla \eta_j \, dx$$

$$\leq C \int_\Omega v_k \, \eta_j^{p-1} |\nabla u|^{p-1} |\nabla \eta_j| \, dx.$$

Young's inequality leads to

$$\int_\Omega v_k \, \eta_j^{p-1} |\nabla u|^{p-1} |\nabla \eta_j| \, dx$$

$$\leq \frac{1}{2} \Big(\beta \int_{\{v_k = u^\beta\}} u^{\beta-1} |\nabla u|^p \eta_j^p \, dx + k\varepsilon \int_{\{v_k = ku^\varepsilon\}} u^{\varepsilon-1} |\nabla u|^p \eta_j^p \, dx$$

(2.102)

$$+ C\beta^{1-p} \int_{\{v_k = u^\beta\}} u^{\beta+p-1} |\nabla \eta_j|^p \, dx$$

$$+ Ck\varepsilon^{1-p} \int_{\{v_k = ku^\varepsilon\}} u^{\varepsilon+p-1} |\nabla \eta_j|^p \, dx.$$

By (2.101) and (2.102) we have

$$\beta \int_{\{v_k = u^\beta\}} u^{\beta-1} |\nabla u|^p \eta_j^p \, dx + k\varepsilon \int_{\{v_k = ku^\varepsilon\}} u^{\varepsilon-1} |\nabla u|^p \eta_j^p \, dx$$

(2.103)
$$\leq C\beta^{1-p} \int_{\{v_k = u^\beta\}} u^{\beta+p-1} |\nabla \eta_j|^p \, dx$$

$$+ Ck\varepsilon^{1-p} \int_{\{v_k = ku^\varepsilon\}} u^{\varepsilon+p-1} |\nabla \eta_j|^p \, dx.$$

The cancellation is justified, as the terms to be canceled are integrable; indeed, all of them are estimated by a constant multiple of $|\nabla u|^p \eta_j^p$ and $|\nabla u|^p$ is integrable on the support of η_j. Since, by Young's inequality

(2.104)
$$|u^{\varepsilon+p-1}| \, |\nabla \eta_j|^p \leq \frac{s-p}{s} u^{\frac{s(p-1+\varepsilon)}{s-p}} + \frac{p}{s} |\nabla \eta_j|^s, \qquad s > p,$$

(or, if $s = p$, the integrability is transparent without any effort) and this term majorizes the remaining ones, the right-hand side of (2.103) is finite. Now, taking into account (2.104) we may let $j \to \infty$ in (2.103) and obtain

$$\beta \int_{\{v_k = u^\beta\}} u^{\beta-1} |\nabla u|^p \eta^p \, dx + k\varepsilon \int_{\{v_k = ku^\varepsilon\}} u^{\varepsilon-1} |\nabla u|^p \eta^p \, dx$$

(2.105)
$$\leq C\beta^{1-p} \int_{\{v_k = u^\beta\}} u^{\beta+p-1} |\nabla \eta|^p \, dx$$

$$+ Ck\varepsilon^{1-p} \int_{\{v_k = ku^\varepsilon\}} u^{\varepsilon+p-1} |\nabla \eta|^p \, dx.$$

Finally, letting $k \to \infty$ we obtain

(2.106)
$$\beta \int_\Omega u^{\beta-1} |\nabla u|^p \eta^p \, dx \leq C\beta^{1-p} \int_\Omega u^{\beta+p-1} |\nabla \eta|^p \, dx$$

where the right-hand side may be infinite, but is finite for $\beta = \varepsilon$. Similarly as in the proof of Lemma 2.111 we utilize Sobolev's inequality and iterate to obtain that u is locally bounded on Ω. Setting $\beta = 1$ in (2.106) we see that $u \in W^{1,p}_{\text{loc}}(\Omega)$. By Young's inequality,

$$|A(x, \nabla u)|^{\frac{s}{s-1}} \leq C|\nabla u|^{\frac{s(p-1)}{s-1}} \leq C|\nabla u|^p u^{\beta-1} + C u^{\frac{s(1-\varepsilon)(p-1)}{s-p}}$$

and thus $A(x, \nabla u) \in L^{s'}_{\text{loc}}$. This allows us pass to limit in

$$\int_\Omega A(x, \nabla u) \cdot \nabla \eta_j \leq 0$$

and obtain

$$\int_\Omega A(x, \nabla u) \cdot \nabla \eta \leq 0.$$

Thus, u is a weak subsolution of (2.69) in Ω. $\square$

The next theorem has the same hypotheses as the previous lemma, except that u is assumed to be bounded below.

2.116 THEOREM. *Let $\Omega' \subset \Omega$ be an open set and $u \in W^{1,p}(\Omega')$ be a weak solution of (2.69) in Ω'. Suppose that $\mathbf{C}_s(\Omega \setminus \Omega') = 0$, $p \leq s \leq n$. If, for some $\varepsilon > 0$,*

$$u \in L^{\frac{s(p-1+\varepsilon)}{s-p}}(\Omega),$$

then u is a weak solution in all of Ω. In particular, if $\mathbf{C}_p(\Omega \setminus \Omega') = 0$, then $\Omega \setminus \Omega'$ is removable for any bounded solution of (2.69).

PROOF. By Lemma 2.109, the function u^- is a subsolution bounded below on Ω' and thus by Lemma 2.115, u^- is a subsolution on Ω. This implies that u is locally lower bounded and thus using Lemma 2.115 again we obtain that u is a subsolution on Ω, and therefore, locally bounded above. The final step consists in third using of Lemma 2.115 to show that $-u$ is a subsolution as well. $\square$

2.3.7 Estimates of supersolutions.

The following two results are the main ingredients in proving that the Wiener condition is sufficient for regularity at the boundary. These same results will be established in Chapter 4 for equations with a general structure.

2.117 LEMMA. *Let u be a nonnegative weak supersolution of (2.69) on $B(4r)$ and $\eta \in \mathcal{C}_c^\infty(B(r))$ be nonnegative with $|\nabla \eta| \leq Cr^{-p}$. Then*

$$(2.107) \qquad \int_{B(r)} |\nabla u|^{p-1} \eta^{p-1} |\nabla \eta| \, dx \leq C r^{n-p} (\inf_{B(r/2)} u)^{p-1}$$

where $C = C(n, p, a, c)$.

PROOF. Choose $\beta \in (1-p, 0)$ such that $(n-p)\beta > -p$. Referring to Lemma 2.112 we have

$$\int_{B(r)} \eta^p u^{\beta-1} |\nabla u|^p \, dx \leq C \int_{B(r)} |\nabla \eta|^p u^{p+\beta-1} \, dx$$

$$\leq C r^{-p} \int_{B(r)} u^{p+\beta-1} \, dx.$$

Since $(n-p)\beta > -p$, we obtain from Theorem 2.113 and Hölder's inequality,

$$\int_{B(r)} |\nabla u|^{p-1} \eta^{p-1} |\nabla \eta| \, dx$$
$$\leq \Big(\int_{B(r)} |\nabla u|^p u^{\beta-1} \eta^p \, dx \Big)^{(p-1)/p} \Big(\int_{B(r)} u^{(1-\beta)(p-1)} |\nabla \eta|^p \, dx \Big)^{1/p}$$
$$\leq C \Big(\int_{B(r)} u^{p+\beta-1} |\nabla \eta|^p \, dx \Big)^{(p-1)/p} \Big(\int_{B(r)} u^{(1-\beta)(p-1)} |\nabla \eta|^p \, dx \Big)^{1/p}$$
$$\leq C \Big(r^{n-p} (\inf_{B(r)} u)^{p+\beta-1} \Big)^{(p-1)/p} \Big(r^{n-p} (\inf_{B(r)} u)^{(1-\beta)(p-1)} \Big)^{1/p}$$
$$\leq C r^{n-p} (\inf_{B(r)} u)^{p-1}. \qquad \square$$

2.118 THEOREM. *Let u be a nonnegative supersolution of (2.69) on $B(4r)$ and $\eta \in \mathcal{C}_c^\infty(B(r))$ be nonnegative with $|\nabla \eta| \leq Cr^{-p}$. Let $k > 0$. Then*

$$(2.108) \qquad \int_{B(r) \cap \{u \leq k\}} |\nabla u|^p \eta^p \, dx \leq C r^{n-p} k (\inf_{B(r)} u)^{p-1},$$

with $C = C(n, p, a, c)$.

PROOF. Write $u_k = \min(u, k)$ and test (2.69) by

$$(k - u_k)\eta^p.$$

We obtain

$$\int_{B(r)} |\nabla u_k|^p \eta^p \, dx \leq C \int_{B(r)} \boldsymbol{A}(x, \nabla u) \cdot \nabla u_k \eta^p \, dx$$

$$\leq C \int_{B(r)} \boldsymbol{A}(x, \nabla u) \cdot (k - u_k) \eta^{p-1} |\nabla \eta| \, dx$$

$$\leq C k \int_{B(r)} |\nabla u_k|^{p-1} \eta^{p-1} |\nabla \eta| \, dx.$$

By Lemma 2.117,

$$\int_{B(r)} |\nabla u_k|^p \eta^p \, dx \leq C r^{n-p} k (\inf_{B(r)} u)^{p-1}.$$

which is the required estimate. □

2.119 DEFINITION. We say that a function u is p-finely continuous on an open set U if for each $z \in U$ and $\varepsilon > 0$, the set $\{\,|u - u(z)| \geq \varepsilon\,\}$ is p-thin at z.

2.120 REMARK. This definition of fine continuity will be superseded later in Definition 2.134.

2.121 THEOREM. *Let u be a supersolution of (2.69) on Ω. Then there exists a representative of u which is lower semicontinuous and p-finely continuous.*

PROOF. Let $B(8R) = B(z, 8R) \subset\subset \Omega$. Choose $l \in \mathbf{R}$ and write $E = \{u \geq l\}$, $u_l = \min\{u, l\}$. Let
$$m_0 = \lim_{r \to 0+} m(r),$$
where
$$m(r) = \inf_{B(r)} u_l, \quad r > 0.$$

We want to show that E is p-thin at z if $m_0 < l$. Fix $r \in (0, R)$ and write $v = u_l - m(8r)$. Find $\eta \in \mathcal{C}_c^\infty(B(2r))$ such that $0 \leq \eta \leq 1$, $\eta = 1$ on $B(r)$ and $|\nabla \eta| \leq Cr^{-1}$. Then by Corollary 2.25, we may test the p-capacity of $B(r) \cap E$ by $(l - m_0)^{-1} v\eta$. Since $u - m(8r)$ is a nonnegative supersolution on $B(8r)$, by Theorem 2.118

(2.109) $$\int_{B(2r)} |\nabla v|^p \eta^p \, dx \leq Cr^{n-p}\bigl[l - m(8R)\bigr]\bigl[m(2r) - m(8r)\bigr]^{p-1}$$

and by Theorem 2.113

(2.110) $$\int_{B(2r)} v^p |\nabla \eta|^p \, dx \leq Cr^{-p}\bigl[l - m(8R)\bigr] \int_{B(2r)} v^{p-1} \, dx$$
$$\leq Cr^{n-p}\bigl[l - m(8R)\bigr]\bigl[m(2r) - m(8r)\bigr]^{p-1}.$$

By (2.109), (2.110) and Lemma 2.5,
$$\gamma_{p;r}\bigl(B(r) \cap E\bigr) \leq C\, \mathbf{c}_p\bigl(B(r) \cap E; B(8r)\bigr)$$
$$\leq C(l - m_0)^{-p} \int_{B(r)} |\nabla[v\eta]|^p \, dx$$
$$\leq Cr^{n-p}(l - m_0)^{-p}\bigl[l - m(8R)\bigr]\bigl[m(2r) - m(8r)\bigr]^{p-1}$$

or

(2.111) $$\left(\frac{\gamma_{p;r}\bigl(B(r) \cap E\bigr)}{r^{n-p}}\right)^{1/(p-1)} \leq K\bigl[m(2r) - m(8r)\bigr],$$

where
$$K^{p-1} = C\bigl[l - m(8R)\bigr]\bigl[l - m_0\bigr]^{p-1}.$$

Integrating over r from $\rho > 0$ to R we obtain

$$\int_\rho^R \left(\frac{\gamma_{p;r}(B(r) \cap E)}{r^{n-p}}\right)^{1/(p-1)} \frac{dr}{r}$$

$$\leq K\left(\int_{2\rho}^{2R} \frac{m(r)\,dr}{r} - \int_{8\rho}^{8R} \frac{m(r)\,dr}{r}\right)$$

$$= K\left(\int_{2\rho}^{8\rho} \frac{m(r)\,dr}{r} - \int_{2R}^{8R} \frac{m(r)\,dr}{r}\right)$$

$$\leq K \ln 4 \left(m_0 - m(8R)\right).$$

Letting $\rho \to 0^+$ proves that E is p-thin at z. We conclude that u_l has a Lebesgue point at z if we define $u_l(z) = m_0$. If u_l is defined in terms of its Lebesgue points, the above observations show that u_l is lower semicontinuous and p-finely continuous. The same must then hold for u. □

2.3.8 Estimates of energy minimizers.

In this section we will consider the equation

(2.112) $$-\operatorname{div} \boldsymbol{A}(x, \nabla u) = \mu,$$

where $\mu \in (W_0^{1,p}(\Omega))^*$ is a nonnegative measure. The estimates obtained here will be used to demonstrate the necessity of the Wiener condition for boundary regularity of solutions of (2.69). Because of the relatively simple structure of this equation, this analysis is simpler than that of the corresponding analysis given in the general setting of Chapter 4.

2.122 THEOREM. *Let u be a nonnegative weak solution of (2.112) on $B(4r)$ and $\eta \in \mathcal{C}_c^\infty(B(r))$ be nonnegative with $|\nabla \eta| \leq Cr^{-p}$. Then*

(2.113) $$\int_{B(r)} \eta^p\,d\mu \leq Cr^{n-p}(\inf_{B(r)} u)^{p-1},$$

where $C = C(n, p, a, c)$.

PROOF. Test (2.112) by η^p. We obtain

$$\int_{B(r)} \eta^p\,d\mu = p\int_{B(r)} \boldsymbol{A}(x, \nabla u) \cdot \eta^{p-1} \nabla \eta\,dx$$

$$\leq C \int_{B(r)} |\nabla u|^{p-1} \eta^{p-1} |\nabla \eta|\,dx.$$

By Lemma 2.117, it follows that

$$\int_{B(r)} \eta^p\,d\mu \leq Cr^{n-p}(\inf_{B(r)} u)^{p-1}.$$

□

2.123 Theorem. *Let u be a nonnegative solution of (2.112) on Ω and $B(8R) = B(z, 8R) \subset \Omega$. Then*

$$\int_0^R \left(\frac{\mu(B(r))}{r^{n-p}}\right)^{1/(p-1)} \frac{dr}{r} \leq Cu(z),$$

where $C = C(n, p, a, c)$.

PROOF. Let
$$m(r) = \inf_{B(r)} u, \quad r > 0.$$

Fix $r \in (0, R)$ and write $v = u - m(8r)$. Find $\eta \in \mathcal{C}_c^\infty(B(2r))$ such that $0 \leq \eta \leq 1$, $\eta = 1$ on $B(r)$ and $|\nabla \eta| \leq Cr^{-1}$. Then by Theorem 2.122,

$$\mu(B(r)) \leq \int_{B(2r)} \eta^p \, d\mu \leq Cr^{n-p} \Big[\inf_{B(2r)} v\Big]^{p-1}$$
$$\leq Cr^{n-p} \big[m(2r) - m(8r)\big]^{p-1}.$$

As in the proof of Theorem 2.121 we conclude that

$$\int_0^R \left(\frac{\mu(B(r))}{r^{n-p}}\right)^{1/(p-1)} \frac{dr}{r}$$
$$\leq C\big[\lim_{r \to 0+} m(r) - m(8(R))\big] \leq Cu(z). \qquad \square$$

2.124 Lemma. *Let u be a solution of of (2.112) on $B(2r) = B(z, 2r)$, $\ell \in \mathbf{R}$ and $\delta > 0$. Let $v = \delta^{-1}(u - \ell)^+$. Suppose that $\gamma \in (p-1, \frac{n(p-1)}{n-p+1})$ and $\tau = \gamma/(p-1)$. Let $\eta \in \mathcal{C}_c^\infty(B(2r))$ be nonnegative, $|\nabla \eta| \leq Cr^{-1}$. Then*

$$(2.114) \quad \int_{B(2r)} (1+v)^{-\tau} |\nabla v|^p \eta^p \, dx \leq Cr^{-p} \int_{B(2r) \cap \{u > \ell\}} (1+v)^\gamma + C\delta^{1-p} \mu(B(2r))$$

with $C = C(n, p, \gamma, a, c)$.

PROOF. We use
$$\Big[1 - (1+v)^{1-\tau}\Big]\eta^p$$
as a test function and obtain

$$(\tau - 1) \int_{B(2r)} \mathbf{A}(x, \nabla u) \cdot (1+v)^{-\tau} \nabla v \, dx$$
$$= -p \int_{B(2r)} \mathbf{A}(x, \nabla u) \cdot \Big[1 - (1+v)^{1-\tau}\Big] \eta^{p-1} \nabla \eta \, dx$$
$$+ \int_{B(2r)} \Big[1 - (1+v)^{1-\tau}\Big] \eta^p \, d\mu.$$

It follows

$$\int_{B(2r)} (1+v)^{-\tau} |\nabla v|^p \eta^p \, dx$$
$$\leq C \int_{B(2r)} |\nabla v|^{p-1} \eta^{p-1} |\nabla \eta| \, dx + C\delta^{1-p} \int_{B(2r)} \eta^p \, d\mu.$$

Using Young's inequality we obtain

$$\int_{B(2r)} (1+v)^{-\tau}|\nabla v|^p \eta^p \, dx$$
$$\leq \frac{1}{2} \int_{B(2r)} (1+v)^{-\tau}|\nabla v|^p \eta^p \, dx + C \int_{B(2r) \cap \{v > \ell\}} (1+v)^{\gamma}|\nabla \eta|^p \, dx$$
$$+ C\delta^{1-p} \int_{B(2r)} \eta^p \, d\mu,$$

so that (2.114) holds. □

2.125 THEOREM. *Let u be a solution of of (2.112) on Ω and $B(R) = B(z, R) \subset \Omega$. Suppose that $\gamma \in (p-1, \frac{n(p-1)}{n-p+1})$. Then*

$$(2.115) \quad u(z) \leq C \left(R^{-n} \int_{B(R) \cap \{u > 0\}} u^\gamma \, dx \right)^{1/\gamma} + C \int_0^R \left(\frac{\mu(B(r))}{r^{n-p}} \right)^{1/(p-1)} \frac{dr}{r},$$

where $C = C(n, p, \gamma, a, c)$.

PROOF. Write
$$\tau = \frac{\gamma}{p-1}, \qquad q = \frac{p\gamma}{p-\tau}.$$

Fix a constant $\kappa \in (0, 1)$ to be specified later and set
$$r_j = R 2^{-j-2}.$$

We define recursively
$$\ell_0 = 0,$$
$$\ell_{j+1} = \ell_j + \left(\frac{1}{\kappa r_j^n} \int_{B(r_j) \cap \{u > \ell_j\}} (u - \ell_j)^\gamma \right)^{1/\gamma}$$

and denote
$$\delta_j = \ell_{j+1} - \ell_j.$$

Since u is lower semicontinuous (Theorem 2.121),

(2.116) $$u(z) \leq \lim_j \ell_j.$$

We claim that

(2.117) $$\delta_j \leq \frac{1}{2}\delta_{j-1} + C \left(\frac{\mu(B(2r_j))}{r_j^{n-p}} \right)^{1/(p-1)}, \qquad j = 1, 2, \ldots.$$

If $\delta_{j-1} \geq 2\delta_j$, then the claim (2.117) holds trivially. Let use suppose that

(2.118) $$\delta_{j-1} \leq 2\delta_j.$$

Let $\eta_j \in C_c^\infty(B(2r_j))$ be a nonnegative cutoff function such that $\eta_j = 1$ on $B(r_j)$ and $|\nabla \eta_j| \leq C r_j^{-1}$. Write
$$L_j = B(2r_j) \cap \{u > \ell_j\},$$
$$v_j = \frac{(u - \ell_j)^+}{\delta_j},$$
$$w_j = (1 + v_j)^{\gamma/q} - 1.$$

Then
$$\nabla w_j = \frac{\gamma}{q}(1+v_j)^{-\tau/p}$$
and thus by Lemma 2.124
$$\int_{L_j} |\nabla w_j|^p \, \eta_j^p \, dx$$
$$= \left(\tfrac{\gamma}{q}\right)^p \int_{L_j} (1+v_j)^{-\tau} |\nabla v_j|^p \eta_j^p \, dx$$
$$\leq C r_j^{-p} \int_{L_j} (1+v_j)^\gamma + C\delta_j^{1-p} \mu(B(2r_j)).$$

Since
$$w_j^p \leq C(1+v)^\gamma \chi_{L_j},$$
we have
$$\int_{L_j} |\nabla[w_j \eta_j]|^p \leq C \int_{L_j} \left(w_j^p |\nabla \eta_j|^p + |\nabla w_j|^p \eta_j^p \right) dx$$
$$\leq C r_j^{-p} \int_{L_j} (1+v_j)^\gamma + C\delta_j^{1-p}\mu(B(2r_j)).$$

Using Sobolev's inequality (Corollary 1.57) we obtain

(2.119)
$$\left(r^{-n} \int_{L_j} w_j^q \eta_j^q \right)^{p/q} \leq C r_j^{p-n} \int_{L_j} |\nabla[w_j \eta_j]|^p$$
$$\leq C r_j^{-n} \int_{L_j} (1+v_j)^\gamma + C r_j^{p-n} \delta_j^{1-p} \mu(B(2r_j)).$$

Since
$$|B(2r_j) \cap \{v_j > 0\}| = |B(2r_j) \cap \{v_{j-1} > 1\}|$$
$$\leq \int_{B(r_{j-1})} v_{j-1}^\gamma \, dx = \kappa r_{j-1}^n = 2^n \kappa r_j^n,$$
we have
$$\kappa r_j^n = \int_{B(r_j)} v_j^\gamma \, dx \leq 2^{-n-1} |B(r_j) \cap \{0 < v_j^\gamma < 2^{-n-1}\}| + C \int_{B(r_j)} w_j^q \, dx$$
$$\leq \tfrac{1}{2} \kappa r_j^n + C \int_{B(2r_j)} w_j^q \eta_j^q \, dx,$$
so that

(2.120)
$$\kappa^{p/q} \leq C \left(r_j^{-n} \int_{B(2r_j)} w_j^q \eta_j^q \right)^{p/q},$$

which estimates the left-hand side of (2.119). Concerning the right-hand side, from (2.118) we deduce that
$$(1+v_j)^\gamma \chi_{L_j} \leq C \left(\frac{\delta_{j-1}}{\delta_j} \right)^\gamma v_{j-1}^\gamma \leq C v_{j-1}^\gamma.$$

It follows

(2.121)
$$r^{-n} \int_{L_j} (1+v_j)^\gamma \, dx \leq C r_{j-1}^{-n} \int_{B(r_{j-1})} v_{j-1}^\gamma = C\kappa.$$

A combination of (2.119), (2.120) and (2.121) yields
$$\kappa^{p/q} \le C_{\text{spec}}\kappa + C\, r_j^{p-n} \delta_j^{1-p} \mu(B(2r_j)).$$

Now we can specify the choice of κ so that
$$\kappa^{p/q} - C_{\text{spec}}\kappa > 0$$

which proves the claim (2.117). Summing up (2.117) over $j = 1, 2, \ldots, i$ we obtain

$$\ell_{i+1} - \ell_1 \le \frac{1}{2}(\ell_i - \ell_1 + \delta_0) + \sum_{j=1}^{i} \left(\frac{\mu(B(2r_j))}{r_j^{n-p}} \right)^{1/(p-1)}$$

$$\le \frac{1}{2}(\ell_{i+1} - \ell_1) + C \left(\int_{B(r_0) \cap \{u>0\}} u^\gamma \, dx \right)^{1/\gamma} + C \int_0^R \left(\frac{\mu(B(r))}{r^{n-p}} \right)^{1/(p-1)} \frac{dr}{r}.$$

Letting $i \to \infty$, by (2.116) we obtain the assertion. $\square$

2.3.9 Dirichlet problem.

The Dirichlet problem consists of finding a solution of a partial differential equation which assumes given boundary data. In this section we will consider the Dirichlet problem corresponding to the Euler-Lagrange equation of the functional

$$\mathbf{J}(u) = \int_\Omega \mathbf{F}(x, \nabla u)\, dx$$

where Ω is a bounded open set. We suppose that $\mathbf{F}$ and $\mathbf{A}$ are as in Theorem 2.96. As seen from the theorem, the solution is obtained naturally by minimizing a functional where the prescribed boundary function is the trace of a $W^{1,p}(\Omega)$ function. However, the traditional form of Dirichlet problem assumes that the boundary function f is merely continuous and nothing more. Accordingly, the classical solution of the Dirichlet problem is then understood as the local minimizer of $\mathbf{J}$ in Ω which satisfies

(2.122) $$\lim_{\substack{x \to z \\ x \in \Omega}} u(x) = f(z)$$

for each $z \in \partial\Omega$. Under suitable assumptions, it turns out that the Dirichlet problem has a unique solution. The solution is not classical, but "generalized" and to what extent (2.122) is satisfied is an object of a further investigation. This question of boundary regularity leads to the Wiener criterion which will be considered later in Chapter 4.

2.126 DEFINITION. With any function $u_0 \in W^{1,p}_{\text{loc}}(\Omega)$ we associate the problem

(2.123) $$\begin{cases} u - u_0 \in W^{1,p}_0(\Omega) \\ \int_{\text{spt}\,\varphi} \mathbf{A}(x, \nabla u) \cdot \nabla \varphi \, dx = 0 \text{ for all } \varphi \in W^{1,p}_c(\Omega), \end{cases}$$

which differs from the problem (2.80), p. 113, by allowing $u, u_0 \notin W^{1,p}(\Omega)$.

Let f be a continuous function on $\partial\Omega$. We say that u solves the Dirichlet problem with boundary function f if there is a $u_0 \in \mathcal{C}(\overline{\Omega}) \cap W^{1,p}_{\text{loc}}(\Omega)$ such that $u_0 \llcorner \partial\Omega = f$ and u solves (2.123).

2.127 LEMMA. *Let u be a supersolution and v a subsolution of (2.69) on Ω. Suppose there exist functions $h \in W^{1,p}(\Omega)$ and $w \in W_0^{1,p}(\Omega) + \mathcal{C}_0(\Omega)$ such that*

$$u \geq h - w \quad \text{and} \quad v \leq h + w.$$

Then

$$v \leq u \quad \text{in } \Omega.$$

PROOF. We may assume $w \geq 0$. First we will investigate the case that $w \in W_0^{1,p}(\Omega)$. Let w_j be sequence of nonnegative functions from $\mathcal{C}_c^\infty(\Omega)$ such that $w_j \to w$ a.e. and $\|w_j - w\|_{1,p} \to 0$. Write

$$E_j = \Omega \cap \{h - w - u + 2w_j > 0\}.$$

We test u by $(h - w - u + 2w_j)^+$, which is obviously in $W_c^{1,p}(\Omega)$, and obtain

$$\int_{E_j} |\nabla u|^p \, dx \leq C \int_{E_j} \mathbf{A}(x, \nabla u) \cdot \nabla u \, dx \leq C \int_{E_j} \mathbf{A}(x, \nabla u) \cdot \nabla [h - w + 2w_j] \, dx$$

$$\leq C \int_{E_j} |\nabla u|^{p-1} |\nabla [h - w + 2w_j]| \, dx$$

$$\leq C \left(\int_{E_j} |\nabla u|^p \, dx \right)^{1/p'} \left(\int_{E_j} |\nabla [h - w + 2w_j]|^p \, dx \right)^{1/p}.$$

It follows

$$\left(\int_{E_j} |\nabla u|^p \, dx \right)^{1/p} \leq C \left(\int_{E_j} |h - w + 2w_j|^p \, dx \right)^{1/p}$$

$$\leq C \left(\int_\Omega |h + w|^p \, dx \right)^{1/p} + C \left(\int_\Omega |2w_j - 2w|^p \, dx \right)^{1/p}.$$

Since

$$\{h + w > u\} \subset \bigcup_{j=1}^\infty \bigcap_{i=j}^\infty E_j,$$

letting $j \to \infty$ be obtain

$$\int_{\{h+w>u\}} |\nabla u|^p \leq C \int_\Omega |\nabla [h + w]|^p \, dx.$$

Similarly

$$\int_{\{h-w<v\}} |\nabla v|^p \leq C \int_\Omega |\nabla [h - w]|^p \, dx.$$

Let

$$\varphi_j = \min\{3w_j, (v - u)^+\}.$$

Then

$$\{\varphi_j = v - u > 0\} \subset \{h + w > v\} \cap \{h - w < v\},$$

and thus

$$\varphi_j \in W_c^{1,p}(\Omega).$$

We test u by φ_j and obtain

$$\int_{\{0<v-u\leq 3w_j\}} \boldsymbol{A}(x,\nabla u) \cdot (\nabla u - \nabla v)\, dx$$

$$\leq 3\int_{\{3w_j<v-u\}} \boldsymbol{A}(x,\nabla u) \cdot \nabla w_j\, dx$$

$$\leq C\int_{\{3w_j<v-u\}} |\nabla u|^{p-1}|\nabla w_j|\, dx$$

$$\leq C\int_{\{3w_j<v-u\}} \left(|\nabla u|^p + |\nabla w_j|^p\right) dx$$

$$\leq C\int_{\{3w_j<v-u\}} \left(|\nabla u|^p + |\nabla w|^p + |\nabla w_j - \nabla w|^p\right) dx.$$

Since $v - u \leq 2w$, the integrals over $\{v - u > 3w_j\}$ tend to 0 and we obtain

$$\int_{\{u<v\}} \boldsymbol{A}(x,\nabla u) \cdot (\nabla u - \nabla v)\, dx \leq 0.$$

Similarly, testing v by φ_j and letting $j \to \infty$ we obtain

$$\int_{\{u<v\}} \boldsymbol{A}(x,\nabla v) \cdot (\nabla v - \nabla u)\, dx \leq 0.$$

Hence

$$\int_{\{u<v\}} \left[\boldsymbol{A}(x,\nabla v) - \boldsymbol{A}(x,\nabla u)\right] \cdot (\nabla v - \nabla u)\, dx \leq 0,$$

and using (2.74) we infer that $\nabla[v - u]^+ = 0$. It follows that $v \leq u$ in Ω. This concludes the proof under the assumption that $w \in W_0^{1,p}(\Omega)$. Now, if $w \in W^{1,p}(\Omega) + \mathcal{C}_0(\Omega)$, then for any $\varepsilon > 0$ there is $w_\varepsilon \in W_0^{1,p}(\Omega)$ such that $|w_\varepsilon - w| \leq \varepsilon$ on Ω. Using the previous part we obtain

$$v \leq u + \varepsilon$$

and letting $\varepsilon \to 0$ we conclude the proof. $\square$

2.128 REMARK. The assumptions of Lemma 2.127 are satisfied, in particular, if $u, v \in W_0^{1,p}(\Omega)$ and

(2.124) $$(v - u)^+ \in W_0^{1,p}(\Omega).$$

This condition has the intuitive meaning that $v \leq u$ on the boundary of Ω in the sense pertaining to the space $W^{1,p}(\Omega)$. Note that Lemma 2.26, p. 75, provides a powerful method of verifying that $v \leq u$ on the boundary of Ω in the sense of (2.124).

2.129 THEOREM. *Let f be a continuous function on $\partial\Omega$. Then there is a unique solution of the Dirichlet problem in Ω with prescribed boundary function f.*

PROOF. In view of Tietze's extension theorem, we may suppose that
$$f \in \mathcal{C}_c(\mathbf{R}^n).$$
By means of mollifying we find a sequence f_j of functions from $\mathcal{C}_c^1(\mathbf{R}^n)$ such that $|f_j - f| \leq 2^{-j}$ on $\overline{\Omega}$. There is a constant M such that $|f_j| \leq M$ on $\overline{\Omega}$ for each j. Let u_j be the unique minimizer of $\mathbf{J}$ in $\{u \in W^{1,p}(\Omega) : u - f_j \in W_0^{1,p}(\Omega)\}$ as found in Theorem 2.96. Then from Lemma 2.127 we deduce that

(2.125) $$|u_i - u_j| \leq 2^{-i} + 2^{-j} \quad \text{on } \Omega$$

and
$$|u_j| \leq M \quad \text{on } \Omega.$$
It follows that the sequence $\{u_j\}$ converges uniformly. We denote the limit by u_∞. First, we show that u_∞ is a local minimizer of $\mathbf{J}$ in Ω. Choose a function $w \in W_c^{1,p}(\Omega)$. Write
$$v_j = \min\{u_j + M, w\}, \quad v = \min\{u_\infty + M, w\}.$$
Since u_j is a local minimizer of (2.79), we have
$$\int_{\text{spt } w} \mathbf{F}(x, \nabla u_j) \, dx \leq \int_{\text{spt } w} \mathbf{F}(x, \nabla[u_j + w - v_j]) \, dx,$$
so that
$$\int_\Omega \mathbf{F}(x, \nabla v_j) \, dx \leq \int_\Omega \mathbf{F}(x, \nabla w) \, dx.$$
By (2.68),
$$\int_\Omega |\nabla v_j|^p \, dx \leq C \int_\Omega \mathbf{F}(x, \nabla w) \, dx.$$
We infer that $\{v_j\}$ is bounded in $W_0^{1,p}(\Omega)$ and thus there is a subsequence (denoted as the whole sequence) which converges weakly in $W_0^{1,p}(\Omega)$. The weak limit can be only v. Using weak lower semicontinuity of $\mathbf{J}$ we infer that

(2.126) $$\int_\Omega \mathbf{F}(x, \nabla v) \, dx \leq \liminf_{j \to \infty} \int_\Omega \mathbf{F}(x, \nabla v_j) \, dx \leq \int_\Omega \mathbf{F}(x, \nabla w) \, dx.$$

Consider $\Omega' \subset\subset \Omega$. We can choose a $w' \in W_c^{1,p}(\Omega)$ such that $w' \geq 2M$ in Ω'. Then $v' := \min\{u_\infty + M, w'\} = u_\infty + M$ in Ω'. This verifies that $u_\infty \in W^{1,p}(\Omega')$. If $\varphi \in W_c^{1,p}(\Omega')$, set
$$w = \begin{cases} M + u_\infty + \varphi^+ & \text{in } \Omega', \\ w' & \text{in } \Omega \setminus \Omega'. \end{cases}$$
By (2.126),
$$\int_{\Omega'} \mathbf{F}(x, \nabla u_\infty) \, dx = \int_{\Omega'} \mathbf{F}(x, \nabla v) \, dx$$
$$\leq \int_{\Omega'} \mathbf{F}(x, \nabla w) \, dx$$
$$= \int_{\Omega'} \mathbf{F}(x, \nabla[u_\infty + \varphi^+]) \, dx.$$
Similarly we show that
$$\int_{\Omega'} \mathbf{F}(x, \nabla u_\infty) \, dx \leq \int_{\Omega'} \mathbf{F}(x, \nabla[u_\infty - \varphi^-]) \, dx.$$

It follows that

$$\int_{\Omega'} \mathbf{F}(x, \nabla u_\infty) \, dx \leq \int_{\Omega' \cap \{\varphi=0\}} \mathbf{F}(x, \nabla u_\infty) \, dx$$
$$+ \int_{\Omega' \cap \{\varphi>0\}} \mathbf{F}(x, \nabla u_\infty) \, dx + \int_{\Omega' \cap \{\varphi<0\}} \mathbf{F}(x, \nabla u_\infty) \, dx$$
$$\leq \int_{\Omega'} \mathbf{F}(x, \nabla [u_\infty + \varphi]) \, dx,$$

which establishes that u_∞ is a local minimizer of $\mathbf{J}$ in Ω. Denote

$$g_j = (u_{j+1} - f_{j+1}) - (u_j - f_j).$$

Then, $g_j \in W_0^{1,p}(\Omega)$ and from (2.125) we obtain

$$\|g_j\|_\infty \leq 2^{-j+2}.$$

We find $\psi_j \in \mathcal{C}_c^\infty(\mathbf{R}^n)$ with support in Ω such that

$$\|\psi_j\|_\infty \leq 2^{-j+2} \quad \text{and} \quad \|\psi_j - g_j\|_{1,p} \leq 2^{-j}.$$

Set

$$\psi = \sum_{j \to \infty} \psi_j.$$

Then $\psi \in \mathcal{C}_c(\Omega)$ and

$$u_\infty - (f + \psi) = u_1 - f_1 + \sum_{j \to \infty} (g_j - \psi_j) \in W_0^{1,p}(\Omega).$$

This proves that $u = u_\infty$ solves (2.123) with $u_0 = f + \psi$ and thus we conclude the existence part of the proof.

Now, for uniqueness. Suppose that u is a solution of (2.123) with $u_0 \in \mathcal{C}(\overline{\Omega}) \cap W_{\text{loc}}^{1,p}(\Omega)$, $u_0 \, \llcorner \, \partial\Omega = f$. For any positive integer j we have

$$u \leq (u_j + 2^{-j}) + (u - u_\infty).$$

Since $u - u_\infty \in W_0^{1,p}(\Omega) + \mathcal{C}_0(\Omega)$, from comparison principle (Theorem 2.127) we infer that

$$u \leq u_j + 2^{-j+1}.$$

Letting $j \to \infty$ we obtain

$$u \leq u_\infty.$$

Similarly, $u \geq u_\infty$ which concludes the proof of uniqueness. $\square$

2.3.10 Application of thinness: the Wiener criterion.

2.130 THEOREM. *Let $E \subset \mathbf{R}^n$ be p-thin at $z \in \mathcal{O} \cap \overline{E} \setminus E$ and $r_0 > 0$. Then there is a nonnegative supersolution u of (2.69) on $B(z, r_0)$ such that its p-finely continuous representative satisfies*

$$u(x) < \liminf_{\substack{y \to x \\ y \in E}} u(y).$$

PROOF. By Lemma 2.48 there is an open set U containing E such that U is thin at z. Hence we may assume that E itself is open. Let use write $B(r) = B(z,r)$ for each $r > 0$ and $\mathbf{c}_p = \mathbf{c}_p(\cdot, B(r_0))$. Fix $\rho, R \in (0, r_0/8)$, $\mathbf{R} < \rho$. Let u_R be the capacitary extremal for $\mathbf{c}_p(E \cap B(R))$ and μ_R be the corresponding capacitary distribution. Let $u \in W_0^{1,p}(B(r_0))$ be the weak solution of

$$-\operatorname{div} \mathbf{A}(x, \nabla u) = \mu_R.$$

First we estimate the p-Dirichlet integral of u. We have

$$\|\nabla u\|_p^p = \int_{B(r_0)} |\nabla u|^p \, dx \leq C \int_{B(r_0)} \mathbf{A}(x, \nabla u) \cdot \nabla u \, dx$$
$$= C \langle \mu_R, u \rangle \leq C \|\nabla u\|_p \|\mu\|_{(W_0^{1,p})^*},$$

so that, by Theorem 2.36

(2.127) $$\int_{B(r_0)} |\nabla u|^p \, dx \leq C \mathcal{E}_{p;\infty}(\mu_R; B(r_0)) = C \mathbf{c}_p(E \cap B(R)).$$

Now we pass to pointwise estimates. By Theorem 2.125,

(2.128) $$u_R(y) \leq C \left(\rho^{-n} \int_{B(y,\rho)} u_R^\gamma \, dx \right)^{1/\gamma} + C \int_0^\rho \left(\frac{\mu_R(B(y,r))}{r^{n-p}} \right)^{1/(p-1)} \frac{dr}{r}$$

for each $y \in B(z, \rho)$. Using Poincaré's inequality (Theorem 1.57) we obtain

(2.129)
$$\left(\rho^{-n} \int_{B(y,\rho)} u_R^\gamma \, dx \right)^{1/\gamma} \leq C \left(\int_{B(r_0)} u_R^\gamma \, dx \right)^{1/\gamma}$$
$$\leq C \left(\int_{B(r_0)} |\nabla u_R|^p \right)^{1/p}$$
$$\leq C \left(\mathbf{c}_p(E \cap B(R)) \right)^{1/p},$$

where the constant C depends on n, p, r_0, ρ and γ, but not on R. Later we will need also a similar estimate,

(2.130)
$$\left(\rho^{-n} \int_{B(y,\rho)} u^\gamma \, dx \right)^{1/\gamma} \leq C \left(\int_{B(r_0)} |\nabla u|^p \right)^{1/p}$$
$$\leq C \left(\mathbf{c}_p(E \cap B(R)) \right)^{1/p}$$

where we have used (2.127). By (2.128) and (2.129) we have

(2.131) $$u_R(y) \leq C_1 \mathbf{c}_p(B(R))^{1/p} + C \int_0^\rho \left(\frac{\mu_R(B(y,r))}{r^{n-p}} \right)^{1/(p-1)} \frac{dr}{r},$$

where C, C_1 depend on n, p, r_0, ρ and γ. For R small enough and $y \in B(z, R) \cap E$ we infer

$$1 \leq 2u_R(y) - C_1 \mathbf{c}_p(B(R))^{1/p} \leq C \int_0^\rho \left(\frac{\mu_R(B(y,r))}{r^{n-p}} \right)^{1/(p-1)}.$$

Using Theorem 2.123 it follows

(2.132) $$1 \leq C u(y), \quad y \in B(z, R) \cap E.$$

Now, we use Theorem 2.125 again to obtain
$$u(z) \leq C \left(\rho^{-n} \int_{B(\rho)} u^\gamma \, dx \right)^{1/\gamma} + C \int_0^\rho \left(\frac{\mu_R(B(r))}{r^{n-p}} \right)^{1/(p-1)} \frac{dr}{r}.$$
Using (2.130) and Theorem 2.45 it follows
$$u(z) \leq C \left(\mathbf{c}_p(E \cap B(R)) \right)^{1/p} + C \int_0^\rho \left(\frac{\mathbf{c}_p(B(R) \cap B(r) \cap E)}{r^{n-p}} \right)^{1/(p-1)} \frac{dr}{r}.$$

Since E is p-thin at z, we may use Lebesgue's dominated convergence theorem to observe that the right part can be made arbitrarily small by choosing R small enough. Comparing with (2.132), we conclude that, with an appropriate choice of R,
$$u(z) < \liminf_{\substack{x \to z \\ x \in E}} u(y). \qquad \square$$

2.131 DEFINITION. We will denote by H_f the weak solution of the Dirichlet problem for (2.69) with boundary function f. A point $z \in \partial\Omega$ is called *regular* for Ω if

(2.133)
$$\lim_{\substack{x \to z \\ x \in \Omega}} H_f(x) = f(z)$$

for any continuous f defined on $\partial\Omega$.

The next result shows that regular points can be characterized in terms of functions $f \in W^{1,p}(\Omega)$.

2.132 LEMMA. *Suppose* (2.133) *holds for every function* $f \in \mathcal{C}_c^\infty(\mathbf{R}^n)$. *Then z is regular.*

PROOF. Let $f \in \mathcal{C}(\partial\Omega)$. We extend f via Tietze's theorem to a function from $\mathcal{C}_c(\mathbf{R}^n)$ denoted again as f. For $\varepsilon > 0$, let $g \in \mathcal{C}_c^\infty(\mathbf{R}^n)$ be such that
$$|f - g| < \varepsilon \quad \text{on } \mathbf{R}^n.$$
Then $H_f - f$ and $H_g - g$ are in $W_0^{1,p}(\Omega) + \mathcal{C}_0(\Omega)$. We have
$$H_f \leq (H_g + \varepsilon) + (H_f - f) + (g - H_g) \quad \text{on } \Omega.$$
The comparison principle (Theorem 2.127) implies that $H_f \leq H_g + \varepsilon$ in Ω. Since
$$\lim_{\substack{x \to z \\ x \in \Omega}} H_g(x) = g(z),$$
we have
$$\limsup_{\substack{x \to z \\ x \in \Omega}} H_f(x) \leq g(z) + \varepsilon \leq f(z) + 2\varepsilon.$$
Together with a similar lower estimate we obtain
$$\lim_{\substack{x \to z \\ x \in \Omega}} H_f(x) = f(z)$$
thus showing that z is regular. $\qquad \square$

2.133 THEOREM (Wiener criterion). *A point $z \in \partial\Omega$ is regular for the equation (2.69) if and only if $\mathbf{R}^n \setminus \Omega$ is not p-thin at z.*

PROOF. Suppose that z is regular for (2.69), but $\mathbf{R}^n \setminus \Omega$ is p-thin at z. Let $B(r_0)$ be a ball containing Ω. By Theorem 2.130 there is a nonnegative supersolution u on $B(r_0)$ such that
$$u(z) < \liminf_{\substack{x \to z \\ x \in \partial\Omega}} u(y).$$
Then there are $a > u(z)$ and $r \in (0, r_0)$ such that $u \geq a$ on $B(z,r) \cap \partial\Omega \setminus \{z\}$. We find a function $f \in C_c^\infty(\mathbf{R}^n)$ such that
$$0 \leq f \leq a, \quad \text{spt } f \subset B(z,r) \quad \text{and} \quad f(z) = a.$$
Denote by h the solution of the Dirichlet problem with boundary data f. Then $f \leq u$ on $\partial\Omega$ in the sense of Remark 2.128, and thus by the comparison principle $h \leq u$ in Ω. Since z is regular,
$$a = \lim_{\substack{x \to z \\ x \in \Omega}} h(x) \leq \liminf_{\substack{x \to z \\ x \in \Omega}} u(x).$$
Hence
$$a \leq \liminf_{x \to z} u(x)$$
which contradicts that u is p-finely continuous and $u(z) < a$.

Conversely, suppose that $\mathbf{R}^n \setminus \Omega$ is not p-thin at z and $f \in C_c^1(\mathbf{R}^n)$. Let h be the continuous representative of the solution of the Dirichlet problem with boundary function f. Choose $k < f(z)$. There is a ball $B(z,r)$ such that $k < f$ on $\overline{B}(z,r) \cap \partial\Omega$. Set
$$u = \begin{cases} \min\{h, k\} & \text{on } B(z,r) \cap \Omega, \\ k & \text{on } B(z,r) \setminus \Omega. \end{cases}$$
As in the proof of Lemma 2.109 we see that u is a supersolution of (2.69) in $B(z,r)$. By Theorem 2.121 there is a representative of u which is lower semicontinuous and p-finely continuous by Theorem 2.121. By Lemma 2.26, this representative (denoted by u in the sequel) is equal to $\min\{h, k\}$ on $B(z,r) \cap \Omega$ and to k p-q.e. on $B(z,r) \setminus \Omega$. Since $\mathbf{R}^n \setminus \Omega$ is not p-thin at z, we infer that $u(z) = k$. Since u is lower semicontinuous at z it follows that
$$\liminf_{\substack{x \to z \\ x \in \Omega}} h(x) \geq \liminf_{\substack{x \to z \\ x \in \Omega}} u(x) = k.$$
Taking into account also the symmetrical estimate we conclude that
$$\lim_{\substack{x \to z \\ x \in \Omega}} h(x) = f(z),$$
so that z is regular. □

2.4 Fine topology

In this section we prove some properties of the p-fine topology associated in a certain sense with the Sobolev space $W^{1,p}$. We avoid some hard methods of the theory of function spaces and nonlinear potentials. However, then we are not able to proceed without the knowledge of nontrivial properties of capacitary extremals which were obtained in the preceding section using the fact that extremals are supersolutions of the p-Laplace equation.

2.134 DEFINITION. We say that a set $A \subset \mathbf{R}^n$ is *p-finely closed* if
$$b_p F \subset F.$$

Complements of p-finely closed sets are termed *p-finely open*. It is immediately seen that the collection of all p-finely open sets forms a topology, called the *p-fine topology*. Notions related to this topology will be attributed by "p-fine" or "p-finely". In particular, continuous function with respect to the p-fine topology will be *p-finely continuous functions*. This terminology is obviously consistent with Definition 2.119. The abstract topological language also gives meaning to the notion of p-fine continuity at a point. Notice that the characterization which will be given in Theorem 2.136 is not immediate since its proof relies on deep results.

2.135 REMARK. Since $b_p E \subset \overline{E}$ for any $E \subset \mathbf{R}^n$, the p-fine topology is finer than the Euclidean one. On the other hand, it is coarser than the density topology in view of Corollary 2.51, p. 86. Any p-polar set is p-finely closed.

2.136 THEOREM. *For any set $E \subset \mathbf{R}^n$, $E \cup b_p E$ is the p-fine closure of E. A function u on a p-fine neighborhood U of z is p-finely continuous at z if and only if for each $\varepsilon > 0$, the set $\{|u - u(z)| \geq \varepsilon\}$ is p-thin at z.*

PROOF. Let F be a p-finely closed set containing E. Then
$$b_p E \subset b_p F \subset F,$$
and thus the p-fine closure of E contains $E \cup b_p E$. Conversely, let z be a point of $\overline{E} \setminus E$ at which E is p-thin. Then by Theorem 2.130 there is a supersolution of the p-Laplace equation u on a neighborhood U of z such that its p-finely continuous representative satisfies
$$u(z) < t < \liminf_{x \to z,\, x \in E} u(x)$$
for some level t. Set
$$F = (\mathbf{R}^n \setminus U) \cup \{x \in U : u(x) \geq t\}.$$

From the p-fine continuity we obtain that $b_p F \subset F$, so that F is p-closed. Since $z \notin F$, z does not belong to the p-fine closure of E. The assertion concerning the characterization of p-fine continuity is then an obvious consequence. □

2.137 THEOREM. *Let $U \subset \mathbf{R}^n$ be a p-finely open set and $z \in U$. Then there exists an upper semicontinuous p-finely continuous function $f \in W^{1,p}(\mathbf{R}^n)$ with compact support in U such that $f(z) > 0$. In particular, there exists a bounded p-finely open set V such that $z \in V$ and $\overline{V} \subset U$.*

PROOF. As in the preceding proof, using Theorem 2.130 we find a p-finely continuous supersolution of the p-Laplace equation u on a ball $B(z,r)$ and $t \in \mathbf{R}$ such that
$$u(z) < t < \inf\{u(x) : x \in B(z,r) \setminus U\}.$$
Let $\eta \in \mathcal{C}_c^\infty(\mathbf{R}^n)$ be a cutoff function such that
$$0 \leq \eta \leq 1, \quad \operatorname{spt}\eta \subset B(z,r) \quad \text{and} \quad \eta(z) = 1.$$
Set
$$f = \eta\,(t - u)^+.$$
Obviously, $f \in W^{1,p}(\mathbf{R}^n)$ and $f(z) > 0$. By Theorem 2.121, u is lower semicontinuous and thus f is upper semicontinuous and so the set $t - u \geq 0$ is closed. Hence the support of f is compact and
$$\operatorname{spt} f \subset \operatorname{spt}\eta \cap \{t - u \geq 0\} \subset U. \qquad \square$$

2.138 LEMMA. *If $\{A_i\}$ is a sequence of sets each of which is thin at x, then there exists a sequence of real numbers $\{r_i\}$ such that*
$$\bigcup_{i=1}^\infty \bigl(A_i \cap B(x,r_i)\bigr)$$
is thin at x.

PROOF. Fix a positive integer i. Consider a sequence of functions
$$f_j(r) = \left(\frac{\mathbf{C}_p\bigl(A_i \cap B(x,1/j) \cap B(x,r)\bigr)}{r^{n-p}}\right)^{1/(p-1)} \frac{1}{r}.$$
Obviously $f_j \to 0$ pointwise on $(0,1)$. Because A_i is thin at x, it follows that f_1 is integrable. By Lebesgue theorem, there exists $r_i \in \{1/j : j = 1, 2 \ldots\}$ such that
$$\int_0^1 \left(\frac{\mathbf{C}_p\bigl(A_i \cap B(x,r_i) \cap B(x,r)\bigr)}{r^{n-p}}\right)^{1/(p-1)} \frac{dr}{r} = \int_0^1 f_{1/r_i}(r)\,dr < 2^{-i}.$$
Since capacity is countably subadditive, the result easily follows. $\qquad \square$

2.139 THEOREM. *Let u be a function defined on a p-finely open set G. Then u is p-finely continuous at $z \in G$ if and only if there exists a set A that is p-thin at z and*

(2.134)
$$\lim_{\substack{x \to z \\ z \in G \setminus A}} u(x) = u(z).$$

PROOF. If u is p-finely continuous at z, then for each $j = 1, 2, \ldots$ there are p-fine neighborhoods U_j of z, $j = 1, 2, \ldots$, such that
$$|u(x) - u(z)| < 1/j \quad \text{for each } x \in U_j.$$
The sets $A_j := \mathbf{R}^n \setminus U_j$ are p-thin at z and thus by Lemma 2.138 there are a radii $r_i \downarrow 0$ such that
$$A := \bigcup_{j=1}^\infty \bigl(A_j \cap B(z,r_j)\bigr)$$

is p-thin at z. Since
$$|u(x) - u(z)| < 1/j \quad \text{for each } x \in G \cap B(z, r_j) \setminus A,$$
we obtain
$$\lim_{\substack{x \to z \\ z \in G \setminus A}} u(x) = u(z).$$
Conversely, (2.134) implies that for each $\varepsilon > 0$, the set
$$\{x \in G \colon |u(x) - u(z)| \geq \varepsilon\}$$
is p-thin at z. By Theorem 2.136, z is a p-fine interior point of $\{x \in G : |u(x) - u(z)| < \varepsilon\}$. □

2.140 THEOREM. *Let $A \subset \mathbf{R}^n$ be an arbitrary set. Then*
$$\mathbf{C}_p(A \cup b_p A) = \mathbf{C}_p(A).$$

PROOF. We may suppose that $\mathbf{C}_p(A) < \infty$. Choose $\varepsilon > 0$ and find an open set $G \supset A$ such that
$$\mathbf{C}_p(G) < \mathbf{C}_p(A) + \varepsilon.$$
Let u be the capacitary extremal for $\mathbf{C}_p(G)$. Then by p-fine continuity of u (Theorem 2.121), $u \geq 1$ on $G \cup b_p(G)$. According to Lemma 2.25, p. 75,
$$\mathbf{C}_p(G \cup b_p(G)) \leq \int_{\mathbf{R}^n} \left(|\nabla u|^p + |u|^p \right) dx \leq \mathbf{C}_p(G),$$
and thus
$$\mathbf{C}_p(A \cup b_p A) \leq \mathbf{C}_p(G \cup b_p G) \leq \mathbf{C}_p(G) \leq \mathbf{C}_p(A) + \varepsilon. \qquad \square$$

2.141 THEOREM (Choquet property). *For any $E \subset \mathbf{R}^n$ and any $\varepsilon > 0$ there is an open set U such that $U \cup b_p E = \mathbf{R}^n$ and $\mathbf{C}_p(E \cap U) < \varepsilon$.*

PROOF. Let $\{B_j\}$ be a sequence of balls with rational centers and rational radii. Choose $\varepsilon > 0$. For any j, let u_j be the capacitary extremal for $\mathbf{c}_p(E \cap B_j, 2B_j)$. Since u_j is p-quasicontinuous, there is an open set V_j with
$$\mathbf{C}_p(V_j) < 2^{-j}\varepsilon$$
such that the set
$$U_j := (B_j \cap \{u_j < 1\}) \cup V_j$$
is open. Set
$$U := \bigcup_{j=1}^{\infty} U_j.$$
We want to prove that $U \cup b_p E = \mathbf{R}^n$. Obviously,
$$\mathbf{R}^n \setminus \overline{E} \subset \bigcup \{B_j : B_j \cap E = \emptyset\} \subset U.$$

Choose a point $z \in \overline{E} \setminus b_p E$. By Theorem 2.130, there is a positive p-finely continuous supersolution v of the p-Laplace equation on an open set Ω containing z such that
$$v(z) < 1 < \liminf_{x \to z, \, x \in E} v(x).$$

We find a ball B_j such that $z \in B_j$, $v > 1$ on $B_j \cap E$ and $2B_j \subset \Omega$. Let us test v and u_j by $(u_j - v)^+$, using that $u_j - (u_j - v)^+ \in \mathcal{Y}(E \cap B_j, 2B_j)$ and v is a supersolution of the p-Laplace equation. We obtain

$$\int_{2B_j \cap \{u_j > v\}} (|\nabla u_j|^{p-2} \nabla u_j - |\nabla v|^{p-2} \nabla v) \cdot (\nabla u_j - \nabla v) \leq 0.$$

Since the integrand is nonnegative, this shows that $v \geq u_j$ a.e. Comparing the p-finely continuous representatives we obtain that $u_j(z) < 1$. It follows that $z \in U_j \subset U$. It remains to prove that $\mathbf{C}_p(E \cap U) < \varepsilon$. For any j, $u_j \geq 1$ p-quasi everywhere on $E \cap B_j$, and thus

$$\mathbf{C}_p(E \cap U_j) \leq \mathbf{C}_p(E \cap B_j \cap \{u_j < 1\}) + \mathbf{C}_p(V_j) < 2^{-j}\varepsilon.$$

Using countable subadditivity of the capacity we obtain the assertion. □

2.142 COROLLARY (Kellogg property). $\mathbf{C}_p(A \setminus b_p A) = 0$ for any $A \subset \mathbf{R}^n$.

2.143 COROLLARY. A set $A \subset \mathbf{R}^n$ is p-polar if and only if $b_p(A) = \emptyset$.

The following theorem says that p-finely open sets are "p-quasiopen."

2.144 THEOREM. Let $G \subset \mathbf{R}^n$ be a p-finely open set. Then for any $\varepsilon > 0$ there is an open set V with $\mathbf{C}_p(V) < \varepsilon$ such that $G \cup V$ is open.

PROOF. By Theorem 2.141, there is an open set U such that $U \cup b_p(\mathbf{R}^n \setminus G) = \mathbf{R}^n$ and $\mathbf{C}_p(U \setminus G) < \varepsilon$. Let V be an open set containing $U \setminus G$ such that $\mathbf{C}_p(V) < \varepsilon$. Since G is p-finely open, $G \subset U$ and thus $G \cup V = U \cup V$ is open. □

2.145 THEOREM. Let u be a function on a p-finely open set G. Then u is p-quasicontinuous on G if and only if u is p-finely continuous at p-quasi every point of G.

PROOF. Suppose that u is p-quasicontinuous in G. Find open sets V_j with $\mathbf{C}_p(V_j) < 2^{-j}$ such that u restricted to $G \setminus V_j$ is continuous. Then by Theorem 2.140,

$$\mathbf{C}_p(V_j \cup b_p V_j) = \mathbf{C}_p(V_j) < 2^{-j}.$$

The set
$$Z := \bigcap_j (V_j \cup b_p V_j)$$

is thus p-polar. If $x \notin Z$, then by Theorem 2.136, x belongs to the p-fine interior of $G \setminus V_k$ for some positive integer k. This means that there is a p-finely open set U_x containing x such that u restricted to U_x is continuous. It follows that u is p-finely continuous at x.

For the converse, assume that u is p-finely continuous at p-quasi every point of G. There is a p-polar set Z such that u is p-finely continuous on the p-finely open set $G' := G \setminus Z$. Choose $\varepsilon > 0$. Let $\{(a_j, b_j)\}$ be the sequence of all intervals with rational endpoints and $U_j = G' \cap \{a_j < u < b_j\}$. By Theorem 2.144, there are open sets V_j with $\mathbf{C}_p(V_j) < 2^{-j}\varepsilon$ and V' with $\mathbf{C}_p(V') < \varepsilon$ such that the $U_j \cup V_j$ and $G \cup V'$ are open. Also, there is an open set $W \supset Z$ with $\mathbf{C}_p(W) < \varepsilon$. Set

$$V := W \cup V' \cup \bigcup_{j=1}^{\infty} V_j.$$

Then $\mathbf{C}_p(V) < 3\varepsilon$ and it is easy to verify that u restricted to $G \setminus V$ is continuous. □

2.146 THEOREM (Quasi-Lindelöf property). *For each family $\mathcal{V}$ of p-finely open sets there is a countable subfamily whose union differs from the union of the whole family in a p-polar set.*

PROOF. Denote by U the union of the whole family. Let $\mathcal{V}_1$ be the family of all countable unions of sets from $\mathcal{V}$ and $\mathcal{B}$ be the family of all balls with rational centers and rational radii. With every ball $B \in \mathcal{B}$ we find a sequence $\{V_{B,j}\}_j$ of sets from $\mathcal{V}_1$ such that

$$\mathbf{C}_p(B \cap U \setminus V_{B,j}) \to \inf\{\mathbf{C}_p(B \cap U \setminus V) : V \in \mathcal{V}_1\}.$$

Set

$$W := \bigcup_{B,j} V_{B,j}.$$

Then $W \in \mathcal{V}_1$ and

(2.135) $$\mathbf{C}_p(B \cap U \setminus W) = \inf\{\mathbf{C}_p(B \cap U \setminus V) : V \in \mathcal{V}_1\}$$

for each $B \in \mathcal{B}$. Assume that $U \setminus W$ is not p-polar. By Corollary 2.142,

$$0 < \mathbf{C}_p(U \setminus W) \leq \mathbf{C}_p(U \setminus W \cap b_p(U \setminus W)) + \mathbf{C}_p(U \setminus W \setminus b_p(U \setminus W))$$
$$= \mathbf{C}_p(U \setminus W \cap b_p(U \setminus W)),$$

so that there exists

$$x \in U \setminus W \cap b_p(U \setminus W).$$

Find $V_0 \in \mathcal{V}$ such that $x \in V_0$. Then the set $\mathbf{R}^n \setminus V_0$ is p-thin at x, whereas $U \setminus W$ is not p-thin at x. It follows that

$$\int_0^{1/2} \left(\frac{\mathbf{C}_p(B(x,2r) \setminus V_0)}{r^{n-p}}\right)\frac{dr}{r} < \infty = \int_0^1 \left(\frac{\mathbf{C}_p(B(x,r) \cap U \setminus W)}{r^{n-p}}\right)\frac{dr}{r}$$

and whence there is $r \in (0, 1/2)$ such that

$$\mathbf{C}_p(B(x,2r) \setminus V_0) < \mathbf{C}_p(B(x,r) \cap U \setminus W).$$

We insert a ball $B := B(z,\rho) \in \mathcal{B}$ between $B(x,r)$ and $B(x,2r)$, i.e.

$$B(x,r) \subset B(z,\rho) \subset B(x,2r).$$

Then

$$\mathbf{C}_p(B \cap U \setminus V_0) \leq \mathbf{C}_p(B \setminus V_0) < \mathbf{C}_p(B \cap U \setminus W),$$

which contradicts (2.135). □

2.147 THEOREM. *Let $\Omega \subset \mathbf{R}^n$ be an open set. A p-quasicontinuous function $u \in W^{1,p}(\mathbf{R}^n)$ belongs to $W_0^{1,p}(\Omega)$ if and only if $u(x) = 0$ at p-quasi every point x of the p-fine boundary of Ω.*

PROOF. Notice that by Theorem 2.145, p-quasicontinuity is the same as p-fine continuity p-q.e. Set
$$v = \begin{cases} u & \text{on } \Omega, \\ 0 & \text{outside } \Omega. \end{cases}$$
Let H denote the p-fine boundary of Ω. First, assume that $u = 0$ at p-quasi every point x of H. If x is such a point, and, in addition, u is p-finely continuous at x, then the p-fine limit of v at x is 0 and thus v is p-finely continuous at x. It is also evident that v is p-finely continuous at p-quasi every point in Ω and outside the p-fine closure of Ω. Therefore, v is p-quasicontinuous on $\mathbf{R}^n$ and thus by Lemma 2.26, $u \in W_0^{1,p}(\Omega)$. Conversely, if $u \in W_0^{1,p}(\Omega)$, by Lemma 2.26, v is p-quasicontinuous on $\mathbf{R}^n$. If x is a point of H at which both v and u are p-finely continuous, then $u(x) = v(x) = 0$.

2.148 REMARK. From Theorems 2.147 and 2.55 we obtain the following criterion which determines when Sobolev functions have the same boundary values: Let $u, v \in W^{1,p}(\Omega)$. Then $u - v \in W_0^{1,p}(\Omega)$ if the zero extension of $u - v$ is weakly differentiable on $\mathbf{R}^n$ and
$$\lim_{r \to 0} r^{-n} \int_{B(x,r) \cap \Omega} |u(y) - v(y)|\, dy = 0$$
at p-quasi every x in the p-fine boundary of Ω. In particular, $u - M \in W_0^{1,p}(\Omega)$ if and only if

(2.136)
$$\lim_{r \to 0} \left(\fint_{B(x,r) \cap \Omega} u(y)\, dy \right) = M$$

for p-quasi all $x \in \partial \Omega$. We will use this to say that $u = M$ on $\partial \Omega$. Similarly, we will say that $u \leq M$ and $u \geq M$ on $\partial \Omega$ if $=$ can be replaced by $\leq$ or $\geq$ in (2.136).

2.5 Fine Sobolev spaces

In this section we study Sobolev spaces on p-finely open sets. The motivation comes from the natural problem of identifying minimizers of the p-Dirichlet integral among all functions in $W^{1,p}(\mathbf{R}^n)$ which coincide with some smooth function outside a general bounded set U. Indeed, the minimizer is the solution of the Dirichlet problem for the p-Laplace equation on the p-fine interior of U. We need Sobolev classes on p-finely open sets if only to describe the meaning of the Dirichlet problem on a p-finely open set.

Throughout this chapter we identify the space of functions with domain Ω with the space of functions with domain $\mathbf{R}^n$ which vanish outside Ω. If a function u is a.e. approximately differentiable, we will denote by Du the approximate derivative.

2.149 DEFINITION. Let $\Omega \subset \mathbf{R}^n$ be a p-finely open set. By definition, a function u belongs to $W^{1,p}(\Omega)$ if $u \in L^p(\Omega)$, u is approximately differentiable a.e. in Ω and there are sequences $\{\Omega_j\}$ of open sets containing Ω and $\{u_j\}$ of functions such that $u_j \in W^{1,p}(\Omega_j)$,
$$\|u_j - u\|_{p;\Omega} \to 0 \quad \text{and} \quad \|\nabla u_j - Du\|_{p;\Omega} \to 0.$$
The space $W_c^{1,p}(\Omega)$ is defined as the set of all functions in $W^{1,p}(\mathbf{R}^n)$ whose support is compact and contained in Ω. The space $W_0^{1,p}(\Omega)$ is the closure of $W_c^{1,p}(\Omega)$ with

respect to the $1,p$-norm. We denote $R_c^p(\Omega) = W_c^{1,p}(\Omega) \cap L^\infty(\Omega)$. The algebra $R^p(\Omega)$ is defined as the family of all a.e. approximately differentiable functions $u \in L^\infty(\Omega)$ such that $Du \in L^p(\Omega)$ and Du is a "weak derivative of u" in the sense that

$$\int_\Omega u \nabla \varphi \, dx = - \int_\Omega \varphi \, Du \, dx$$

for each $\varphi \in R_c^p(\Omega)$. The space $W_{p\text{-loc}}^{1,p}(\Omega)$ is defined as the collection of all measurable functions on Ω such that there is a family $\{G_\alpha\}$ of p-finely open set which covers Ω up to a p-polar set with $u \llcorner G_\alpha \in R^p(G_\alpha)$. Finally, we introduce $W_i^{1,p}(\Omega)$ as the space of all functions u from $W_{p\text{-loc}}^{1,p}(\Omega)$ such that $Du \in L^p(\Omega)$. We use

$$\|u\|_{1,p} = \left(\int_\Omega [|u|^p + |Du|^p] \, dx \right)^{1/p}$$

as the norm for $W^{1,p}(\Omega)$, $W_i^{1,p}(\Omega)$ and $W_0^{1,p}(\Omega)$.

2.150 REMARK. It is immediate that the "new" spaces $W^{1,p}(\Omega)$ and $W_0^{1,p}(\Omega)$ coincide with the "old" ones if the set Ω is open. We will show later that the spaces $W_i^{1,p}$ and $W^{1,p}$ are the same. On the other hand, $W_{p\text{-loc}}^{1,p}(\Omega)$ differs from $W_{\text{loc}}^{1,p}(\Omega)$. Indeed, if $z_0 \in \Omega$ and $p(1+\alpha) \geq n$, then the function

$$x \mapsto |x - x_0|^{-\alpha}$$

belongs to $W_{p\text{-loc}}^{1,p}(\Omega)$, but not to $W_{\text{loc}}^{1,p}(\Omega)$.

2.151 LEMMA. *Let $u \in R^p(\Omega)$ and $\eta \in R_c^p(\Omega)$. Then $u\eta \in R_c^p(\Omega)$ and*

$$\nabla(u\eta) = u \nabla \eta + \eta \, Du.$$

PROOF. Let $\varphi \in \mathcal{C}_c^\infty(\mathbf{R}^n)$. Then $\varphi \eta \in R_c^p(\Omega)$ and we have

$$\int_\Omega u\eta \nabla \varphi \, dx = \int_\Omega u \nabla(\eta \varphi) \, dx - \int_\Omega u\varphi \nabla \eta \, dx = - \int_\Omega \eta \varphi Du \, dx - \int_\Omega u\varphi \nabla \eta \, dx.$$

Hence $u \nabla \eta + \eta \, Du$ is the weak derivative of $u\eta$. Obviously, $u\eta \in L^\infty(\Omega)$ and $\nabla(u\eta) \in L^p(\Omega)$.

2.152 LEMMA. *Let $u \in W_{p\text{-loc}}^{1,p}(\Omega)$. Then u is almost everywhere approximately differentiable and has a p-quasicontinuous representative in Ω.*

PROOF. There is a system G_α of p-finely open sets covering Ω up to a p-polar set such that $u \llcorner G_\alpha \in R^p(G_\alpha)$ for each α. If $z \in G_\alpha$, by Theorem 2.137 there is a p-finely continuous function $\eta_z \in R_c^p(G_\alpha)$ such that $\eta_z = 1$ on a p-fine neighborhood V_z of z. By Lemma 2.151, $u\eta_z \in R_c^p(\Omega) \subset W^{1,p}(\mathbf{R}^n)$. Let v_z be a p-quasicontinuous representative of $u\eta_z$. Then $v_z = u$ a.e. in V_z. By the quasi-Lindelöf property (Theorem 2.146), there is a sequence z_j such that the collection $\{V_{z_j}\}$ covers Ω up to a polar set Z_0. By Theorem 2.23, for each couple i,j there is p-polar set $Z_{i,j}$ such that $v_{z_i} = v_{z_j}$ on $V_{z_i} \cap V_{z_j} \setminus Z_{i,j}$. Hence outside the p-polar set

$$Z := Z_0 \cup \bigcup_{i,j} Z_{i,j},$$

it is clear how to define a p-quasicontinuous representative of u. Since u is approximately differentiable a.e. in each V_{z_j}, it is approximately differentiable a.e. in Ω. $\square$

The following lemma is a substitute for a partition of unity.

2.153 LEMMA. *Let $u \in R_c^p(\Omega)$ be nonnegative. Let U_j, $j = 1, 2, \ldots$, be p-finely open subsets of Ω covering Ω up to a p-polar set Z. Then there are $m(k) \in \mathbf{N}$ and nonnegative functions $u_{k,j} \in R_c^p(U_j)$, $k = 1, 2, \ldots$, $j = 1, 2, \ldots, m(k)$, such that*

$$\sum_{j=1}^{m(k)} u_{k,j} \leq u$$

and

$$\lim_{k \to \infty} \|u - \sum_{j=1}^{m(k)} u_{k,j}\|_{1,p} = 0.$$

PROOF. By Theorem 2.144, for any $j = 0, 1, \ldots$ there are open sets $G_{j,k}$, $k = 1, 2, \ldots$, such that $\mathbf{C}_p(G_{j,k}) \leq 2^{-k-j}$, $Z \subset G_{k,0}$ and the sets $G_{k,j} \cup U_j$ ($j \geq 1$) are open. If we set

$$G_k = \bigcup_{j=0}^{\infty} G_{k,j},$$

then $Z \subset G_k$, the sets G_k and $U_j \cup G_k$ are open,

$$\Omega \subset \bigcup_{j=1}^{\infty} (G_k \cup U_j),$$

and still

$$\mathbf{C}_p(G_k) \to 0.$$

We find functions $\varphi_k \in W^{1,p}(\mathbf{R}^n)$ such that $0 \leq \varphi_k \leq 1$, $\varphi_k = 1$ on G_k and $\|\varphi_k\|_{1,p}^p \leq 2\mathbf{C}_p(G_k)$, and denote $\psi_k = 1 - \varphi_k$. Then using Lebesgue's dominated convergence theorem we easily show that $\|u\psi_k - u\|_{1,p} \to 0$. Now we are going to find a representation

(2.137) $$u\psi_k = \sum_{j=1}^{m(k)} u_{k,j}.$$

Let K be the support of u. Fix $k \in \mathbf{N}$. We find open balls $B_{k,j,i}$, $i, j = 1, 2, \ldots$, such that

$$\overline{B}_{k,j,i} \subset U_j \cup G_k \quad \text{and} \quad \bigcup_{j=1}^{\infty} U_j \cup G_k = \bigcup_{i,j=1}^{\infty} B_{k,j,i}.$$

Since K is compact, from the covering $\{B_{k,j,i}\}_{j,i}$ of K we can select a finite one, so that there is $m = m(k) \in \mathbf{N}$ such that

$$K \subset \bigcup_{i,j=1}^{m} B_{k,j,i}.$$

We find nonnegative functions $\omega_{k,j,i} \in \mathcal{C}_c^\infty(\mathbf{R}^n)$, $i,j = 1, 2, \ldots, m(k)$, such that

$$\{\omega_{k,j,i} > 0\} = B_{k,j,i}.$$

We set

$$u_{k,j} = u\psi_k \frac{\sum_{i=1}^{m(k)} \omega_{k,j,i}}{\sum_{i,j=1}^{m(k)} \omega_{k,j,i}}.$$

Then $u_{k,j} \in R^p(\Omega)$ and the support of $u_{k,j}$ is contained in

$$(K \setminus G_k) \cap \bigcup_{i=1}^{m(k)} \overline{B}_{k,j,i} \subset U_j.$$

Since (2.137) holds, the proof is complete. □

2.154 THEOREM. $W_i^{1,p}(\Omega) \cap L^\infty(\Omega) \subset R^p(\Omega)$.

PROOF. Using the definition of $W_i^{1,p}(\Omega)$ and the quasi-Lindelöf property (Theorem 2.146), we find a sequence $\{U_j\}$ of p-finely open subsets of Ω such that $\{U_j\}$ covers Ω up to a p-polar set and $u \in R^p(U_j)$. Let $\eta \in R_c^p(\Omega)$ be a nonnegative "test function". By Lemma 2.153, there are $m(k) \in \mathbf{N}$ and nonnegative functions $\eta_{k,j} \in R_c^p(\Omega)$, $k = 1, 2, \ldots$, $j = 1, 2, \ldots, m(k)$, such that $\eta_{k,j}$ have support in U_j,

$$\sum_{j=1}^{m(k)} \eta_{k,j} \leq \eta$$

and

$$\lim_{k \to \infty} \|\eta - \sum_{j=1}^{m(k)} \eta_{k,j}\|_{1,p} = 0.$$

By Lemma 2.151, for each $\eta_{k,j}$ we have

$$u\eta_{k,j} \in R_c^p(\Omega)$$

and

$$\nabla(u\eta_{k,j}) = \eta_{k,j} Du + u\nabla \eta_{k,j}.$$

Hence

$$\int_\Omega \eta_{k,j} Du \, dx = -\int_\Omega u\nabla \eta_{k,j} \, dx.$$

Summing for $j = 1, \ldots, m(k)$ and passing to the limit for $k \to \infty$ we infer that

$$\int_\Omega \eta Du \, dx = -\int_\Omega u\nabla \eta \, dx.$$

Hence $u \in R^p(\Omega)$. □

2.155 LEMMA. $W_i^{1,p}(\Omega) \subset W^{1,p}(\Omega)$.

PROOF. Let $u \in W^{1,p}(\Omega)$. We may assume that u is bounded since an unbounded function can be approximated by its truncations. Also we may suppose that u vanishes outside intersection of Ω with some ball. By Theorem 2.154, $u \in R^p(\Omega)$. Since Ω is p-finely open, by Theorem 2.144 there are open sets G_j such that $\mathbf{C}_p(G_j) \to 0$ and the sets $\Omega \cup G_j$ are open. We find $\varphi_j \in W^{1,p}(\mathbf{R}^n)$ such that $0 \le \varphi_j \le 1$, $\varphi_j = 1$ on G_j and

$$\|\varphi_j\|_{1,p} \to 0.$$

Set $\Omega_j = \Omega \cup G_j$, $u_j = u(1-\varphi_j)$. We easily verify that u_j is weakly differentiable and that $\|u\|_{1,p} < \infty$; hence $u_j \in W^{1,p}(\Omega_j)$. Using Lebesgue dominated convergence theorem we obtain that $u_j \to u$ in $W^{1,p}(\Omega)$. $\square$

2.156 REMARK. In view of Theorem 1.45 we may suppose that the approximating functions in the preceding theorem are smooth.

2.157 LEMMA. *Let Ω_j be a sequence of open sets containing Ω, and u_j a sequence of functions, $u_j \in R^p(\Omega_j)$. Let $u \in L^p(\Omega)$ and $g \in L^p(\Omega; \mathbf{R}^n)$. Suppose that $u_j \to u$ a.e.,*

(2.138) $$\int_\Omega |\nabla u_j - g|^p \, dx \to 0$$

and

$$\sup_j \|u_j\|_\infty \le k < \infty.$$

Then $u \in R^p(\Omega)$ and $Du = g$ a.e.

PROOF. We fix $j \in \mathbf{N}$ and $\eta \in R_c^p(\Omega)$. We observe that $u_j \eta$ is weakly differentiable in Ω_j and in $\mathbf{R}^n \setminus K$, where K denotes the support of η. Since $K \subset \Omega_j$, this implies that $u_j \eta$ is weakly differentiable in $\mathbf{R}^n$ and it is easy to see that $\eta \nabla u_j + u_j \nabla \eta$ is the weak derivative of $u_j \eta$ in $\mathbf{R}^n$. We have

$$\nabla(u_j \eta) - (\eta g + u \nabla \eta) = \eta(\nabla u_j - g) + \nabla \eta (u_j - u).$$

Now we pass to the limit. We get

$$\int_{\mathbf{R}^n} |\eta \nabla u_j - \eta g|^p \, dx$$
$$\le \|\eta\|_\infty^p \int_\Omega |\nabla u_j - g|^p \, dx \to 0$$

and

(2.139) $$\lim_{j \to \infty} \int_{\mathbf{R}^n} |\nabla \eta|^p |u_j - u|^p \, dx = 0$$

in view of Lebesgue's dominated convergence theorem. We see that the sequence $\{\eta u_j\}$ is bounded in $W^{1,p}(\mathbf{R}^n)$ and

$$\|\nabla(\eta u_j) - \nabla(\eta u)\|_p \to 0.$$

Using Theorem 1.61 we observe that also

$$\|\eta u_j - \eta u\|_p \to 0.$$

Hence $u\eta \in W^{1,p}(\mathbf{R}^n)$ and

$$\nabla(u\eta) = \eta\,g + u\,\nabla\eta.$$

By Theorem 1.71, u is approximately differentiable and $Du = g$ a.e. on $\{\eta = 1\}$. Using Theorem 2.152 we conclude that $u \in R^p(\Omega)$ and $Du = g$ a.e. $\square$

2.158 THEOREM. *Let $u_j \in W_i^{1,p}(\Omega)$, $u \in L^p(\Omega)$ and $g \in L^p(\Omega; \mathbf{R}^n)$. Suppose that*

$$\|u_j - u\|_p \to 0 \quad \text{and} \quad \|Du_j - g\|_p \to 0.$$

Then $u \in W_i^{1,p}(\Omega)$ and $Du = g$ a.e.

PROOF. In view of the density result of Lemma 2.155, we may assume that $u_j \in W^{1,p}(\Omega_j)$, where Ω_j are open sets containing Ω. We will denote by T_k the truncation operator

$$T_k u = \begin{cases} k, & u > k, \\ u, & |u| \le k, \\ -k, & u < -k. \end{cases}$$

The proof will be done in three steps.

Step 1. Consider $k > 0$ such that $\{|u| = k\}$ has zero measure. We will show that $T_k u$ belongs to $R^p(\Omega)$ and $DT_k u = g\chi_{\{|u|\le k\}}$.

Since $|T_k u_j - T_k u| \le |u_j - u|$, we immediately obtain

$$\lim_{j\to\infty} \int_\Omega |T_k u_j - T_k u|^p\, dx = 0.$$

By the truncation property of Sobolev spaces,

$$\nabla(T_k u_j) = \nabla u_j \text{ a.e. in } \{|u_j| \le k\}$$

and

$$\nabla(T_k u_j) = 0 \text{ a.e. in } \{|u_j| \ge k\}.$$

It follows

$$\int_\Omega |\nabla(T_k u_j) - g\chi_{\{|u|\le k\}}|^p)\, dx$$

$$\le \int_{\{|u_j|\le k\}\cap\{|u|\le k\}} |\nabla u_j - g|^p\, dx$$

$$+ \int_{\{|u_j|>k\}\cap\{|u|\le k\}} |g|^p\, dx$$

$$+ \int_{\{|u_j|\le k\}\cap\{|u|>k\}} |\nabla u_j|^p\, dx.$$

Since $\nabla u_j = 0$ a.e. on $\{u_j = k\}$, $|\{|u| = k\}| = 0$ and $|\nabla u_j|^p \leq 2^{p-1}(|\nabla u_j - g|^p + |g|^p)$, we obtain

(2.140)
$$\int_\Omega |\nabla(T_k u_j) - g\chi_{\{|u|\leq k\}}|^p) \, dx$$
$$\leq 2^{p-1} \int_\Omega |\nabla u_j - g|^p \, dx$$
$$+ \int_{\{|u_j|>k\}\cap\{|u|<k\}} |g|^p \, dx$$
$$+ 2^{p-1} \int_{\{|u_j|<k\}\cap\{|u|>k\}} |g|^p \, dx.$$

Denote
$$E_j = \{|u_j| < k < |u|\} \cup \{|u_j| > k > |u|\}.$$
Using the Lebesgue dominated convergence theorem we infer that

(2.141)
$$\lim_{j\to\infty} \int_{E_j} |g|^p \, dx = \lim_{j\to\infty} \int_\Omega |g|^p \chi_{E_j} \, dx = 0.$$

More precisely, the limit in (2.141) vanishes by the above argument for any subsequence of $\{u_j\}$ converging a.e., and thus it is zero for the whole sequence. By Lemma 2.157, $T_k u \in R^p(\Omega)$ and $DT_k u = g\chi_{\{|u|\leq k\}}$ a.e.

Step 2. We will show that $u \in W_{p\text{-loc}}(\Omega)$.

Let $\eta \in R_c^p(\Omega)$, $0 \leq \eta \leq 1$, and U be a p-finely open set contained in $\{\eta = 1\}$. Consider $k > 0$ with $|\{|u| = k\}| = 0$. Then the function $\eta T_k u$ belongs to $W^{1,p}(\mathbf{R}^n)$ and hence it has a quasicontinuous representative v_k. Let V_k denote the p-fine interior of the set $U \cap \{|v_k| < k\}$. Similar to the proof of Theorem 2.152 we show that there is a quasicontinuous representative of u on $\bigcup V_k$. Now, we want to show that $U \setminus \bigcup V_k$ is p-polar. Using Corollary 2.25 we obtain

(2.142)
$$\mathbf{C}_p(U \setminus V_k) \leq \int_{\mathbf{R}^n} \left(|\frac{1}{k} v_k|^p + |\nabla(\frac{1}{k} v_k)|^p\right) dx$$
$$\leq Ck^{-p}\left(\int_{\mathbf{R}^n} |T_k u|^p |\eta|^p \, dx + \int_{\{u\leq k\}} |g|^p |\eta|^p \, dx + \int_{\{u\leq k\}} |g|^p |\nabla\eta|^p \, dx\right).$$

We will pass to limit for $k \to \infty$. Notice that the condition $|\{|u| = k\}| = 0$ is satisfied for every $k > 0$ except a countable number of values. We see that

(2.143)
$$\int_{\mathbf{R}^n} |T_k u|^p |\eta|^p \, dx + \int_{\{u\leq k\}} |g|^p |\eta|^p \, dx$$
$$\leq k^{-p} \int_\Omega (|u|^p + |g|^p) \, dx \to 0.$$

Since $u \in L^p(\mathbf{R}^n)$, it is finite a.e. and thus $k^{-p}|T_k u|^p \to 0$ a.e. Using the Lebesgue dominated convergence theorem it follows that

(2.144)
$$\int_{\mathbf{R}^n} k^{-p} |T_k u|^p |\nabla\eta|^p \, dx \to 0.$$

By (2.142), (2.143) and (2.144),
$$\mathbf{C}_p(U \setminus V_k) \to 0,$$

in particular

(2.145) $$\mathbf{C}_p(U \setminus \bigcup_k V_k) = 0.$$

By Lemma 2.137, for p-quasi every point $y \in \Omega$ there is a p-fine neighborhood $U = U(y)$ of y and a function $\eta_y \in W_c^{1,p}(\mathbf{R}^n)$ such that $0 \leq \eta_y \leq 1$, η_y has its support in Ω and $\eta_y = 1$ in $U(y)$. By the above part of the proof, for p-q.e. point $z \in U(y)$ there is its p-fine neighborhood $V(z)$ and $k \in \mathbf{R}^n$ such that $\eta_y T_k u = u$ a.e. in $V(z)$ and $\eta_y T_k u \in R^p(V(z))$. Hence $u \in W_{p\text{-loc}}^{1,p}(\Omega)$.

Step 3. It remains to notice that $Du = g$ a.e. and that the $1,p$-norm of u is finite, which is now obvious. □

2.159 THEOREM. $W_i^{1,p}(\Omega) = W^{1,p}(\Omega)$.

PROOF. The inclusion $W^{1,p}(\Omega) \subset W_i^{1,p}(\Omega)$ immediately follows from Theorem 2.158. The opposite inclusion results from Lemma 2.155. □

2.160 LEMMA. *Suppose that $u \in W^{1,p}(\Omega)$ has compact support in Ω. Then $u \in W_c^{1,p}(\Omega)$.*

PROOF. By employing a truncation argument, we are free to assume that u is bounded. There are open sets $\Omega_j \supset \Omega$ and functions $u_j \in W^{1,p}(\Omega_j)$, $j = 1, 2, \ldots$, such that
$$u_j \to u \text{ in } L^p(\Omega),$$
$$\nabla u_j \to \nabla u \text{ in } L^p(\Omega; \mathbf{R}^n), \text{ and}$$
$$\|u_j\|_\infty \leq \|u\|_\infty.$$
Since Ω is p-quasiopen, there are open sets G_k, $k = 1, 2, \ldots$, such that every $\Omega \cup G_k$ is open and $\mathbf{C}_p(G_k) < 1/k$. We find $\varphi_k \in W^{1,p}(\mathbf{R}^n)$ such that
$$\varphi_k = 1 \text{ on } G_k \text{ and } \int_{\mathbf{R}^n} (|\varphi_k|^p + |\nabla \varphi_k|^p)\, dx < 2/k.$$

We denote
$$\psi_k = 1 - \varphi_k.$$
For fixed j and k, the function
$$v_{j,k} = \begin{cases} u_j \psi_k & \text{on } \Omega_j \\ 0 & \text{on } G_k \end{cases}$$
is both in $W^{1,p}(G_k)$ and $W^{1,p}(\Omega_j)$, thus it is in $W^{1,p}(\Omega \cup G_k)$, as $\Omega \cup G_k \subset \Omega_j \cup G_k$. Recall that $\Omega \cup G_k$ is an open set. Now, let us pass to a limit for $j \to \infty$. We denote
$$g_j = \begin{cases} \nabla u_j & \text{in } \Omega, \\ 0 & \text{outside } \Omega. \end{cases}$$
Obviously
$$v_{j,k} \to v\psi_k \text{ in } L^p(\Omega \cup G_k)$$
and
$$\|\nabla v_{j,k} - (v \nabla \psi_k + g\psi_k)\|_p$$
$$\leq \|(v_j - v)\nabla \psi_k\|_p + \|(g_j - g)\psi_k\|_p \to 0.$$

We infer that $v\psi_k \in W^{1,p}(\Omega \cup G_k)$ and
$$\nabla(v\psi_k) = v\nabla\psi_k + g\psi_k.$$
On the other hand, $v\psi_k \in W^{1,p}(\mathbf{R}^n \setminus K)$, where K is the support of v, as $v\psi_k$ vanishes outside K. Since $\mathbf{R}^n = (\Omega \cup G_k) \cup (\mathbf{R}^n \setminus K)$, $v\psi_k \in W^{1,p}(\mathbf{R}^n)$. Letting $k \to \infty$ we get that $v\psi_k \to v$ in $L^p(\mathbf{R}^n)$ as
$$\|v\psi_k - v\|_p = \|v\varphi_k\|_p \leq \|v\|_\infty \|\varphi_k\|_p \to 0.$$
We have
$$\|\nabla(v\psi_k) - g\|_p \leq \|v\nabla\varphi_k\|_p + \|g\varphi_k\|_p \to 0.$$
It follows that $v \in W^{1,p}(\mathbf{R}^n)$ and $\nabla v = g$. □

We conclude this section by showing that Lemma 2.26 and Theorem 2.147 hold also for Sobolev spaces on p-finely open sets.

2.161 THEOREM. *Let $u \in W^{1,p}(\Omega)$ and $\tilde{u}$ be a p-quasicontinuous representative of u in Ω. Let v be a function on $\mathbf{R}^n$ which coincides with $\tilde{u}$ on Ω and vanishes outside Ω. Then the following assertions are equivalent:*
  (i) $u \in W_0^{1,p}(\Omega)$,
  (ii) $v \in W_0^{1,p}(\Omega')$ *for every open set* $\Omega' \supset \Omega$,
  (iii) v *is p-quasicontinuous in* $\mathbf{R}^n$.

PROOF. (i) $\implies$ (ii): Suppose that $u \in W_c^{1,p}(\Omega)$. Then $v \in W^{1,p}(\mathbf{R}^n)$. If $\Omega' \supset \Omega$ is an open set, then the support of v is contained in Ω' and thus $v \in W_0^{1,p}(\Omega')$. Since the set of all $u \in W^{1,p}(\Omega)$ for which (ii) is true is closed in $W^{1,p}(\Omega)$, we infer that (ii) holds for any $u \in W_0^{1,p}(\Omega)$.

(ii) $\implies$ (iii): We find open sets G_k, $k = 1, 2, \ldots$, such that the sets $\Omega \cup G_k$ are open and $\mathbf{C}_p(G_k) \to 0$. Fix $k \in \mathbf{N}$ and $\varepsilon > 0$. Assuming (ii), we have $v \in W_0^{1,p}(\Omega \cup G_k)$. We find a function $w_k \in \mathcal{C}_c(\Omega \cup G_k)$ such that $\|w - v\|_{1,p} \leq 1/k$. Let $\tilde{v}$ be a p-quasicontinuous representative of v in $\mathbf{R}^n$. (It exists, as by (ii) $v \in W^{1,p}(\mathbf{R}^n)$.) Then, using Corollary 2.25 and letting $k \to \infty$,
$$\mathbf{C}_p(\{|v - \tilde{v}| > \varepsilon\} \setminus \Omega) \leq \mathbf{C}_p(\{|w_k - \tilde{v}| > \varepsilon\} \cup G_k)$$
$$\leq \mathbf{C}_p(\{|w_k - \tilde{v}| > \varepsilon\}) + \mathbf{C}_p(G_k) \leq \varepsilon^{-p} \|w_k - \tilde{v}\|_{1,p}^p + \mathbf{C}_p(G_k) \to 0.$$
It follows that $\tilde{v} = v = 0$ p-q.e. outside Ω. By Corollary 2.23, $\tilde{v} = v$ p-q.e. in Ω. Since $\tilde{v}$ is p-quasicontinuous in $\mathbf{R}^n$, we deduce that v is p-quasicontinuous in $\mathbf{R}^n$ as well.

(iii) $\implies$ (i): We we may assume that u is bounded and the support of u is bounded. There exist open sets G_k, $k = 1, 2, \ldots$, such that $\mathbf{C}_p(G_k) \to 0$, the sets $\Omega \cup G_k$ are open and the restrictions $v \bigsqcup \mathbf{R}^n \setminus G_k$ are continuous. We find functions $\varphi_k \in W^{1,p}(\mathbf{R}^n)$ such that $0 \leq \varphi_k \leq 1$, $\varphi_k = 1$ on G_k and $\|\varphi_k\|_{1,p}^p \leq 2\mathbf{C}_p(G_k)$, and denote $\psi_k = 1 - \varphi_k$. We easily show that $v\psi_k \to v$ in $W^{1,p}(\Omega)$. It is enough to prove that each $v\psi_k$ belongs to $W_0^{1,p}(\Omega)$. Fix $k \in \mathbf{N}$ and write $G = G_k$, $\psi = \psi_k$. For each positive integer j, set
$$v_j = \max(v\psi - \frac{1}{j}, \min(0, v\psi + \frac{1}{j})).$$
Each function v_j is in $W^{1,p}(\Omega)$. Let z be a point of the closure of $\{v_j \neq 0\}$ in $\mathbf{R}^n$. There are $z_i \to z$ such that $v_j(z_i) \neq 0$. By the definition of v_j, $z_i \in \Omega \setminus G$ and

$|v(z_i)| \geq |v(z_i)\psi(z_i)| \geq 1/j$. Since G is open and $v \, \llcorner \, \mathbf{R}^n \setminus G$ is continuous, it follows that $z \notin G$ and $|v(z)| \geq 1/j$. Hence $z \in \Omega$. This proves that the support of v_j is contained in Ω, so that $v_j \in W_c^{1,p}(\Omega)$. Using the Lebesgue dominated convergence theorem we easily get that $v_j \to v\psi$ in $W^{1,p}(\Omega)$, so that $v\psi \in W_0^{1,p}(\Omega)$. This concludes the proof. $\square$

2.162 COROLLARY. *A p-quasicontinuous function $u \in W^{1,p}(\mathbf{R}^n)$ belongs to $W_0^{1,p}(\Omega)$ if and only if $u(x) = 0$ at p-quasi every point x of the p-fine boundary of Ω.*

2.6 Historical notes

Capacity as developed in this chapter has its roots in the work of Aronszajn and Smith [**AS1**], [**AS2**], Aronszajn, Mulla and Szeptycki [**AMS**] and Choquet [**Ch1**]. A capacity that preceded our $(p;r)$-capacity was introduced by Loewner for the case $p = n$, [**Lo**], which has proved to be indispensable in the development of the theory of n-dimensional quasiconformal mappings. Significant developments to this theory include Maz'ya [**Maz1**], [**Maz6**], Maz'ya and Havin [**HM2**], and Meyers [**Me1**].

The concept of quasicontinuous and finely continuous representatives for $W^{1,2}$ functions was introduced by Deny [**De**]. Further development included work by Deny and Lions, [**DL**], Aronszajn and Smith, [**AS1**] [**AS2**], and Fuglede, [**Fu1**], [**Fu2**], [**Fu3**]. Malý [**Mal1**] proved that a function $u \in W^{1,p}$ can be approximated by a Hölder continuous $v \in W^{1,p}$ with arbitrarily small $\mathbf{C}_q(\{u \neq v\})$, where $q < p$.

Theorem 2.28 is one of the first results in the book [**MaSh**] where multipliers are thoroughly studied.

The observations about energy are well known and their classical versions go back to Gauss and other founders of potential theory.

The relationship between Hausdorff measure and capacity has been investigated extensively. Contributors to this area of investigation include, but is certainly not limited to, Kametani [**Ka**], Erohin [**Er**], Carleson [**Car**], Taylor [**Ta**], Reshetnyak [**Re1**], Maz'ya and Havin [**HM2**] and Martio [**Mart**].

The presentation of capacity in this chapter is a modification of the development in [**FZ**]. There were many contributors to the development of capacity defined in terms of a general kernel and its relation to the variational capacity, cf. Fuglede [**Fu1**], Meyers [**Me1**], Havin and Maz'ya [**HM1**], [**HM2**], Reshetnyak [**Re1**]. A general theory of capacities was developed by Choquet [**Ch1**]. A capacity can be also associated with a monotone operator, as shown in the recent work by Dal Maso and Skrypnik [**DMS**]. An interesting survey on various aspect of applications of capacity theory to partial differential equations was written by Frehse [**Fr**].

The results in Section 2.1.2 (Polar Sets) are due to Littman [**Lit**]. That Sobolev functions $W^{1,p}$ possess Lebesgue points p-quasi-everywhere was proved in [**FZ**]. The extension of this result to the spaces $W^{k,p}$ and Bessel potentials was also established by Bagby and Ziemer [**BZ**], Calderón, Fabes and Riviere [**CFR**] and Meyers [**Me2**].

The history of potential theory begins in 17th century. Its development can be traced to such greats as Newton, Euler, Laplace, Lagrange, Fourier, Green, Gauss, Poisson, Dirichlet, Riemann, Weierstrass, and Poincaré. We refer to the book by Kellogg [**Kel2**] for references to some of the old works. The mean-value property of harmonic functions was known to Gauss. Weyl proved that distributional solutions of the Laplacian are harmonic [**We**]. Superharmonic functions were introduced by Riesz [**Rif1**] and their lower approximation by smooth function was proved in [**Rif2**]. Existence of a Green's function for a general simply connected domain in the plane is due to Osgood [**Os**]. A development of potential theory based on Green's potentials is done in the book by Helms [**Hel**]. N. Wiener was probably first in introducing the idea of thin sets, [**Wie1**] and among the early developers of thin sets were Brelot [**Br2**], Choquet [**Ch2**], G. C. Evans [**Evgc**], and O. Frostman [**Fro**]. The fine topology was introduced by Cartan [**Cart**]. The Perron-Wiener-Brelot method of solving the Dirichlet problems is based on an idea by Perron [**Pe**], but discontinuous upper functions were considered by Brelot [**Br1**]. Wiener [**Wie2**] proved that the Dirichlet problem is solvable for continuous boundary functions. It was Zaremba [**Za**] who observed that domains with deleted isolated points are examples of open sets for which the classical Dirichlet problem is not solvable. A more sophisticated example was constructed by Lebesgue [**Le2**]. Kerimov [**Ke**] investigated the behavior of the solution of a uniformly elliptic equation near a "singular" boundary point and established a sufficient condition for regularity there in terms of a Wiener sum. Barriers as a criterion for regularity were used already by Poincaré and systematically introduced by Lebesgue [**Le3**]. The main goal of Section 2 is Wiener's criterion proved first in the paper [**Wie1**] by Wiener. For further introduction to fine topics in the classical potential theory we refer to the books by Brelot [**Br3**] and [**Br3**].

Section 3 gives an overview of higher regularity of solutions to variational problems. An important step is that provided by Theorem 2.103 due to Nirenberg [**Ni1**], which implies that derivatives of the solution of the variational problem are solutions of a linear elliptic equation with L^∞ coefficients. The proof in subsection 2.3.4 is based on the work by De Giorgi [**DeG**] while that in subsection 2.3.5 is due to Moser, [**Mos1**], [**Mos2**]. Sections 2.3.6–2.3.10 are simplified versions of developments in Chapters 4 and 6 in which we also present historical comments to the material.

The potential theory for the equation (2.69) has been developed by Granlund, Lindqvist and Martio [**GLM**], Lindqvist [**Lin1**], Lindqvist and Martio [**LM**], Heinonen and Kilpeläinen [**HK1**], [**HK2**], [**HK3**], Heinonen, Kilpeläinen and Martio [**HKM1**], Kilpeläinen [**Ki1**], [**Ki2**], Kilpeläinen and Malý [**KM1**], [**KM2**], [**KM3**], [**KM4**]. Many results are collected in the book by Heinonen, Kilpeläinen and Martio [**HKM2**], where the situation is generalized to a weighted case. Theorem 2.130 was proved by Kilpeläinen and Malý in [**KM4**]. The fact that the p-fine topology is the coarsest topology making all $\mathcal{A}$-superharmonic functions continuous was proved by Heinonen, Kilpeläinen and Martio in [**HKM1**], i.e. before it was known that Theorem 2.130 holds.

The quantity

$$\mathbf{W}^\mu_{m,p}(x) = \int_0^{r_0} \left(\frac{\mu(B(x,r))}{r^{n-pm}}\right)^{1/(p-1)} \frac{dr}{r}$$

used with $m = 1$ in estimates of solutions of the equation (2.112), is a nonlinear potential, called the Wolff potential. The study of such potentials was initiated by Adams and Meyers [**AM3**], Hedberg [**He1**] and Hedberg and Wolff [**HW**]. The work of Adams and Meyers was part of a larger collaborative effort, [**AM1**], [**AM2**].

The **W**-potentials provide a useful alternative to **V**-potentials in generalizing newtonian potentials to the nonlinear case. The $\mathbf{V}_p$-potentials defined by

$$\mathbf{V}_{m,p}^{\mu} = k_m * (k_m * \mu)^{1/(p-1)},$$

where k_m is either the bessel kernel g_m or the Riesz kernel I_m, were studied by Maz'ya and Havin [**HM1**], [**HM2**] and Meyers [**Me1**]. The current state of development of nonlinear potential theory is presented in the monograph by Adams and Hedberg [**AH**].

The results on the Wiener criterion and estimates in terms of Wolff potentials of solutions with a measure on the right side were generalized to situation of weighted nonlinear potential theory by Mikkonen, [**Mik**]. Further notes on the history of results connected to the Wiener criterion will be discussed in the Historical Remarks of Chapters 4 and 5.

Fine topology in nonlinear potential theory has been studied by Adams and Meyers [**AM3**], Hedberg [**He1**], Meyers [**Me3**], Hedberg and Wolff [**HW**], Adams and Lewis [**AL**], Heinonen, Kilpeläinen and Malý [**HKMy**] and others. The connection between quasitopologies and fine topologies has been studied by Fuglede [**Fu2**]. A survey on fine topologies is written by Lukeš and Malý [**LMy**]. The quasi-Lindelöf property (Theorem 2.146) was established in the linear case by Doob [**Do**], while an abstract result applicable to the nonlinear theory was proved by Fuglede [**Fu4**]. The linear case of Theorem 2.145 is due to Brelot [**Br4**] and Deny [**De**]. A general form of it which allows application to the nonlinear theory was proved by Fuglede [**Fu4**]. In classical potential theory, the Choquet property first appeared in [**Ch2**]. The Kellogg property was conjectured and proven in the planar case by Kellogg [**Kel1**], [**Kel2**] and the case of general dimension is due to Evans [**Evgc**]. Both results are generalized to the nonlinear case by Hedberg and Wolff [**HW**]. See Dal Maso and Defranceschi [**DD**] for a Kellogg-type result involving the Laplacian and a measure on the right-hand side. Theorem 2.147 was first proved by Havin [**Ha**] in the case $p = 2$ and by Bagby [**Ba**] for all $p > 1$. Theorem 2.147 has a version valid for $W^{k,p}$ functions. Its proof offered considerable difficulty due in part to the fact that, unlike the situation in $W^{1,p}$, functions in $W^{k,p}$ are not closed under truncation. Hedberg [**He3**] proved this extension of Theorem 2.147 under the restriction $p > 2 - \frac{1}{n}$. Later, Hedberg and Wolff [**HW**] established the result for all $p > 1$.

Recent expositions of trace theory for Sobolev functions appear in [**EG**] and [**Z5**].

Our development in Section 2.5 is based on the approach by Kilpeläinen and Malý [**KM3**], where fine Sobolev spaces are introduced and the p-fine Dirichlet problem is studied. The idea of Sobolev spaces on finely open sets appears also in earlier works by Feyel and de La Pradelle, [**FP1**], [**FP2**] and by Fuglede [**Fu6**]. The fine Dirichlet problem for linear equations was introduced and solved by Fuglede; we refer to his monograph [**Fu5**]. Theorem 2.161 is a fine version of the above mentioned result by Bagby [**Ba**]. Related papers have been written by Latvala [**Lat1**], [**Lat2**].

The algebra $R^p(\Omega)$ is related to the Royden's algebra of continuous functions with derivatives in L^p, cf. [**Roy**], [**Lewl**].

CHAPTER 3

Quasilinear Equations

3.1 Basic properties of weak solutions

The objective of Chapters 3, 4, and 5 is to discuss results relating to the pointwise behavior of solutions and supersolutions of quasilinear elliptic equations of type

$$(3.1) \qquad -\operatorname{div} \boldsymbol{A}(x, u, \nabla u) + \boldsymbol{B}(x, u, \nabla u) = 0.$$

This will include investigation of Wiener-type criteria and comparison with nonlinear potentials. We will also consider problems with a measure on the right hand side, the Dirichlet problem and obstacle problems.

The model equation for (3.1) is

$$(3.2) \qquad -\operatorname{div}(|\nabla u|^{p-2}\nabla u) + \lambda |u|^{p-2}u = 0$$

with $\lambda \in \mathbf{R}$.

Recall that u is a *solution* (frequently termed "weak solution") of (3.1) in open set $\Omega \subset \mathbf{R}^n$ if $u \in W^{1,p}_{\text{loc}}(\Omega)$ and

$$(3.3) \qquad \int_\Omega \boldsymbol{A}(x, u, \nabla u) \cdot \nabla \varphi + \boldsymbol{B}(x, u, \nabla u)\, \varphi\, dx = 0$$

for all bounded "test functions" $\varphi \in W^{1,p}_c(\Omega)$.

Similarly we define *subsolutions* and *supersolutions*. Namely, we test by non-negative bounded functions $\varphi \in W^{1,p}_c(\Omega)$ and require

$$(3.4) \qquad \int_\Omega \boldsymbol{A}(x, u, \nabla u) \cdot \nabla \varphi + \boldsymbol{B}(x, u, \nabla u)\, \varphi\, dx \begin{cases} \geq 0 & \text{for supersolutions,} \\ \leq 0 & \text{for subsolutions} \end{cases}$$

for each test function.

We define solutions, supersolutions and subsolutions also on p-finely open sets. Nevertheless, unless specified otherwise, we will suppose in the sequel that Ω is open.

3.1 DEFINITION. We redefine the symbol $U \subset\subset \Omega$ in situations where U and Ω are p-finely open sets. Namely, we write $U \subset\subset \Omega$ if there is $\varphi \in W^{1,p}_c(\Omega)$ such that $\varphi \geq 1$ on U. Note that this concept agrees with the usual notation when U and Ω are open.

We introduce $W^{1,p}_{\text{loc}}(\Omega)$ as the family of all function u such that

$$U \subset\subset \Omega \implies u \llcorner U \in W^{1,p}(U).$$

Besides the class $W_c^{1,p}(\Omega)$ we consider a more restrictive class $W_{cc}^{1,p}(\Omega)$ which is defined as the family of all functions from $W_c^{1,p}(\Omega)$ which vanish outside some $U \subset\subset \Omega$. The classes $W_c^{1,p}(\Omega)$ and $W_{cc}^{1,p}(\Omega)$ coincide if Ω is open.

Now we are ready to introduce solutions (subsolutions, supersolutions, respectively) on p-finely open sets analogous to those on open sets. Namely, we require that $u \in W_{\text{loc}}^{1,p}(\Omega)$ and test as above in (3.3), (3.4) by nonnegative bounded functions from $W_{cc}^{1,p}(\Omega)$.

Throughout we suppose that the functions $\boldsymbol{A} \colon \mathbf{R}^n \times \mathbf{R} \times \mathbf{R}^n \to \mathbf{R}^n$ and $\boldsymbol{B} \colon \mathbf{R}^n \times \mathbf{R} \times \mathbf{R}^n \to \mathbf{R}$ are given. We suppose that $\boldsymbol{A}(\cdot, \zeta, \xi), \boldsymbol{B}(\cdot, \zeta, \xi)$ are Lebesgue measurable, $\boldsymbol{A}(x, \cdot, \cdot), \boldsymbol{B}(x, \cdot, \cdot)$ are Borel measurable, and the following structure conditions are satisfied:

(3.5)
$$\begin{aligned}|\boldsymbol{A}(x,\zeta,\xi)| &\leq a_1|\xi|^{p-1} + a_2|\zeta|^{p-1} + a_3,\\ |\boldsymbol{B}(x,\zeta,\xi)| &\leq b_0|\xi|^p + b_1|\xi|^{p-1} + b_2|\zeta|^{p-1} + b_3,\\ \boldsymbol{A}(x,\zeta,\xi)\cdot\xi &\geq c_1|\xi|^p - c_2|\zeta|^p - c_3.\end{aligned}$$

Here c_1 is a positive constant, a_1 and b_0 are nonnegative constants whereas the remaining coefficients are nonnegative functions assumed to lie in suitable function spaces, namely multiplier spaces and Morrey spaces. Specifically, we make the following assumptions:

(3.6)
$$\begin{aligned}a_2^{1/(p-1)}, a_3^{1/(p-1)} &\in M(W^{1,p}(\mathbf{R}^n), L^p(\mathbf{R}^n)),\\ b_1^p, b_2, b_3, c_2, c_3 &\in \mathcal{M}^{n/(p-\epsilon)}(\mathbf{R}^n),\end{aligned}$$

where $\epsilon \in (0,1)$. Throughout the symbol ϵ will be used exclusively for its purpose in (3.6). It should be distinguished from ε which will generally be used to designate a small positive number.

It is a mere technicality that $\boldsymbol{A}(\cdot, \zeta, \xi)$ and $\boldsymbol{B}(\cdot, \zeta, \xi)$ are also defined outside Ω; if not, we can extend them by setting $\boldsymbol{A}(x, \zeta, \xi) = |\xi|^{p-2}\xi$ and $\boldsymbol{B}(x, \zeta, \xi) = 0$ for $x \notin \Omega$.

The model example $\boldsymbol{A}(x, \zeta, \xi) = |\xi|^{p-2}\xi$, $\boldsymbol{B}(x, \zeta, \xi) = \lambda |\zeta|^{p-2}\zeta$ leads to (3.2).

Unless otherwise specified, reference to a solution (subsolution, supersolution) of (3.1) will carry with it the assumption that either (3.5) and (3.6) are in force.

We will assume without loss of generality that $c_1 = 1$ and $b_2 = c_2$. We will frequently consider solutions defined in a ball $B(r)$. It is convenient to simplify the structure of (3.5) by introducing

(3.7) $\quad \boldsymbol{k} = \boldsymbol{k}(r) = \left(\|a_3^{1/(p-1)}\|^{p-1} + r^\epsilon \|b_3 + b_0 c_3\|\right)^{1/(p-1)} + (r^\epsilon \|c_3\|)^{1/p},$

where the norms are taken in the respective spaces (3.6) over $B(r)$. More precisely, we use norm in $\mathcal{M}^{n/(p-\epsilon)}(B(r))$ for $\|b_3 + b_0 c_3\|$, $\|c_3\|$ and $\|\ldots\|_{M(p,p);(B(r)}$ for $\|a_2^{1/(p-1)}\|$, $\|a_3^{1/(p-1)}\|$, where

$$\|f\|_{M(p,p);B(r)} := \sup\{\|fv\|_{p;B(r)} \colon v \in W_0^{1,p}(B(r)), \|\nabla v\|_{p;B(r)} \leq 1\}.$$

Then,

(3.8)
$$|A(x,u,\nabla u)| \le a_1 |\nabla u|^{p-1} + \bar{a}(|u|+k)^{p-1}$$
$$|B(x,u,\nabla u)| \le b_0|\nabla u|^p + b_1 |\nabla u|^{p-1} + \bar{b}(|u|+k)^{p-1}$$
$$A(x,u,\nabla u) \cdot \nabla u \ge |\nabla u|^p - \bar{b}(|u|+k)^p$$

with $\bar{a} = a_2 + k^{1-p}a_3$, $\bar{b} = b_2 + k^{1-p}(b_3 + b_0 c_3) + k^{-p}c_3$. Moreover, for $r=1$,

$$\|\bar{a}^{1/(p-1)}\|^{p-1} \le \|a_2^{1/(p-1)}\|^{p-1} + 1, \quad \|\bar{b}\| \le \|b_2\| + 2.$$

3.2 REMARK. By considering the transformation $x \mapsto xr$, we see the effect of scaling on various parameters. Specifically, suppose u is a weak subsolution on $B(r)$ and let $\tilde{u}(x) = u(rx)$ for $x \in B(1)$. Then $\tilde{u}$ satisfies the equation $-\operatorname{div} \tilde{A} + \tilde{B} = 0$ where

$$\tilde{A}(x,\zeta,\xi) = r^{p-1} A(rx,\zeta,\xi/r) \quad \tilde{B}(x,\zeta,\xi) = r^p B(rx,\zeta,\xi/r).$$

Moreover, $\tilde{A}$ and $\tilde{B}$ satisfy the same structure conditions as in (3.5) with the exception that coefficients now become

(3.9)
$$\tilde{a}_1 = a_1, \quad \tilde{a}_2(x) = r^{p-1} a_2(rx), \quad \tilde{a}_3(x) = r^{p-1} a_3(rx),$$
$$\tilde{b}_0 = b_0, \quad \tilde{b}_1(x) = r b_1(rx), \quad \tilde{b}_2(x) = r^p b_2(rx), \quad \tilde{b}_3(x) = r^p b_3(rx),$$
$$\tilde{c}_2(x) = r^p c_2(rx), \quad \tilde{c}_3(x) = r^p c_3(rx).$$

For $k(r)$ we have

(3.10)
$$k(r) = \left(\|\tilde{a}_3^{1/(p-1)}\|^{p-1} + \|\tilde{b}_3 + \tilde{b}_0 \tilde{c}_3\| \right)^{1/(p-1)} + \|\tilde{c}_3\|^{1/p}$$

where the norms are in the respective spaces over $B(1)$.

3.3 REMARK. Recall that the space $M(W_0^{1,p}(\Omega), L^p(\Omega))$ of multipliers is defined as the space of all functions g such that the operator

$$u \mapsto gu$$

is bounded from $W_0^{1,p}(\Omega)$ to $L^p(\Omega)$ and endowed with the operator norm; see Theorem 2.28, p. 77. If $p < n$, then $L^n(\Omega) \subset M(W_0^{1,p}(\Omega), L^p(\Omega))$, which follows easily from the Sobolev imbedding theorem. For $p = n$ we need slightly better integrability to get a sufficient condition.

The Morrey space $\mathcal{M}^{n/(p-\epsilon)}(\Omega)$ consists of all functions f such that

$$\int_{\Omega \cap B(r)} f \, dx \le K r^{n-p+\epsilon}$$

for some $K > 0$ and all balls $B(r)$. A typical application of Adams' theorem Corollary 1.93, p. 56 and Corollary 1.95 says that

$$\{|f|^{1/p} : f \in \mathcal{M}^{n/(p-\epsilon)}(\Omega)\} \subset M(W_0^{1,p}(\Omega), L^p(\Omega)) \quad \text{if } |\Omega| < \infty$$

and, even more,

(3.11)
$$\int_\Omega |u|^p f \, dx \le C \|f\|_{\mathcal{M}^{n/(p-\epsilon)}(\Omega)} \left(\delta \|\nabla u\|_p^p + C \delta^{1-p/\epsilon} \|u\|_p^p \right)$$

for any $f \in \mathcal{M}^{n/(p-\epsilon)}(\Omega)$, $u \in W_0^{1,p}(\Omega)$ and $\delta > 0$.

The following lemma allows reduction to the case $b_0 = 0$ in most situations.

3.4 LEMMA. *Let u be a supersolution of (3.1) with $0 \leq u \leq M$. Then u is a supersolution of*
$$-\operatorname{div} \boldsymbol{A}'(x, u, \nabla u) + \boldsymbol{B}'(x, u, \nabla u) = 0$$
where
$$\boldsymbol{A}'(x, \zeta, \xi) \cdot \xi \geq c_1'|\xi|^p - c_2'|\zeta|^p - c_3'$$
$$|\boldsymbol{A}'(x, \zeta, \xi)| \leq a_1'|\xi|^{p-1} + a_2'|\zeta|^{p-1} + a_3'$$
$$|\boldsymbol{B}'(x, \zeta, \xi)| \leq b_1'|\xi|^{p-1} + b_2'|\zeta|^{p-1} + b_3'$$
and

$a_1' = e^{b_0 M/c_1} a_1, \quad a_2' = e^{b_0 M/c_1} a_2, \qquad\qquad a_3' = e^{b_0 M/c_1} a_1,$
$b_1' = e^{b_0 M/c_1} b_1, \quad b_2' = e^{b_0 M/c_1}(b_2 + b_0 M c_2/c_1), \quad b_3' = e^{b_0 M/c_1}(b_3 + b_0 c_3/c_1)$
$c_1' = c_1, \qquad\quad c_2' = e^{b_0 M/c_1} c_2, \qquad\qquad\qquad c_3' = e^{b_0 M/c_1} c_3.$

Similarly, if u is a subsolution of (3.1), then
$$\int_\Omega \left(\boldsymbol{A}''(x, u, \nabla u) \cdot \nabla \varphi + \boldsymbol{B}''(x, u, \nabla u) \varphi \right) \leq 0$$
holds for all $\varphi \geq 0$ with $\{\varphi > 0\} \subset \{u > 0\}$. Here,
$$\boldsymbol{A}''(x, \zeta, \xi) \cdot \xi \geq c_1'|\xi|^p - c_2'|\zeta|^p - c_3'$$
$$|\boldsymbol{A}''(x, \zeta, \xi)| \leq a_1'|\xi|^{p-1} + a_2'|\zeta|^{p-1} + a_3'$$
$$|\boldsymbol{B}''(x, \zeta, \xi)| \leq b_1'|\xi|^{p-1} + b_2'|\zeta|^{p-1} + b_3'$$
and a_i', b_i', c_i' are as in the preceding situation.

PROOF. Write $b = b_0/c_1$. We may assume that $c_1 = 1$. Let u be a supersolution of (3.1). Choose a nonnegative function $\varphi \in W_0^{1,p}$. Testing (3.4) with $e^{-bu}\varphi$ we obtain

$$\int_\Omega \boldsymbol{A}(x, u, \nabla u)\, e^{-bu} \nabla \varphi \, dx$$
$$+ \int_\Omega \left[-b\boldsymbol{A}(x, u, \nabla u) \cdot \nabla u\, e^{-bu} + \boldsymbol{B}(x, u, \nabla u)\, e^{-bu} \right]^+ \varphi \, dx$$
$$\geq \int_\Omega \boldsymbol{A}(x, u, \nabla u)\, e^{-bu} \nabla \varphi \, dx$$
$$+ \int_\Omega \left[-b\boldsymbol{A}(x, u, \nabla u) \cdot \nabla u\, e^{-bu} + \boldsymbol{B}(x, u, \nabla u)\, e^{-bu} \right] \varphi \, dx \geq 0.$$

Set
$$\boldsymbol{A}'(x, \zeta, \xi) = e^{b(M-u(x))} \boldsymbol{A}(x, \zeta, \xi),$$
$$\boldsymbol{B}'(x, \zeta, \xi) = e^{b(M-u(x))} \left[\boldsymbol{B}(x, \zeta, \xi) - b\boldsymbol{A}(x, \zeta', \xi) \cdot \xi \right]^+,$$
where
$$\zeta' = \min\{\zeta^+, M\}.$$

Then
$$A'(x,\zeta,\xi)\cdot\xi \geq |\xi|^p - e^{bM}(c_2|\zeta|^p + c_3),$$
$$|A'(x,\zeta,\xi)| \leq e^{bM}\left(a_1|\xi|^{p-1} + a_2|\zeta|^{p-1} + a_3\right),$$
$$0 \leq B'(x,\zeta,\xi) \leq e^{b(M-u(x))}\left(b|\xi|^p + b_1|\xi|^{p-1} + b_2|\zeta|^{p-1} + b_3\right.$$
$$\left. - b|\xi|^p + c_2 b|\zeta'|^p + c_3 b\right)$$
$$\leq e^{bM}\left(b_1|\xi|^{p-1} + (b_2 + c_2 bM)|\zeta|^{p-1} + b_3 + c_3 b\right).$$

Similarly we prove the statement concerning subsolutions with
$$A''(x,\zeta,\xi) = e^{bu(x)}A(x,\zeta,\xi),$$
$$B''(x,\zeta,\xi) = -e^{bu(x)}\Big[B(x,\zeta,\xi) + bA(x,\zeta',\xi)\cdot\xi\Big]^{-}.$$

3.5 PROPOSITION. *Suppose that $b_0 = 0$. If (3.4) holds with "≤ 0" for all nonnegative $\varphi \in \mathcal{C}_c^\infty(\Omega)$, then u is a weak subsolution of (3.1).*

PROOF. From the assumptions it follows that
$$A(x,u,\nabla u) \in L_{\text{loc}}^{p'}(\Omega) \quad \text{and} \quad B(x,u,\nabla u) \in \big[W_0^{1,p}(\Omega')\big]^*$$
for each $\Omega' \subset\subset \Omega$. For example, if $\psi \in W_0^{1,p}(\Omega')$ and $\eta \in \mathcal{C}_c^\infty(\Omega)$ with $\chi_{\Omega'} \leq \eta \leq \chi_\Omega$, then
$$\int_{\Omega'} b_2|u|^{p-1}|\psi|\,dx$$
$$\leq \left(\int_\Omega b_2|u|^p\eta^p\,dx\right)^{(p-1)/p}\left(\int_{\Omega'} b_2|\psi|^p\,dx\right)^{1/p}$$
$$\leq C\|u\eta\|_{1,p;\Omega}^{(p-1)/p}\|\psi\|_{1,p;\Omega'}.$$
If $\varphi \in W_c^{1,p}(\Omega)$, then there is $\Omega' \subset\subset \Omega$ such that $\operatorname{spt}\varphi \subset \Omega'$. By mollifying we obtain a sequence $\varphi_j \in \mathcal{C}_c^\infty(\Omega)$ such that $\varphi_j \geq 0$, $\operatorname{spt}\varphi_j \subset \Omega'$ and $\varphi_j \to \varphi$ in $W_0^{1,p}(\Omega')$. By assumptions we have
$$\int_\Omega A(x,u,\nabla u)\cdot\nabla\varphi_j\,dx + \int_\Omega B(x,u,\nabla u)\cdot\varphi_j\,dx \geq 0$$
and we can pass to the limit to obtain the assertion. □

3.6 REMARK. Suppose that $|\Omega| < \infty$. If u is a weak subsolution in Ω and $u - f \in W_0^{1,p}(\Omega)$ for some given $f \in W^{1,p}(\mathbf{R}^n)$, it is easily seen that the argument above leads to the conclusion that (3.4) holds for all nonnegative $\varphi \in W_0^{1,p}(\Omega)$. This is true also for Ω p-finely open.

3.7 REMARK. Statements analogous to Proposition 3.5 and Remark 3.6 hold also for supersolutions and solutions.

3.8 THEOREM. *Let u be a weak subsolution of (3.1) with $b_0 = 0$ in Ω. Assume that $|\Omega| < \infty$ and $b_3 + a_3^{p'} + c_3 \in L^1(\Omega)$. Suppose that there is $\eta \in W_0^{1,p}(\Omega)$ such that $u \leq \eta$ in Ω. Then $u^+ \in W_0^{1,p}(\Omega)$. The domain Ω is allowed to be p-finely open.*

PROOF. Let η_j be a sequence of nonnegative functions from $W_c^{1,p}(\Omega)$ such that $\eta_j \uparrow \eta$ and
$$\|\eta_j - \eta\|_{1,p} \to 0.$$
(If Ω is p-finely open, we take $\eta_j \in W_{cc}^{1,p}(\Omega)$; for their construction we refer to Theorem 2.137 and Lemma 2.153.) Set
$$\varphi_j = (u + \eta_j - \eta)^+,$$
$$\Omega_j = \{\varphi_j > 0\}.$$
Then φ_j is a legitimate test function, because $\operatorname{spt} \varphi_j \subset \operatorname{spt} \eta_j$. We have

$$\int_{\Omega_j} |\nabla \varphi_j|^p \, dx \leq 2^{p-1} \int_{\Omega_j} \left(|\nabla u|^p + |\nabla(\eta_j - \eta)|^p \right) dx$$
$$\leq C + 2^{p-1} \int_{\Omega_j} \boldsymbol{A}(x, u, \nabla u) \cdot \nabla \varphi_j \, dx + 2^{p-1} \int_{\Omega_j} \boldsymbol{A}(x, u, \nabla u) \cdot \nabla(\eta - \eta_j) \, dx$$
$$+ 2^{p-1} \int_{\Omega_j} \left(c_2 |u|^p + c_3 \right) dx$$
$$\leq C - 2^{p-1} \int_{\Omega_j} \boldsymbol{B}(x, u, \nabla u) \varphi_j \, dx + 2^{p-1} \int_{\Omega_j} \boldsymbol{A}(x, u, \nabla u) \cdot \nabla(\eta - \eta_j) \, dx$$
$$+ 2^{p-1} \int_{\Omega_j} \left(c_2 \eta^p + c_3 \right) dx$$
$$\leq C + 2^{p-1} \int_{\Omega_j} \left(b_1 |\nabla u|^{p-1} + b_2 \eta^{p-1} + b_3 \right) \varphi_j \, dx$$
$$+ 2^{p-1} \int_{\Omega_j} \left(a_1 |\nabla u|^{p-1} + a_2 \eta^{p-1} + a_3 \right) |\nabla(\eta - \eta_j)| \, dx$$
$$+ 2^{p-1} \int_{\Omega_j} \left(c_2 \eta^p + c_3 \right) dx.$$

Choose $\delta > 0$. Using Young's inequality and Corollary 1.95 we obtain

$$\int_\Omega |\nabla \varphi_j|^p \, dx \leq C + C \int_{\Omega_j} \Big(\delta |\nabla u|^p + \delta^{1-p} b_1^p \varphi_j^p$$
$$+ (b_2 + c_2 + \delta a_2^{p'}) \eta^p + (b_2 + b_3) |\varphi_j|^p$$
$$+ b_3 + \delta a_3^{p'} + c_3 + \delta^{1-p} |\nabla \eta - \nabla \eta_j|^p \Big) dx$$
$$\leq C(\delta + \delta^{1-p}) + C \int_{\Omega_j} \left(\delta |\nabla \varphi_j|^p + \delta^{1-p/\epsilon} |\varphi_j|^p \right) dx.$$

Choosing δ small enough we obtain

$$\int_\Omega |\nabla \varphi_j|^p \, dx \leq C + C \int_\Omega |\varphi_j|^p \leq C + \int_\Omega |\eta|^p \, dx \leq C.$$

We see that the sequence $\{\varphi_j\}$ is bounded in $W_0^{1,p}(\Omega)$, and the only candidate for its weak limit is u^+. It follows that $u^+ \in W_0^{1,p}(\Omega)$. □

3.1.1 Upper bounds for weak solutions.

With the Sobolev inequality as the main tool, it is shown here that a weak subsolution is locally bounded above by its integral average. In case of a more restricted structure, an upper bound is obtained in terms of its boundary values.

3.9 THEOREM. *Let u be a weak subsolution of* (3.1) *defined in some ball $B(r) \subset \Omega$. Assume that $b_0 = 0$ in* (3.5). *Then for $0 < \sigma < 1$,*

$$\text{(3.12)} \qquad \sup_{B(\sigma r)} u^+ \leq \frac{C}{(1-\sigma)^{n/p}} \left[\left(\fint_{B(r)} (u^+)^p \, dx \right)^{1/p} + \boldsymbol{k}(r) \right]$$

where

$$C = C(p, n, a_1, \|a_2^{1/(p-1)}\|, r^\epsilon \|b_1^p\|, r^\epsilon \|b_2\|)$$

and $\boldsymbol{k}(r)$ is as in (3.7). *The norms are taken in the respective spaces* (3.6) *over $B(r)$. Similarly, if u is a weak supersolution, then* (3.12) *holds with u^+ replaced by u^-.*

PROOF. We will assume that u is a subsolution and first consider the case $r = 1$, $\sigma = 1/2$. Set
$$\bar{u} = u^+ + \boldsymbol{k}$$
where $\boldsymbol{k} = \boldsymbol{k}(1)$. For $q \geq 1$ and $l > k$, we define functions F and G by

$$\text{(3.13)} \qquad F(u) = \min\{\bar{u}^q, l^{q-1}\bar{u}\} \qquad G(u) = \min\{\bar{u}^\beta, l^{\beta-1}\bar{u}\} - k^\beta,$$

where q and β are related by $pq = p + \beta - 1$. Finally, with $\eta \in C_c^\infty(B(1))$, we define a test function
$$\varphi(x) = \eta(x)^p G(u(x)).$$
Employing φ in (3.4), the structure conditions (3.8) and the fact that $\nabla u = \nabla \bar{u}$ a.e. on spt φ yield

$$\int_{B(1)} \eta^p |\nabla v|^p \, dx \leq a_1 p \int_{B(1)} |v \nabla \eta| \, |\eta \nabla v|^{p-1} \, dx$$
$$+ q^{p-1} p \int_{B(1)} \bar{a} \, |v \nabla \eta| \, (\eta v)^{p-1} \, dx$$
$$+ \int_{B(1)} b_1 \, \eta v \, |\eta \nabla v|^{p-1} \, dx$$
$$+ (1+\beta) q^{p-1} \int_{B(1)} \bar{b} \, (\eta v)^p \, dx$$

where $v = v(x) = F(u)$. Young's inequality in the form
$$ab^{p-1} \leq \frac{1}{p} \delta^p a^p + \frac{1}{p'} \delta^{-p'} b^p, \quad \delta > 0,$$
along with the fact that $(1+\beta) q^{p-1} \leq (p+1) q^p$ leads to

$$\text{(3.14)} \qquad \int_{B(1)} \eta^p |\nabla v|^p \, dx \leq C(a_1, p) q^p \left(\int_{B(1)} |v \nabla \eta|^p \, dx \right.$$
$$\left. + \int_{B(1)} \bar{a} \, |v \nabla \eta| \, (\eta v)^{p-1} + \int_{B(1)} (\eta v)^p f \, dx \right)$$

where

(3.15) $$f = b_1^p + \bar{b}.$$

Choose $\delta > 0$. The term on the right involving $\bar{a}$ can be estimated as follows. We have by Hölder's and Young's inequality and the multiplier property

(3.16) $$\int_{B(1)} \bar{a} \, |v\nabla\eta| \, (\eta v)^{p-1} \, dx \leq \frac{\delta}{p'} \|\bar{a}^{1/(p-1)}\eta v\|_p^p + \frac{\delta^{1-p}}{p} \|v\nabla\eta\|_p$$
$$\leq \frac{\delta}{p'} \|\bar{a}^{1/(p-1)}\|_{M(p,p)}^p \|\nabla(\eta v)\|_p^p + \frac{\delta^{1-p}}{p} \|v\nabla\eta\|_p.$$

As for the third term on the right of (3.14), we use Corollary 1.95 and obtain

(3.17) $$\int_{B(1)} f(\eta v)^p \, dx \leq \delta \|\nabla(\eta v)\|_p^p + \delta^{1-p/\epsilon} \|\eta v\|_p^p.$$

Choosing $\delta > 0$ sufficiently small, reference to (3.14) leads to

$$\|\eta \nabla v\|_p \leq C q^{p/\epsilon} \|v(\eta + |\nabla\eta|)\|_p$$

where $C = C(p, n, a_1, \|a_2^{1/(p-1)}\|, \|b_1^p\|, \|b_2\|)$. This, along with the Sobolev's inequality, Corollary 1.57, implies

(3.18) $$\|\eta v\|_{\chi p} \leq C q^{p/\epsilon} \|v(\eta + |\nabla\eta|)\|_p$$

where $\chi \in (1, \infty)$, $(n-p)\chi \leq n$. Now let h and h' be real numbers satisfying $h' < h \leq 1$ and choose η so that $\eta = 1$ on $B(h')$, $0 \leq \eta \leq 1$ on $B(h)$ and identically zero outside $B(h)$. Then we obtain

$$\|v\|_{\chi p; B(h')} \leq C q^{p/\epsilon} (h - h')^{-1} \|v\|_{p; B(h)}.$$

Letting $l \to \infty$ in this inequality yields

$$\|\overline{u}^q\|_{\chi p; B(h')} \leq C q^{p/\epsilon} (h - h')^{-1} \|\overline{u}^q\|_{p; B(h)}.$$

This is valid irrespective of the finiteness of either norm. However, the right side is finite if $q = 1$ and thus by iteration, it is clear that $\|\overline{u}^q\|_p < \infty$ for any $q \geq 1$. With the notation

$$\alpha := pq = p + \beta - 1,$$

this can be written as

$$\|\overline{u}\|_{\chi\alpha; B(h')} \leq [C \cdot (\alpha/p)^{p/\epsilon} (h - h')^{-1}]^{p/\alpha} \|\overline{u}\|_{\alpha; B(h)}.$$

For each nonnegative integer i, let

$$\alpha_i = \chi^i p, \qquad h_i = 1/2 + 2^{-i-1}, \qquad h'_i = h_{i+1}$$

so that we obtain

$$\|\overline{u}\|_{\alpha_{i+1}; B(h_{i+1})} \leq C^{1/\chi^i} K^{i/\chi^i} \|\overline{u}\|_{\alpha_i; B(h_i)}$$

where $K = 2\chi^{p/\epsilon}$. Iteration of this inequality leads to

$$\|\overline{u}\|_{\infty; B(r/2)} \leq C^{\sum \frac{1}{\chi^i}} K^{\sum \frac{i}{\chi^i}} \|\overline{u}\|_{p; B(1)}.$$

Since $\bar{u} = u^+ + \boldsymbol{k}$, with a new constant C we have

$$\|u^+\|_{\infty;B(r/2)} \leq C\left[\|u^+\|_{p;B(1)} + \boldsymbol{k}\right].$$

The case of general r is obtained by a change of variables $x \mapsto rx$; see also Remark 3.2.

Now, given $\sigma \in (0, 1/2)$ we have

$$\begin{aligned}\|u^+\|_{\infty;B(\sigma r)} &\leq \|u^+\|_{\infty;B(r/2)} \\ &\leq C\, r^{-n/p}\left[\|u^+\|_{p;B(r)} + \boldsymbol{k}(r)\right] \\ &\leq C\, r^{-n/p}(1-\sigma)^{-n/p}\left[\|u^+\|_{p;B(r)} + \boldsymbol{k}(r)\right].\end{aligned}$$

If $1/2 < \sigma < 1$, for each ball $B(z, (1-\sigma)r) \subset B(\sigma r)$ we have

$$\begin{aligned}\|u\|_{\infty;B(z,(1-\sigma)r)} &\leq Cr^{-n/p}(1-\sigma)^{-n/p}\left[\|u^+\|_{p;B(z,2(1-\sigma)r)} + \boldsymbol{k}(r)\right] \\ &\leq Cr^{-n/p}(1-\sigma)^{-n/p}\left[\|u^+\|_{p;B(r)} + \boldsymbol{k}(r)\right],\end{aligned}$$

which implies

$$\|u\|_{\infty;B(\sigma r)} \leq Cr^{-n/p}(1-\sigma)^{-n/p}\left[\|u^+\|_{p;B(r)} + \boldsymbol{k}(r)\right].$$

The proof with u a supersolution proceeds in a similar way, after defining $\bar{u} = u^- + \boldsymbol{k}$. $\square$

3.10 COROLLARY. *With the same hypotheses as in the previous theorem, inequality (3.12) can be improved as follows: If u is a subsolution, for any $0 < q \leq p$ we have*

$$(3.19) \qquad \sup_{B(\sigma r)} u^+ \leq \frac{C}{(1-\sigma)^{n/q}}\left[\left(\fint_{B(r)} (u^+)^q\, dx\right)^{1/q} + \boldsymbol{k}(r)\right]$$

where C depends in addition on q.

Similarly, if u is a weak supersolution, then (3.19) holds with u^+ replaced by u^-.

PROOF. We write $v = u^+ + \boldsymbol{k}(r)$ is u is a subsolution and $v = u^- + \boldsymbol{k}(r)$ is u is a supersolution. Set

$$M(\sigma) = \sup_{B(\sigma r)} v,$$

$$S = \sup_{0 < \sigma < 1}\left\{(1-\sigma)^{\kappa}\left(\fint_{B(\sigma r)} v^p\, dx\right)^{1/p}\right\},$$

where $\kappa = \frac{n}{q} - \frac{n}{p}$. Given $\sigma \in (0,1)$, denote $\sigma' = (1+\sigma)/2$ and notice that $1 - \sigma' = (1-\sigma)/2$. By Theorem 3.9,

$$M(\sigma) \leq C(1-\sigma)^{-n/p}\left(\fint_{B(\sigma' r)} v^p\, dx\right)^{1/p} \leq CS(1-\sigma)^{-n/q}.$$

We have

$$(1-\sigma)^\kappa \left(\fint_{B(\sigma r)} v^p \, dx\right)^{1/p} \leq C(1-\sigma)^\kappa \left(M(\sigma)^{p-q} \fint_{B(\sigma r)} v^q \, dx\right)^{1/p}$$

$$\leq C(1-\sigma)^\kappa \left[S(1-\sigma)^{-n/q}\right]^{(p-q)/p} \left(\fint_{B(\sigma r)} v^q \, dx\right)^{1/p}$$

$$\leq CS^{(p-q)/p} \left(\fint_{B(r)} v^q \, dx\right)^{1/p}.$$

Passing to the supremum on the left-hand side we obtain

$$S \leq CS^{(p-q)/p} \left(\fint_{B(r)} v^q \, dx\right)^{1/p},$$

so that

$$S \leq C \left(\fint_{B(r)} v^q \, dx\right)^{1/q}.$$

Hence, for arbitrary $0 < \sigma < 1$,

(3.20)
$$(1-\sigma)^{-n/p} \left(\fint_{B(\sigma r)} v^p \, dx\right)^{1/p}$$
$$\leq C(1-\sigma)^{-n/q} \left(\fint_{B(r)} v^q \, dx\right)^{1/q}.$$

Combining (3.12) and (3.20) leads to the desired result (3.19). □

The previous results can be extended to obtain uniform estimates of the solution over the entire domain.

3.11 THEOREM. *Let u be a weak subsolution of (3.1) in Ω with $|\Omega| < \infty$ and suppose there is a real number L and $\eta \in W_0^{1,p}(\Omega)$ such that $u \leq L + \eta$. Assume (3.6) and that $b_0 = 0$. Suppose that $a_3^{p'} + b_1^p + b_2 + b_3 + c_3 \in L^1(\Omega)$. Then*

$$\sup_\Omega u \leq L + C'|L| + C\left[|\Omega|^{-1/p} \left(\int_\Omega ((u-L)^+)^p \, dx\right)^{1/p} + k\right]$$

where

$$C = C(p, n, \epsilon, \fint_\Omega f \, dx, |\Omega|^{\epsilon/n} \|b_1\|, |\Omega|^{\epsilon/n} \|b_2\|),$$
$$k = (|\Omega|^{\epsilon/n} \|b_3\|)^{1/(p-1)} + (|\Omega|^{\epsilon/n} \|c_3\|)^{1/p},$$
$$f := b_1^p + \bar{b},$$
$$\bar{b} = b_2 + k^{1-p} b_3 + k^{-p} c_3.$$

The constant C' has the same dependence as C. The domain Ω is allowed to be p-finely open.

3.1 BASIC PROPERTIES OF WEAK SOLUTIONS

PROOF. We will first assume that $L = 0$, thus by Theorem 3.8 implying $u^+ \in W_0^{1,p}(\Omega)$. We will also assume initially that $|\Omega| = 1$. Now define

$$\overline{u} = u^+ + \boldsymbol{k}$$

and let

$$F(u) = \min\{\overline{u}^q, l^{q-1}\overline{u}\} \qquad \varphi(x) = G(u(x)) = \min\{\overline{u}^\beta, l^{\beta-1}\overline{u}\} - \boldsymbol{k}^\beta,$$

where $q \geq 1$ is arbitrary and $pq = p+\beta-1$. Notice that $F(u)$ and $G(u)$ are as in the proof of Theorem 3.9. We have $\varphi \in W_0^{1,p}(\Omega)$, $\varphi \geq 0$, $\boldsymbol{A}(x,u,\nabla u)\chi_{\{\varphi>0\}} \in L^{p'}(\Omega)$ and $\boldsymbol{B}(x,u,\nabla u)\chi_{\{\varphi>0\}} \in [W^{1,p}(\Omega)]^*$. Similarly as in the proof of Theorem 3.8 we show that φ is an admissible test function in (3.4). Furthermore, note that on $\{\varphi > 0\}$ we have

$$|\boldsymbol{B}(x,u,\nabla u)| \leq b_1 |\nabla u|^{p-1} + \overline{b}|\overline{u}|^{p-1}$$
$$\boldsymbol{A}(x,u,\nabla u) \cdot \nabla u \geq |\nabla u|^p - \overline{b}|\overline{u}|^{p-1}.$$

With $v := F(u)$, we obtain

$$\int_\Omega |\nabla v|^p \, dx \leq \int_\Omega b_1 v |\nabla v|^{p-1} \, dx$$
$$+ (1+\beta)q^{p-1} \int_\Omega \overline{b} v^p \, dx,$$

which, by Young's inequality, leads to

(3.21) $$\int_\Omega |\nabla v|^p \, dx \leq Cq^p \left(\int_\Omega v^p f \, dx \right),$$

where $f := b_1^p + \overline{b} \in \mathcal{M}^{n/(p-\epsilon)}(\Omega)$. Since $v - \boldsymbol{k}^q \in W_0^{1,p}(\Omega)$, we can apply Sobolev's inequality and Corollary 1.95 to obtain

$$\|v - \boldsymbol{k}^q\|_{\chi p} \leq C\|\nabla v\|_p,$$
$$\int_\Omega f(v - \boldsymbol{k}^q)^p \leq C\delta \|\nabla v\|_p^p + C\delta^{1-p/\epsilon}\|v\|_p^p, \quad \delta > 0,$$

with $1 < \chi < n/(n-p)$. This along with (3.21) yields

$$\|v\|_{\chi p} \leq C\|\nabla v\|_p + \boldsymbol{k}^q|\Omega|^{1/\chi p}$$
$$\leq C\left[q^{p/\epsilon}\|v\|_p + q\boldsymbol{k}^q\left(\int_\Omega f \, dx\right)^{1/p} + \boldsymbol{k}^q|\Omega|^{1/\chi p}\right]$$
$$\leq Cq^{p/\epsilon}\|v\|_p + Cq\boldsymbol{k}^q$$
$$\leq Cq^{p/\epsilon}\|v\|_p$$

since $|\Omega| = 1$ and $\fint_\Omega f \, dx \leq C$. As in the proof of Theorem 3.9, let $l \to \infty$ and iterate the resulting inequality. Then, by replacing u by $(u - L)^+$ and rescaling, the desired conclusion is reached. $\square$

In the event of a more restrictive structure, we obtain a bound on the solution only in terms of its boundary values. We will need this result in the sequel.

3.12 THEOREM. *Let u be a weak subsolution of (3.1) in Ω with $|\Omega| < \infty$. Assume $b_2 = c_2 = 0$, b_1 is constant in (3.5) and $b_3, c_3 \in L^1(\Omega)$. Suppose that there is a real number L and $\eta \in W_0^{1,p}(\Omega)$ such that $u \leq L + \eta$. Let*

$$k := \left(|\Omega|^{(p-n)/n}\|b_3\|_{1;\Omega}\right)^{1/(p-1)} + \left(|\Omega|^{(p-n)/n}\|c_3\|_{1;\Omega}\right)^{1/p}$$
$$+ \left(|\Omega|^{\epsilon/n}\|b_3\|_{\mathcal{M}^{n/(p-\epsilon)}(\Omega)}\right)^{1/(p-1)} + \left(|\Omega|^{\epsilon/n}\|c_3\|_{\mathcal{M}^{n/(p-\epsilon)}(\Omega)}\right)^{1/p}.$$

Then we have

$$\sup_\Omega u \leq Ck + L$$

where $C = C(n, p, b_1)$. In case u is a solution, u can be replaced by $|u|$ in the above inequality. The domain Ω is allowed to be p-finely open.

PROOF. We will assume initially that $u \leq 0$ on $\partial\Omega$ and that $|\Omega| = 1$. We will also assume without loss of generality that $k > 0$, for if not we can let k tend to zero. By Theorem 3.8, $u^+ \in W_0^{1,p}(\Omega)$. Let

$$\overline{u} = u^+ + k$$
$$M = \sup_\Omega \overline{u}$$
$$w = \log\left(\frac{M}{M - \overline{u} + k}\right)$$
$$v = 1 + w^q, \quad q \geq (p-1)/p,$$
$$\varphi = \frac{w^\beta}{[M - \overline{u} + k]^{p-1}} - \frac{\ell}{M^{p-1}}, \quad \beta = pq - p + 1,$$

where

$$\ell = \begin{cases} 1, & \beta = 0, \\ 0, & \beta > 0. \end{cases}$$

Taking into account that u is bounded above, by Theorem 3.11 we observe that $\varphi \in W_0^{1,p}(\Omega)$. Notice that

$$\nabla v = \frac{qw^{q-1}}{M - \overline{u} + k}\nabla u$$
$$\nabla \varphi = \frac{(\beta w^{pq-p} + (p-1)w^\beta)\nabla u}{(M - \overline{u} + k)^p}.$$

Using φ as a test function in (3.4), we obtain

(3.22)
$$\int_\Omega \frac{\beta w^{pq-p} + (p-1)w^\beta}{(M - \overline{u} + k)^p} \left(|\nabla u|^p - c_3\right) dx$$
$$\leq \int_\Omega A(x, u, \nabla u) \cdot \nabla \varphi \, dx$$
$$\leq -\int_\Omega B(x, u, \nabla u)\varphi \, dx$$
$$\leq \int_\Omega \frac{w^\beta}{[M - \overline{u} + k]^{p-1}} \left(b_1|\nabla u|^{p-1} + b_3\right) dx.$$

3.1 BASIC PROPERTIES OF WEAK SOLUTIONS

If $\beta = 0$, we deduce

$$(p-1)\int_\Omega \frac{|\nabla u|^p}{(M-\overline{u}+k)^p}\,dx$$
$$\leq b_1 \int_\Omega \frac{|\nabla u|^{p-1}}{(M-\overline{u}+k)^{p-1}}\,dx$$
$$+ (p-1)\int_\Omega \frac{c_3}{(M-\overline{u}+k)^p}\,dx$$
$$+ \int_\Omega \frac{b_3}{(M-\overline{u}+k)^{p-1}}\,dx$$
$$\leq b_1 \int_\Omega \frac{|\nabla u|^{p-1}}{(M-\overline{u}+k)^{p-1}}\,dx + p.$$

Using Young's inequality, we then obtain

(3.23) $$\int_\Omega |\nabla w|^p\,dx \leq C(n, p, b_1).$$

If $\beta \geq 1$, we have

$$\beta w^{\beta-1} + (p-1)w^\beta \leq pq\,v^p,$$
$$w \leq v$$

and (3.22) yields

$$\beta q^{-p}\int_\Omega |\nabla v|^p\,dx = \beta \int_\Omega \frac{w^{pq-p}|\nabla u|^p}{(M-\overline{u}+k)^p}\,dx$$
$$\leq b_1 \int_\Omega \frac{w^\beta |\nabla u|^{p-1}}{[M-\overline{u}+k]^{p-1}}\,dx$$
$$+ \int_\Omega \frac{c_3[\beta w^{\beta-1} + (p-1)w^\beta]}{(M-\overline{u}+k)^p}\,dx$$
$$+ \int_\Omega \frac{b_3 w^\beta}{(M-\overline{u}+k)^{p-1}}\,dx$$
$$\leq b_1 q^{1-p}\int_\Omega v|\nabla v|^{p-1}\,dx$$
$$+ \int_\Omega pq(k^{-p}c_3 + k^{1-p}b_3)v^p\,dx.$$

It follows that

$$\int_\Omega |\nabla v|^p\,dx \leq Cq^p \int_\Omega g\,(w^{pq}+1)\,dx$$

with

$$\|g\|_{\mathcal{M}^{n/(p-\epsilon)}} \leq C, \qquad \|g\|_1 \leq C,$$

and thus by Corollary 1.95 we obtain

(3.24) $$\int_\Omega |\nabla v|^p\,dx \leq Cq^p + Cq^{1/\epsilon}\int_\Omega w^{pq}\,dx.$$

Let $\chi \in [1, n/(n-p))$. By Sobolev's inequality we have

$$\|1+w\|_{\chi pq}^q \leq C(1 + \|w^q\|_{\chi p}) \leq C(1 + \|\nabla v\|_p).$$

We start the iteration with

$$\|1+w\|_{\chi p} \leq 1 + C\|\nabla w\|_p \leq C$$

which holds by (3.23). Then we continue with $q = \chi^i p$, $i = 1, 2, \ldots$. By (3.24), we have

$$\|1+w\|_{\chi pq} \leq C(1 + q^{p/\epsilon}\|w^q + 1\|_p)^{1/q} \leq Cq^{p/(\epsilon q)}\|1+w\|_{pq}.$$

Iterating this inequality we obtain that

$$\sup w \leq C.$$

Hence, $M \leq Ck$, which establishes our conclusion when $L = 0$. The general case follows by replacing u^+ by $(u-L)^+$ and rescaling. □

3.1.2 Weak Harnack inequality.

The weak Harnack inequality, which states roughly that the infimum of a nonnegative weak supersolution is bounded above by its integral average, is one of the main tools in our subsequent analysis. The essential ingredients in its proof is the use of the Sobolev and John-Nirenberg inequalities in Moser's iteration scheme.

At the end of this subsection we are at a stage that brings us close to obtaining interior Hölder estimates for the solution. We postpone this to Chapter 4, because the version of Harnack's inequality obtained there is better suited for this purpose.

3.13 THEOREM. *Let u be a weak supersolution of (3.1) defined in Ω. Assume $0 \leq u \leq M \leq \infty$ in some ball $B(r) \subset \Omega$, where $M = \infty$ is allowed if $b_0 = 0$. Then for any $\sigma, \tau \in (0,1)$ and $\gamma \in \big(0, n(p-1)/(n-p)\big)$, there is a constant C such that*

$$(3.25) \qquad \left(\fint_{B(\sigma r)} u^\gamma \, dx\right)^{1/\gamma} \leq C\big[\inf_{B(\tau r)} u + k(r)\big].$$

Here

$$C = C(p, n, \gamma, \sigma, \tau, \epsilon, b_0 M, a_1, \|a_2^{1/(p-1)}\|, r^\epsilon \|b_1\|, r^\epsilon \|b_2\|)$$

where the norms are taken relative to $B(r)$. In case $p = n$, (3.25) holds for any $\gamma > 0$.

PROOF. It suffices to consider the case $r = 1$ since rescaling will yield the general result. We will assume throughout that $u \geq \varepsilon > 0$ and then let $\varepsilon \to 0$ to establish the general result. Let $G\colon (0, \infty) \to \mathbf{R}$ be a smooth nonincreasing function and let $\eta \in C_c^\infty(B(1))$. Now set $\bar{u} = u + k$, and define a test function φ as

$$\varphi = G(u)\,\eta^p.$$

Then

$$\nabla\varphi = G'(u)\,\nabla u\,\eta^p + p\,G(u)\,\eta^{p-1}\nabla\eta$$

and we obtain

$$\int_{B(1)} \boldsymbol{A}(x,u,\nabla u) \cdot \nabla u \, G'(u) \, \eta^p \, dx$$
(3.26)
$$+ p \int_{B(1)} \boldsymbol{A}(x,u,\nabla u) \cdot \nabla \eta \, G(u) \, \eta^{p-1} \, dx$$
$$+ \int_{B(1)} \boldsymbol{B}(x,u,\nabla u) \, G(u) \, \eta^p \, dx$$
$$\geq 0.$$

Taking the structure into account, we get

(3.27)
$$-\int_{B(1)} \boldsymbol{A}(x,u,\nabla u) \cdot \nabla u \, G'(u) \, \eta^p \, dx$$
$$\geq \int_{B(1)} \left(|\nabla u|^p - \overline{b}\overline{u}^p \right) |G'(u)| \, \eta^p \, dx ,$$

(3.28)
$$p \int_{B(1)} \boldsymbol{A}(x,u,\nabla u) \cdot \nabla \eta \, G(u) \, \eta^{p-1} \, dx$$
$$\leq p \int_{B(1)} \left(a_1 |\nabla u|^{p-1} + \overline{a}\overline{u}^{p-1} \right) G(u) \, \eta^{p-1} |\nabla \eta| \, dx ,$$

and

(3.29)
$$\int_{B(1)} \boldsymbol{B}(x,u,\nabla u) \, G(u) \, \eta^p \, dx$$
$$\leq \int_{B(1)} \left(b_1 |\nabla u|^{p-1} + \overline{b}\overline{u}^{p-1} \right) G(u) \, \eta^p \, dx.$$

From (3.26)–(3.29) we obtain

$$\int_{B(1)} |G'(u)| \, |\nabla u|^p \, \eta^p \, dx$$
$$\leq \int_{B(1)} \overline{u}^p \, |G'(u)| \, \eta^p \overline{b} \, dx$$
(3.30)
$$+ \int_{B(1)} \overline{u}^{p-1} \, G(u) \eta^{p-1} (\overline{b}\eta + p\overline{a} |\nabla \eta|) \, dx$$
$$+ \int_{B(1)} |\nabla u|^{p-1} \, G(u) \eta^{p-1} (p a_1 |\nabla \eta| + b_1 \eta) \, dx.$$

Set

$$G(u) := \overline{u}^\beta, \quad \beta < 0$$

and obtain

$$
(3.31) \quad \begin{aligned}
|\beta| \int_{B(1)} &|\nabla u|^p \, \overline{u}^{\beta-1} \eta^p \, dx \\
&\leq |\beta| \int_{B(1)} \overline{b} \overline{u}^{\beta-1+p} \eta^p \\
&\quad + \int_{B(1)} \overline{u}^{\beta-1+p} \eta^{p-1} (\overline{b} \eta + p \overline{a} |\nabla \eta|) \\
&\quad + \int_{B(1)} |\nabla u|^{p-1} \overline{u}^\beta \eta^{p-1} (p a_1 |\nabla \eta| + b_1 \eta).
\end{aligned}
$$

Now define

$$
v := \begin{cases} \overline{u}^q & \text{where } pq = p + \beta - 1, \ \beta \neq 1 - p \\ \log \overline{u} & \text{if } \beta = 1 - p. \end{cases}
$$

If $\beta \neq 1 - p$, we use Young's inequality to obtain

$$
\begin{aligned}
\int_{B(1)} \eta^p |\nabla v|^p \, dx &\leq C |q|^{p^2} (1 + |\beta|^{-1})^p \int_{B(1)} |\nabla \eta|^p \, v^p \, dx \\
&\quad + C |q|^p (1 + |\beta|^{-p}) \int_{B(1)} f \eta^p v^p \, dx
\end{aligned}
$$

where

$$
f = b_1^p + \overline{b} \quad \text{and} \quad C = C(p, n, a_1, \|a_2^{1/(p-1)}\|, b_0 M).
$$

We apply Corollary 1.95 as in (3.17) to obtain

$$(3.32) \quad \|\eta \nabla v\|_p \leq C |q|^{p/\epsilon} (1 + |\beta|^{-p})^{1/\epsilon} \|(\eta + |\nabla \eta|) v\|_p.$$

Also, when $\beta = 1 - p$ we have

$$\int_{B(1)} \eta^p |\nabla v|^p \, dx \leq C \int_{B(1)} (f \eta^p + |\nabla \eta|^p) \, dx$$

and by Corollary 1.95, this yields

$$\int_{B(1)} \eta^p |\nabla v|^p \, dx \leq C \int_{B(1)} (\eta^p + |\nabla \eta|^p) \, dx.$$

Thus, we have

$$(3.33) \quad \|\eta \nabla v\|_p \leq \begin{cases} C |q|^{p/\epsilon} (1 + |\beta|^{-p})^{1/\epsilon} \|(\eta + |\nabla \eta|) v\|_p & \text{for } \beta \neq 1 - p \\ C \|\eta + |\nabla \eta|\|_p & \text{for } \beta = 1 - p \end{cases}$$

where $C = C(p, n, \epsilon, a_1, \|a_2^{1/(p-1)}\|, b_0 M)$.

If $\beta \neq 1 - p$, $\beta < 0$, Sobolev's inequality yields

$$(3.34) \quad \|\eta v\|_{\chi p} \leq C |q|^{p/\epsilon} (1 + |\beta|^{-p})^{1/\epsilon} \|(\eta + |\nabla \eta|) v\|_p,$$

where $p^*/p \geq \chi > 1$ for $p < n$ and $\chi > 1$ for $p = n$. Also, we will need later that

$$(3.35) \quad \gamma/\chi < p - 1$$

which is a restriction on γ. Let η be a cutoff function such that $\eta \equiv 1$ on $B(h')$ and identically zero on the complement of $B(h)$, then

$$\|v\|_{\chi p, B(h')} \leq C |q|^{p/\epsilon}(1+|\beta|^{-p})^{1/\epsilon}(h-h')^{-1}\|v\|_{p,B(h)}.$$

With $\alpha := pq = p + \beta - 1$ and $\alpha > 0$, this can be written as

(3.36) $$\|\bar{u}\|_{\chi\alpha, B(h')} \leq [C\alpha^{p/\epsilon}(1+|\beta|^{-p})^{1/\epsilon}(h-h')^{-1}]^{p/\alpha}\|\bar{u}\|_{\alpha, B(h)}$$

while

(3.37) $$\|\bar{u}\|_{\chi\alpha, B(h')} \geq [C(-\alpha)^{p/\epsilon}(h-h')^{-1}]^{p/\alpha}\|\bar{u}\|_{\alpha, B(h)}$$

if $\alpha < 0$. In this latter case note that $|\beta|^{-1} < (p-1)^{-1}$, and that, for notational convenience, we will extend the usual notation for $\|u\|_p$ to include negative values of p.

These inequalities will be iterated. Let $\rho \in (0,1)$ be any number such that $\max(\sigma, \tau) < \rho$ and let a positive integer j be fixed. We set $\alpha_i = \chi^{i-j-1}\gamma$, $i = 0, \ldots, j+1$. Then for any $i = 0, \ldots, j$, the corresponding value of $|\beta|$ will be lower bounded by $p - 1 - \gamma/\chi$ providing by (3.35) an upper bound on the constant $(1+|\beta|^{-p})$ that appears in (3.36). Define

$$h_i = \sigma + 2^{-i}(\rho - \sigma), \quad h'_i = h_{i+1}, \quad i = 0 \ldots j.$$

Hence, from (3.36) we have

$$\|\bar{u}\|_{\gamma, B(\sigma)} \leq C \|\bar{u}\|_{\alpha_0, B(\rho)}$$

for any $\alpha_0 \in \{\chi^{-j-1}\gamma: j = 1, 2, \ldots\}$. An easy application of Hölder's inequality extends the estimate to any $\alpha_0 > 0$.

Iteration of (3.37) for $\alpha < 0$ will yield a lower bound for $\inf \bar{u}$. For this, let

$$h_i = \rho + 2^{-i}(\tau - \rho), \quad h'_i = h_{i+1}, \quad i = 0, 1, 2 \ldots,$$

and obtain

$$\|\bar{u}\|_{-\alpha_0, B(\rho)} \leq C \|\bar{u}\|_{-\infty, B(\tau)}.$$

The proof is concluded by showing the existence of $\alpha_0 > 0$ such that

$$\|\bar{u}\|_{\alpha_0, B(\rho)} \leq C \|\bar{u}\|_{-\alpha_0, B(\rho)}.$$

Referring to the case $\beta = 1 - p$, $v = \ln u$, in (3.33), and to Theorem 1.66, there exists $\alpha_0 > 0$ such that

$$\int_{B(\rho)} e^{\alpha_0 |v - v_0|} \leq C$$

where

$$v_0 = \fint_{B(\rho)} v \, dx.$$

Thus,

$$\int_{B(\rho)} e^{\alpha_0 v} \, dx \int_{B(\rho)} e^{-\alpha_0 v} \, dx \leq C e^{\alpha_0 v_0} e^{-\alpha_p v_0} = C.$$

The final result is obtained by rescaling with $x \mapsto xr$. $\square$

By Lemma 3.4, Corollary 3.10 also holds when $b_0 > 0$ and $0 \leq u \leq M$. Combining Theorem 3.13 and Corollary 3.10 we obtain Harnack's inequality.

3.14 THEOREM. *Let u be a weak solution of (3.1) in some ball $B(r) \subset \Omega$ and $\sigma \in (0,1)$. Suppose that $0 \leq u \leq M \leq \infty$, where $M = \infty$ is allowed if $b_0 = 0$. Then*

$$\sup_{B(\sigma r)} \leq C\Big[\inf_{B(\sigma r)} u + k(r)\Big]$$

with C and $k(r)$ as above.

3.1.3 Removable sets for weak solutions.

In this section we will prove that sets of capacity zero are removable for solutions of (3.1). This will provide some motivation for the proof of the Wiener condition for regularity at the boundary.

Notice that removing a p-polar set from an open set Ω results in a p-finely open set.

We will suppose that $b_0 = 0$ in (3.5).

3.15 THEOREM. *Let $\Omega' \subset \Omega$ be a p-finely open set and u be a lower bounded weak subsolution of (3.1) in Ω'. Suppose that $\mathbf{C}_s(\Omega \setminus \Omega') = 0$, $p \leq s \leq n$. If, for some $\varepsilon > 0$,*

$$u \in L^{\frac{s(p-1+\varepsilon)}{s-p}}(\Omega),$$

then u is a weak subsolution in all of Ω. In particular, if $\mathbf{C}_p(\Omega \setminus \Omega') = 0$, then $\Omega \setminus \Omega'$ is removable for any bounded subsolution of (3.1).

PROOF. We will proceed as in the proof of Theorem 2.116. It is sufficient to prove the theorem for a ball B centered at any point of Ω. Assume that $B = B(1) = \Omega$, $u \geq 0$ and $\varepsilon < 1$. Define

$$\overline{u} = u + k \quad \text{where } k := 1 + k(1).$$

Denote $E = \Omega \setminus \Omega'$. Let $\{\varphi_j\}$ be a sequence of functions from $W^{1,s}(\mathbf{R}^n)$ such that $0 \leq \varphi_j \leq 1$, $\{\varphi_j = 1\}^\circ \supset E$ and $\|\varphi_j\|_{1,s} \to 0$. Let $\eta \in C_c^\infty(\Omega)$ be nonnegative. Each

$$\eta_j := (1 - \varphi_j)\eta$$

belongs to $W_c^{1,s}(\Omega')$. Let $\beta \geq \varepsilon$ and q, γ are determined by $\beta = pq - p + 1$, $\varepsilon = \gamma p - p + 1$. For $l > k$, let

$$v_l = \min\{\overline{u}^q, l^{q-\gamma}\overline{u}^\gamma\},$$
$$w_l = \min\{\overline{u}^\beta, l^{\beta-\varepsilon}\overline{u}^\varepsilon\} - k^\beta.$$

Using $w_l \, \eta_j^p$ as a test function we obtain (as in the proof of Theorem 3.9),

$$\|\eta_j \nabla v_l\|_p \leq C(q+1)^{p/\varepsilon} \|v_l\,(\eta_j + |\nabla \eta_j|)\|_p.$$

Since, by Young's inequality

(3.38)
$$|v_l^p|\,(\eta^p + |\nabla \eta_j|^p) \leq C|u^{\varepsilon+p-1}|\,(\eta^p + |\nabla \eta_j|^p)$$
$$\leq C u^{\frac{s(p-1+\varepsilon)}{s-p}} + C(\eta^p + |\nabla \eta_j|^p)^{s/p}, \quad s > p,$$

(or, if $s = p$, the integrability is transparent without any effort), the sequence $\{v_l \eta_j\}_j$ is bounded in $W_0^{1,p}(\Omega)$ and $v_l \eta$ as the only candidate for the weak limit belongs to $W_0^{1,p}(\Omega)$ with

(3.39)
$$\|\eta \nabla v_l\|_p \leq C(q+1)^{p/\varepsilon} \|v_l\,(\eta + |\nabla \eta|)\|_p.$$

This, along with the Sobolev's inequality, Corollary 1.57, implies

(3.40) $$\|\eta v_l\|_{\chi p} \leq C(q+1)^{p/\epsilon} \|v_l(\eta + |\nabla \eta|)\|_p$$

where $\chi \in (1, \infty)$, $(n-p)\chi \leq n$. With a special choice of η for the while we obtain

$$\|v_l\|_{\chi p; B(h')} \leq C(q+1)^{p/\epsilon}(h-h')^{-1} \|v_l\|_{p; B(h)}.$$

Letting $l \to \infty$ in this inequality yields

$$\|\overline{u}^q\|_{\chi p; B(h')} \leq C(q+1)^{p/\epsilon}(h-h')^{-1} \|\overline{u}^q\|_{p; B(h)}.$$

This is valid irrespective of the finiteness of either norm. However, the right side is finite if $q = \gamma$ and thus by iteration, it is clear that $\|\overline{u}^q\|_p < \infty$ for any $q \geq \gamma$. As in the proof of Theorem 3.9 we conclude that u is locally bounded. Letting $q = 1$ and $l \to \infty$ in (3.39) we see that $u \in W^{1,p}_{\text{loc}}(\Omega)$. Since $\eta_j \to \eta$ in $W^{1,p}(\Omega)$ and spt $\eta_j \subset$ spt η, we can pass to the limit in

$$\int_\Omega \Big(A(x, u, \nabla u) \cdot \nabla \eta_j + B(x, u, \nabla u)\, \eta_j \Big) dx \leq 0$$

and verify (3.4). Thus, u is a weak subsolution of (3.1). □

3.16 THEOREM. *Let $\Omega' \subset \Omega$ be a p-finely open set and $u \in W^{1,p}(\Omega')$ be a weak solution of (3.1) in Ω'. Suppose that $\mathbf{C}_s(\Omega \setminus \Omega') = 0$, $p \leq s \leq n$. If, for some $\varepsilon > 0$,*

$$u \in L^{\frac{s(p-1+\varepsilon)}{s-p}}(\Omega),$$

then u is a weak solution in all of Ω. In particular, if $\mathbf{C}_p(\Omega \setminus \Omega') = 0$, then $\Omega \setminus \Omega'$ is removable for any bounded solution of (3.1).

PROOF. Similarly, as in the preceding proof, we obtain estimates for u^+ and u^- to show that u is locally bounded. Then we may use the preceding theorem to conclude the proof. □

3.17 REMARK. We will illustrate the sharpness of the previous result, at least for integer values of s. Let E denote an $(n-s)$-dimensional plane in $\mathbf{R}^n$ and observe that $\mathbf{C}_s(E) = 0$, cf. Theorem 2.52. It is easily verified that

(3.41) $$u = \begin{cases} \rho^{(p-s)/(p-1)} & p < s \\ \log \rho & p = s, \end{cases}$$

where ρ denotes the distance to E, is a weak solution of $\text{div}(|\nabla u|^{p-2} \nabla u) = 0$ outside E. If $p < s$, then $u \in L^{\frac{s(p-1-\varepsilon)}{s-p}}$ for any $\varepsilon \in (0, p-1)$ whereas $u \in L^q$ for any $q < \infty$ if $p = s$. Nevertheless, u is not a solution in $\mathbf{R}^n$, as $|\nabla u|^p$ is not locally integrable.

3.2 Higher regularity of equations with differentiable structure

In this section some regularity results concerning equations with differentiable structure will be discussed. The main thrust of this book is the treatment of equations with non-smooth structure, and therefore no attempt is made to give a comprehensive survey with proofs relating to higher regularity. Rather, we will state only those results that will be needed later in Chapters 5 and 6.

We will consider equations of type

(3.42) $$\text{div}\, \boldsymbol{A}(x,u,\nabla u) = \boldsymbol{B}(x,u,\nabla u)$$

where $\boldsymbol{A} = (\boldsymbol{A}^1, \boldsymbol{A}^2, \ldots, \boldsymbol{A}^n)$ and $\boldsymbol{B}$ satisfy $\boldsymbol{A}^i \in \mathcal{C}^{1,\gamma}(\overline{\Omega} \times \mathbf{R} \times \mathbf{R}^n)$ and $\boldsymbol{B} \in \mathcal{C}^{0,\gamma}(\overline{\Omega} \times \mathbf{R} \times \mathbf{R}^n)$, $0 < \gamma < 1$. The equations with which we will be concerned are those that are not necessarily uniformly elliptic; indeed, we will focus on those that lose ellipticity when the gradient of the solution vanishes. Thus, with

$$a^{ij}(x,\zeta,\xi) := \frac{\partial \boldsymbol{A}^i}{\partial \xi_j}(x,\zeta,\xi)$$

we assume

(3.43)
$$\sum_{i,j=1}^n a^{ij}(x,\zeta,\xi) h_i h_j \geq \lambda |\xi|^{p-2} |h|^2 \quad \text{for all} \quad h \in \mathbf{R}^n,$$

$$|a^{ij}(x,\zeta,\xi)| \leq \Lambda |\xi|^{p-2},\, i,j = 1,2,\ldots,n,$$

$$\left|\frac{\partial \boldsymbol{A}}{\partial \zeta}(x,\zeta,\xi)\right|, |\nabla \boldsymbol{A}(\cdot,\zeta,\xi)| \leq \Lambda |\xi|^{p-1},$$

$$|\boldsymbol{B}(x,\zeta,\xi)| \leq \Lambda(1 + |\xi|^p)$$

where $0 < \lambda \leq \Lambda < \infty$. Whenever we refer to a weak solution of (3.42), it is assumed to be relative to (3.43).

These equations are degenerate elliptic and the model equations

$$\text{div}(|\nabla u|^{p-2} \nabla u) = 0, \quad \text{div}(|\nabla u|^{p-2} \nabla u) = |\nabla u|^p$$

serve as prototypes.

As we shall see, solutions to these equations are $\mathcal{C}^{1,\alpha}_{\text{loc}}$. In addition, all of the results in Chapters 3 and 4 also apply as results from the following observation.

3.18 LEMMA. *If $u \in W^{1,p}(\Omega)$ is a weak solution with structure (3.43), then it is also a weak solution with structure (3.5), p.162.*

PROOF. Using the first condition of (3.43), we have

$$\sum_{i=1}^n [\boldsymbol{A}^i(x,u,\nabla u) - \boldsymbol{A}^i(x,u,0)] D_i u = \sum_{i=1}^n \left[\int_0^1 \frac{\partial}{\partial t} \boldsymbol{A}^i(x,u,t\nabla u)\, dt\right] D_i u$$

$$= \sum_{i,j=1}^n \int_0^1 \frac{\partial}{\partial \xi_j} \boldsymbol{A}^i(x,u,t\nabla u)\, D_i u D_j u\, dt$$

$$\geq \int_0^1 \lambda t^{p-2} |\nabla u|^p\, dt$$

$$\geq \frac{\lambda}{p-1} |\nabla u|^p.$$

Since by our assumptions $A(\cdot, \cdot, 0)$ is constant on each component and A is differentiated in (3.42), we may assume that $A(\cdot, \cdot, 0) \equiv 0$. Thus the above calculation establishes the third (ellipticity) condition of (3.5), p. 162.

Also, the second condition of (3.43) implies

$$|A^i(x, u, \nabla u) - A^i(x, u, 0)| \leq \int_0^1 |\sum_{i=1}^n \frac{\partial}{\partial \xi_j} A^i(x, u, t\nabla u) D_j u| \, dt$$

$$\leq \Lambda \int_0^1 t^{p-2} |\nabla u|^{p-1} \, dt$$

$$\leq \frac{\Lambda}{p-1} |\nabla u|^{p-1}. \qquad \square$$

Several authors contributed to the development of higher regularity of degenerate elliptic equations. Ural'tseva [**Ur**] and Uhlenbeck [**Uh**] were among the first to obtain $C^{1,\alpha}_{\text{loc}}$ results for homogeneous equations and systems. Then Evans [**Ev**] contributed a relatively simple and short proof for solutions of the p-Laplacian in the case $p \geq 2$. Lewis [**Lew2**] extended Evans' result to include $p > 1$. Later, DiBenedetto [**DiB**], Tolksdorf [**To**], and Manfredi [**Man1**] obtained higher regularity for equations with the general structure (3.43). A result concerning higher integrability of the gradient for solutions of systems was obtained in [**DM**]. In the planar case $n = 2$, it was shown by Iwaniec and Manfredi [**IM**] that the p-Laplacian exhibits an unusual type of higher regularity. Indeed, they have shown that a solution belongs to $C^{k,\alpha}_{\text{loc}}(\Omega)$ where $k + \alpha \to \infty$ as $p \downarrow 1$.

Later in Chapter 5, we will need the following special consequence of the regularity theory, cf. [**DiB**] or [**Man1**].

3.19 THEOREM. *Let $\overline{B}(x_0, R) \subset \Omega$. Then, for some $0 < \alpha < 1$, the solution of* $\operatorname{div} A(\nabla u) = 0$ *in $B(R)$ satisfies*

$$(3.44) \qquad \sup_{B(x_0, R/2)} |\nabla v|^p \leq C \fint_{B(x_0, R)} |\nabla v|^p \, dx$$

$$(3.45) \qquad \operatorname*{osc}_{B(x_0, r)} |\nabla v| \leq C \sup_{B(x_0, R/2)} |\nabla v| \left(\frac{r}{R}\right)^\alpha$$

for all $r \leq R/4$. Here $C = C(n, p, \Lambda/\lambda)$ and $0 < \alpha = \alpha(n, p, \Lambda/\lambda) < 1$.

In Chapter 6 we will need a standard regularization process to transform (3.43) into a uniformly elliptic structure. Thus, for each open set $\Omega' \subset\subset \Omega$, we employ the usual regularization methods to obtain C^2 functions A_ε and B_ε that converge in the $C^{1,\gamma}$-norm to A and B on $\Omega' \times \mathbf{R} \times \mathbf{R}^n$. Consequently, we obtain the following structure conditions for $\varepsilon > 0$ sufficiently small and $(x, \zeta, \xi) \in \Omega' \times \mathbf{R} \times \mathbf{R}^n$:

$$\sum_{i,j=1}^n a_\varepsilon^{ij}(x, \zeta, \xi) h_i h_j \geq \lambda/2 [\varepsilon + |\xi|^2]^{(p-2)/2} |h|^2 \quad \text{for all} \quad h \in \mathbf{R}^n$$

$$(3.46) \qquad |a_\varepsilon^{ij}(x, \zeta, \xi)| \leq 2\Lambda [\varepsilon + |\xi|^2]^{(p-2)/2} |h|^2, \quad i, j = 1, 2, \ldots, n$$

$$|\frac{\partial A_\varepsilon}{\partial \zeta}(x, \zeta, \xi)|, |\nabla A_\varepsilon(\cdot, \zeta, \xi)| \leq 2\Lambda [\varepsilon + |\xi|^2]^{(p-1)/2}$$

$$|B_\varepsilon(x, \zeta, \xi)| \leq 2\Lambda [1 + |\xi|^2]^{p/2}.$$

Classical regularity theory states that solutions u_ε of (3.46) are $\mathcal{C}^{2,\alpha}_{\text{loc}}$ where α depends on ε. However, the regularity theory cited immediately above yields $\mathcal{C}^{1,\alpha}_{\text{loc}}$ regularity where α is now independent of ε. In Chapter 6, we will use the following result of Lieberman [**Li1**] which gives boundary regularity for degenerate equations. We denote the norm in $\mathcal{C}^{1,\beta}(\overline{\Omega})$ as $\|\ldots\|_{1+\beta}$.

3.20 THEOREM. *Let u be a bounded weak solution ($|u| \leq M_0$) of*

(3.47)
$$\operatorname{div} \boldsymbol{A}_\varepsilon = \boldsymbol{B}_\varepsilon$$

with structure (3.46) on a bounded domain Ω with $\mathcal{C}^{1,\alpha}$ boundary. Let $u = f$ on $\partial\Omega$ where $f \in \mathcal{C}^{1,\alpha}(\partial\Omega)$ with $\|f\|_{1+\alpha} \leq M_1$. Then there exists a positive number $\beta = \beta(\alpha, \Lambda/\lambda, p, n)$ such that $u \in \mathcal{C}^{1,\beta}(\overline{\Omega})$ with

$$\|u\|_{1+\beta} \leq C(\alpha, \gamma, \Lambda/\lambda, M_0, M_1, p, n, \Omega).$$

3.3 Historical notes

As stated at the beginning of this chapter, one of the central results of this book is the Hölder continuity of weak solutions of the equation

(3.48)
$$(a^{ij}(x)u_{x_j})_{x_i} = 0, \quad a^{ij} \in L^\infty$$

which was proved independently by De Giorgi [**DeG**] and Nash [**Na**]. De Giorgi's work was extended to a more general class of linear equations by Morrey [**Mo3**] and Stampacchia [**Sta**]. Ladyzhenskaya and Ural'tseva [**LU1**] extended the work to quasilinear equations in divergence form.

Later, Moser [**Mos1**] [**Mos2**] gave a completely different development which made critical use of the John-Nirenberg result, [**JN**] and includes a proof of the Harnack inequality for weak solutions of (3.48).

Moser's technique was adapted by Serrin [**Se1**], [**Se3**] and Trudinger [**T1**] to a large variety of divergence type equations that will be treated in subsequent chapters. The De Giorgi method was used extensively in the book by Ladyzhenskaya and Ural'tseva, [**LU2**].

The results in this chapter are due primarily to Trudinger [**T1**]. The structure (3.5) has been generalized as to allow certain coefficients to lie in Morrey spaces, and more generally, in multiplier spaces. This generalization is made possible through the results of Adams, see Theorem 1.92, Corollary 1.93, p. 56 [**Ad1**] and Maz'ya and Shaposhnikova [**MaSh**], Theorem 2.28, p. 77.

The principle results are the sup bound of the solution in terms of its integral average (Theorem 3.12), the weak Harnack inequality, Theorem 3.13, and the Harnack inequality Theorem 3.14. The Harnack inequality for linear equations was first established by Moser [**Mos2**]. Later, Serrin proved Harnack's inequality for quasilinear equations in divergence form [**Se1**].

Inequalities such as (3.14), in which the L^p norm of the gradient of u can be estimated in terms of the L^p norm of u, are traditionally called Caccioppoli inequalities in recognition of his pioneering work in regularity theory, [**Cac1**], [**Cac2**], where the "reverse Poincaré's inequality" served as a step in the development.

Theorem 3.9 and especially Theorem 3.13 are of great importance in our development. They are due to Trudinger, [**T1**]. Corollary 3.10 was first observed by DiBenedetto and Trudinger [**DT**] and Theorem 3.12 is found in [**GT**]. Theorem 2.116, which supplies some motivation for the need of capacity in dealing with boundary regularity, was proved by J. Serrin, [**Se4**].

A review by Chiarenza [**Chi**], which has many references, concerns various results which focus on the function spaces used for coefficients.

CHAPTER 4

Fine Regularity Theory

4.1 Basic energy estimates

In this section we prove energy estimates, continuity of weak solutions and p-fine continuity of supersolutions. More generally, we will work with supersolutions with a constant upper obstacle l. This will apply later to more sophisticated obstacle problems and to boundary regularity. The class of "supersolutions up to all levels" introduced below strictly contains the family of all supersolutions. We will show later in Remark 4.32 that a function is superharmonic if and only it is a properly chosen representative of a supersolution of the Laplace equation up to all levels. The same correspondence holds for $\mathcal{A}$-superharmonic function in the sense of [**HKM2**].

We start this section with two energy estimates, Theorems 4.4 and 4.7, that are crucial in our subsequent development of regularity at the boundary and regularity of variational inequalities.

The statement and proof of the weak Harnack inequality, Theorem 4.5, are similar to Theorem 3.13, but now we estimate infimum of $u - m$ where m is the infimum of u on a larger ball. Since the structure is not invariant under addition of constants, the results, even for ordinary supersolutions, do not follow directly from those in Chapter 3.

The ultimate consequence of energy estimates is Theorem 4.8 which implies that u is both lower semicontinuous and p-finely continuous. As a consequence we obtain Hölder continuity for solutions.

4.1 DEFINITION. *Let u be a function on Ω and $l \in \mathbf{R}$. We say that u is a supersolution of (3.1) up to level l if $u_l := \min(u, l) \in W^{1,p}_{\mathrm{loc}}(\Omega)$ and*

$$(4.1) \qquad \int_\Omega \Big(A(x, u_l, \nabla u_l) \cdot \nabla \varphi + B(x, u_l, \nabla u_l)\varphi\Big)\, dx \geq 0$$

holds for each "test function" $\varphi \in W^{1,p}_c(\Omega)$ with

$$(4.2) \qquad\qquad 0 \leq \varphi \leq l - u_l \quad \text{on } \Omega.$$

We say that u is *a supersolution of* (3.1) *up to all levels* if u is finite almost everywhere and u is a supersolution up to level l for each $l \in \mathbf{R}$.

Our motivation for this definition follows from the fact that if $k < l$ and if $\{u < k\}$ were an open set and u a supersolution up to level l, then u would be a weak supersolution on $U_k := \{u < k\}$ in the sense of (3.4). Also if $\{u < k\}$ were not open, we may say that u is a supersolution on the p-fine interior of $\{u < k\}$ in the sense of Definition (3.1) on p. 161. Indeed, if $\varphi \in W^{1,p}_0(U_k)$ is nonnegative and bounded, then there exists $\varepsilon > 0$ such that $\varepsilon\varphi \leq l - k \leq l - u_k$. Then, $\varepsilon\varphi$ could be used in (4.1), and thus, φ as well.

Observe also that if u were a solution to the upper obstacle problem with l as the upper obstacle, as in Definition 5.1, then u would be a supersolution up to level l on Ω.

4.2 REMARK. If u is a supersolution of (3.1) up to all levels, then by Theorem 1.71, each truncation u_l is weakly differentiable and thus approximately differentiable almost everywhere. That is, for almost every x_0 there exists a linear mapping $L: \mathbf{R}^n \to \mathbf{R}$ and a measurable set E such that x_0 is a point of density of E and

$$\lim_{\substack{x \to x_0 \\ x \in E}} \frac{|u(x) - u(x_0) - L(x - x_0)|}{|x - x_0|} = 0.$$

It follows that u is approximately differentiable at all points of approximate differentiability of u_l which are density points for $\{u < l\}$. Using the Lebesgue density theorem, we infer that u is approximately differentiable a.e. in Ω. We will denote by Du the approximate derivative (gradient) of u. Thus, in this sense a function u may be differentiable while u may not be weakly differentiable. Since each test function satisfying (4.2) vanishes on $\{u \geq l\}$, we have

$$(4.3) \qquad \int_\Omega \Big(\boldsymbol{A}(x, u, Du) \cdot \nabla\varphi + \boldsymbol{B}(x, u, Du)\,\varphi\Big)\, dx \geq 0$$

for any $\varphi \in W_c^{1,p}(\Omega)$ with

$$0 \leq \varphi \leq (l - u)^+ \quad \text{on } \Omega.$$

Throughout this section, we will work with a more complicated version of (3.8). Consider a ball $B(r)$. Let $\boldsymbol{k}(r)$ be as in (3.7) and

$$(4.4) \qquad \begin{aligned} \boldsymbol{h}(r) &= \|a_2^{1/(p-1)}\|_{M(p,p);B(r)} + \Big(r^\epsilon \|b_2\|_{\mathcal{M}^{n/(p-\epsilon)}(B(r))}\Big)^{1/p}, \\ \boldsymbol{\kappa}(r) &= \boldsymbol{k}(r) + m\,\boldsymbol{h}(r). \end{aligned}$$

We redefine

$$\begin{aligned} \overline{a}(r, \cdot) &:= \big(1 + \boldsymbol{h}^{-1}(r)\big)^{p-1} a_2 + \boldsymbol{k}^{1-p}(r) a_3, \\ \overline{b}(r, \cdot) &:= \big(1 + \boldsymbol{h}^{-1}(r)\big)^p b_2 + \boldsymbol{k}^{1-p}(b_3 + b_0 c_3) + \boldsymbol{k}^{-p} c_3. \end{aligned}$$

For notational convenience and when there is no danger of confusion, we will omit the dependence of r in expressions $\overline{a}(r, \cdot)$, $\overline{b}(r, \cdot)$, $\boldsymbol{k}(r)$, $\boldsymbol{h}(r)$ and $\boldsymbol{\kappa}(r)$. Also we will use $\|\ldots\|$ without subscripts when the norms are $\|\ldots\|_{M(p,p);B(r)}$ for $a_2^{1/(p-1)}$, $a_3^{1/(p-1)}$ or $\|\ldots\|_{\mathcal{M}^{n/(p-\epsilon)}(B(r))}$ for b_1^p, b_2, b_3, c_3. Since

$$\|(1 + \boldsymbol{h}^{-1})a_2^{1/(p-1)}\| \leq \|a_2^{1/(p-1)}\| + \|a_2^{1/(p-1)}\|/\boldsymbol{h} \leq \|a_2^{1/(p-1)}\| + 1$$

and

$$\begin{aligned} \|\overline{a}^{1/(p-1)}\varphi\|_p^{p-1} &= \|\overline{a}|\varphi|^{p-1}\|_{p'} \leq \|(1 + \boldsymbol{h}^{-1})^{p-1} a_2 |\varphi|^{p-1}\|_{p'} + \|\boldsymbol{k}^{1-p} a_3 |\varphi|^{p-1}\|_{p'} \\ &= \|(1 + \boldsymbol{h}^{-1})a_2^{1/(p-1)}\varphi\|_p^{p-1} + \|\boldsymbol{k}^{-1} a_3^{1/(p-1)}\varphi\|_p^{p-1} \\ &\leq \|(1 + \boldsymbol{h}^{-1})a_2^{1/(p-1)}\|^{p-1}\|\varphi\|_{1,p}^{p-1} + \|\boldsymbol{k}^{-1} a_3^{1/(p-1)}\|^{p-1}\|\varphi\|_{1,p}^{p-1} \end{aligned}$$

for any $\varphi \in W_0^{1,p}(\Omega)$, we have

$$\|\overline{a}^{1/(p-1)}\|^{p-1} \leq \left(1 + \|a_2^{1/(p-1)}\|\right)^{p-1} + 1.$$

Similarly

$$r^\epsilon\|\overline{b}\| \leq \left(1 + \left(r^\epsilon\|b_2\|\right)^{1/p}\right)^p + 2.$$

4.3 LEMMA. *If u is a supersolution of (3.1) up to a level l, then u is locally bounded below on Ω.*

PROOF. We may assume that $l \geq 0$. Then we may use the same test functions for estimating u^- as in the proof of Theorem 3.9. □

In the sequel we will assume the following hypotheses and notation to be in force.

(H) $\begin{cases} \text{Let } 0 \leq m < l \leq \infty \text{ be subject to the convention that } b_0 = 0 \text{ if} \\ l = \infty. \text{ Let } 0 < r < r_0 \text{ and } u \geq m \text{ be a supersolution on } B(x_0, r) \\ \text{of (3.1) up to level } l \text{ if } l < \infty, \text{ or up to all levels if } l = \infty. \text{ For a} \\ \text{finite } k \in [m, l] \text{ and } u_k := \min(u, k), \text{ let} \\ \qquad \overline{u} = u_k - m + \kappa(r) \\ \text{where } \kappa \text{ is as in (4.4).} \end{cases}$

We introduce the symbol

(4.5) $\qquad \mathfrak{S} := (n, p, \epsilon, b_0 l, a_1, \|a_2^{1/(p-1)}\|, r^\epsilon\|b_1^p\|, r^\epsilon\|b_2\|)$

as a convenience in referring to the dependence of constants.

4.4 THEOREM. *Assume the conditions of (H). Let $\eta \in C_c^\infty(B(r))$, $0 \leq \eta \leq 1$. Then, for $1 < p \leq n$ and $\beta \leq \beta_0 < 0$,*

(4.6) $\qquad \int_{B(r)} |\nabla \overline{u}|^p \overline{u}^{\beta-1} \eta^p \, dx \leq C|\beta|^{-p+p^2/\epsilon} \int_{B(r)} (r^{-p}\eta^p + |\nabla\eta|^p) \overline{u}^{p+\beta-1} \, dx$

where $C = C(\beta_0, \mathfrak{S})$.

PROOF. In view of Lemma 3.4 we may assume that $b_0 = 0$. We may assume that $k < l$. Indeed, if our result holds for every $k \in (0, l)$, then a routine passage to the limit yields our result for l as well. Under these assumptions, the functions $\overline{u}$ and $\overline{u}^\beta$ are bounded, $\overline{u} \in W^{1,p}_{\text{loc}}(\Omega)$ and $\nabla \overline{u} = Du$ on $\{u < k\}$. We have

$$u_k = (u_k - m) + m \leq \overline{u} + \frac{\overline{u}}{h},$$

and thus
$$a_2 u_k^{p-1} + a_3 \leq (1+h^{-1})^{p-1} a_2 \bar{u}^{p-1} + k^{1-p} a_3 \bar{u}^{p-1}$$
$$\leq \bar{a}\,\bar{u}^{p-1},$$
$$b_2 u_k^{p-1} + b_3 \leq (1+h^{-1})^{p-1} b_2 \bar{u}^{p-1} + k^{1-p} b_3 \bar{u}^{p-1}$$
$$\leq (1+h^{-1})^p b_2 \bar{u}^{p-1} + k^{1-p} b_3 \bar{u}^{p-1}$$
$$\leq \bar{b}\bar{u}^{p-1},$$
$$c_2 u_k^p + c_3 \leq (1+h^{-1})^p c_2 \bar{u}^p + k^{-p} c_3 \bar{u}^p$$
$$\leq \bar{b}\bar{u}^p.$$

With the assurance that $\bar{u} > 0$ on $B(r)$, we may set
$$G(u) := \bar{u}^\beta - (k - m + \kappa(r))^\beta, \quad \beta < 0.$$

Define a test function φ as
$$\varphi := G(u)\eta^p$$

and observe that φ is a nonnegative, bounded element of $W_0^{1,p}(B(r))$ that vanishes on $\{u \geq k\}$. Hence, there exists $\varepsilon > 0$ such that $\varepsilon\varphi + u \leq l$, thus ensuring that φ is an admissible test function for (4.3). Similar to the proof of Theorem 3.13 we obtain

(4.7)
$$|\beta| \int_{B(r)} |\nabla \bar{u}|^p \, \bar{u}^{\beta-1} \eta^p \, dx$$
$$\leq p \int_{B(r)} (a_1 |\nabla \bar{u}|^{p-1} + \bar{a}\bar{u}^{p-1}) \bar{u}^\beta \, \eta^{p-1} \, |\nabla \eta| \, dx$$
$$+ (1 + |\beta|) \int_{B(r)} \bar{b}\bar{u}^{\beta-1+p} \eta^p \, dx$$
$$+ \int_{B(r)} b_1 |\nabla \bar{u}|^{p-1} \, \bar{u}^\beta \eta^p \, dx.$$

Choose $\varepsilon > 0$. By Young's inequality,

(4.8)
$$\int_{B(r)} \left(pa_1 |\nabla \eta| + b_1 \eta \right) |\nabla \bar{u}|^{p-1} \, \bar{u}^\beta \, \eta^{p-1} \, dx$$
$$\leq \varepsilon|\beta| \int_{B(r)} |\nabla \bar{u}|^p \bar{u}^{\beta-1} \eta^p \, dx$$
$$+ \varepsilon^{1-p} |\beta|^{1-p} \int_{B(r)} \left(p^p a_1^p |\nabla \eta|^p + b_1^p \eta^p \right) \bar{u}^{p+\beta-1} \, dx.$$

For the rest of the proof, we treat two cases separately. Let $\beta \neq 1 - p$. Define q by the equality $pq = p + \beta - 1$ and set $v = \bar{u}^q$. Using the multiplier property of $\bar{a}$ we

4.1 BASIC ENERGY ESTIMATES

obtain

$$p \int_{B(r)} \overline{a} u^{p+\beta-1} \eta^{p-1} |\nabla \eta| \, dx = p \int_{B(r)} \overline{a} v^p \eta^{p-1} |\nabla \eta| \, dx$$

$$\leq p\varepsilon |\beta|^{1-p} \int_{B(r)} \overline{a}^{p'} v^p \eta^p \, dx$$

$$+ p\varepsilon^{1-p} |\beta|^{(p-1)^2} \int_{B(r)} v^p |\nabla \eta|^p \, dx$$

(4.9)
$$\leq C\varepsilon |\beta|^{1-p} \int_{B(r)} \left(|\nabla v|^p \eta^p + v^p |\nabla \eta|^p \right) dx$$

$$+ p\varepsilon^{1-p} |\beta|^{(p-1)^2} \int_{B(r)} v^p |\nabla \eta|^p \, dx$$

$$= C\varepsilon |\beta|^{1-p} \int_{B(r)} \left(|q|^p |\nabla \overline{u}|^p \overline{u}^{\beta-1} \eta^p + \overline{u}^{p+\beta-1} |\nabla \eta|^p \right) dx$$

$$+ p\varepsilon^{1-p} |\beta|^{(p-1)^2} \int_{B(r)} \overline{u}^{p+\beta-1} |\nabla \eta|^p \, dx.$$

With

(4.10) $$f := b_1^p + \overline{b} \in \mathcal{M}^{n/(p-\epsilon)}, \qquad r^\epsilon \|f\| \leq C,$$

we apply Corollary 1.95 as in (3.17) to obtain

(4.11)
$$|\beta|^p \varepsilon^{-p} r^{-\epsilon} \int_{B(r)} (r^\epsilon f) v^p \eta^p \, dx$$

$$\leq C \int_{B(r)} \left(|\nabla v|^p \eta^p + v^p |\nabla \eta|^p \right) dx$$

$$+ C\varepsilon^{-p^2/\epsilon} |\beta|^{p^2/\epsilon} r^{-p} \int_{B(r)} v^p \eta^p$$

$$= C \int_{B(r)} |q|^p |\nabla \overline{u}|^p \overline{u}^{\beta-1} \, dx$$

$$+ C\varepsilon^{-p^2/\epsilon} |\beta|^{p^2/\epsilon} r^{-p} \int_{B(r)} \overline{u}^{p+\beta-1} \eta^p$$

$$+ C \int_{B(r)} \overline{u}^{p+\beta-1} |\nabla \eta|^p \, dx.$$

Estimating the asymptotic behavior of β and q for $\beta \to -\infty$ and choosing ε appropriately, by (4.8), (4.9) and (4.11) we obtain (4.6). Also if $\beta = 1 - p$, by similar reasons we have

(4.12)
$$p \int_{B(r)} \overline{a} \, \eta^{p-1} |\nabla \eta| \, dx$$

$$\leq p \int_{B(r)} \overline{a}^{p'} \eta^p \, dx + p \int_{B(r)} |\nabla \eta|^p \, dx$$

$$\leq C \int_{B(r)} |\nabla \eta|^p \, dx$$

and

(4.13)
$$\int_{B(r)} f\eta^p \, dx \leq C \int_{B(r)} |\nabla \eta|^p \, dx + Cr^{-p} \int_{B(r)} \eta^p.$$

Thus we have proved (4.6) in both cases. □

4.5 THEOREM. *Under the assumptions of (H), there is a constant C such that for any $\sigma, \tau \in (0,1)$,*

(4.14)
$$\left(\fint_{B(\sigma r)} \overline{u}^\gamma \, dx \right)^{1/\gamma} \leq C \inf_{B(\tau r)} \overline{u}$$

where $0 < \gamma < n(p-1)/(n-p)$ and
$$C = C(\gamma, \sigma, \tau, \mathfrak{S}).$$

In case $p = n$, the same conclusion holds for any $\gamma > 0$.

PROOF. As in the previous proof, we may initially assume that $k < l$. Let $\rho \in (0,1)$ be any number such that $\max(\sigma, \tau) < \rho$. Consider h, h' with $\rho \leq h' < h < 1$. Let η be a cutoff function such that $\eta \equiv 1$ on $B(h'r)$ and is identically zero on the complement of $B(hr)$. With the usual substitution

$$v := \overline{u}^q \text{ where } pq = p + \beta - 1, \beta \neq 1 - p$$

it is readily verified from Theorem 4.4 that

$$\int_{B(r)} \eta^p |\nabla v|^p \, dx \leq C|\beta|^{p^2/\epsilon} \int_{B(r)} (r^{-p}\eta + |\nabla \eta|^p) v^p \, dx,$$

whenever $\beta \neq 1 - p$, $\beta \leq \beta_0 < 0$. Applying Sobolev's inequality, we obtain

$$\left(\fint_{B(r)} (v\eta)^{\chi p} \right)^{1/(\chi p)} \leq C|\beta|^{p/\epsilon} r \left(\fint_{B(r)} (r^{-p}\eta + |\nabla \eta|^p) v^p \, dx \right)^{1/p}$$

where $p^*/p \geq \chi > \max\{1, \gamma/(p-1)\}$ for $p < n$ and $\chi > \max\{1, \gamma/(p-1)\}$ for $p = n$. With $\alpha := pq = p + \beta - 1$ and $\alpha > 0$, this can be written as

(4.15)
$$\left(\fint_{B(h'r)} \overline{u}^{\chi\alpha} \, dx \right)^{1/(\chi\alpha)} \leq C(h-h')^{-p/\alpha} \left(\fint_{B(hr)} \overline{u}^\alpha \, dx \right)^{1/\alpha}$$

while

(4.16)
$$\left(\fint_{B(h'r)} \overline{u}^{\chi\alpha} \, dx \right)^{1/(\chi\alpha)} \geq C \left(\frac{(p-1-\alpha)^{-p/\epsilon}}{h-h'} \right)^{p/\alpha} \left(\fint_{B(hr)} \overline{u}^\alpha \, dx \right)^{1/\alpha}$$

if $\alpha < 0$. These inequalities will be iterated as in the proof of Theorem 3.13. We obtain

$$\left(\fint_{B(\sigma r)} \overline{u}^\gamma \, dx \right)^{1/\gamma} \leq C \left(\fint_{B(\rho r)} \overline{u}^{\alpha_0} \, dx \right)^{1/\alpha_0}$$

for any $\alpha_0 > 0$ and $\gamma < \chi(p-1)$, and

$$\left(\fint_{B(\sigma r)} \overline{u}^{\alpha_0} \, dx \right)^{1/\alpha_0} \leq C \inf_{B(\tau r)} u.$$

The proof will be concluded by showing that

(4.17) $$\|\overline{u}\|_{\alpha_0, B(\rho)} \leq C \|\overline{u}\|_{-\alpha_0, B(\rho)}$$

for some $\alpha_0 > 0$. For this we examine the case $\beta = 1 - p$. By Theorem 4.4

$$\int_{B(y,s)} |\nabla(\ln \overline{u})|^p \, dx \leq Cs^{n-p}$$

whenever $B(y, 2s) \subset B(r)$. Thus, we may appeal to the John-Nirenberg inequality to obtain (4.17) as in the proof of Theorem 3.13. □

4.6 LEMMA. *Assume the conditions of (H) are in force and $\sigma, \tau, \rho \in (0, 1)$, $\sigma < \rho$. Let η be a cutoff function equal to 1 on $B(\sigma r)$ supported in $B(\rho r)$. Then, with $1 < p \leq n$,*

(4.18)
$$\int_{B(r)} (b_1 \eta + a_1 |\nabla \eta|) |\nabla \overline{u}|^{p-1} \eta^{p-1} \, dx$$
$$+ \int_{B(r)} \overline{a} \overline{u}^{p-1} \eta^{p-1} |\nabla \eta| \, dx + \int_{B(r)} \overline{b} \overline{u}^{p-1} \eta^p \, dx$$
$$\leq C r^{n-p} (\inf_{B(\tau r)} \overline{u})^{p-1}$$

where

$$C = C(\sigma, \tau, \rho, \mathfrak{S}).$$

PROOF. Choose

(4.19) $$\beta = \frac{\epsilon(1-p)}{\epsilon + p(n-p)}.$$

Notice that $0 > \beta > 1 - p$ and $(n-p)\beta > -p$. Let $\rho \in (0, 1)$, $\rho > \max(\sigma, \tau)$. Let η be a cutoff function equal to 1 on $B(\sigma r)$ and with support in $B(\rho r)$. Let $v = \overline{u}^q$, where $q = (p + \beta - 1)/p$, and $f = b_1^p + \overline{b}$. Then

$$\overline{u}^{(p-1)(1-\beta)} = v^{(p-1)(1-\beta)/q}$$

and by Corollary 1.93, Theorem 4.4 and (4.10),

(4.20)
$$r^\epsilon \int_{B(r)} f \overline{u}^{(p-1)(1-\beta)} \eta^p \, dx \leq \int_{B(r)} (r^\epsilon f)(v\eta)^{(p-1)(1-\beta)/q} \, dx$$
$$\leq C \Big(\int_{B(r)} (|\nabla v|^p \eta^p + v^p |\nabla \eta|^p) \, dx \Big)^{(p-1)(1-\beta)/pq}$$
$$\leq C \Big(\int_{B(r)} (r^{-p} \eta^p + |\nabla \eta^p|) v^p \, dx \Big)^{(p-1)(1-\beta)/pq}$$
$$\leq C \Big(r^{-p} \int_{B(\rho r)} \overline{u}^{p+\beta-1} \Big)^{(p-1)(1-\beta)/pq}.$$

By Theorem 4.4,

(4.21)
$$\int_{B(r)} |\nabla \overline{u}|^p \, \overline{u}^{\beta-1} \eta^p \, dx \leq C \int_{B(r)} (r^{-p} \eta^p + |\nabla \eta|^p) \overline{u}^{p+\beta-1} \, dx$$
$$\leq C r^{-p} \int_{B(\rho r)} \overline{u}^{p+\beta-1} \, dx$$

From Theorem 4.5 we obtain

(4.22) $$r^{-p}\int_{B(\rho r)} \bar{u}^{p+\beta-1}\,dx \leq Cr^{n-p}(\inf_{B(\tau r)}\bar{u})^{p+\beta-1}$$

and

(4.23) $$\int_{B(r)} \bar{u}^{(1-\beta)(p-1)}|\nabla\eta|^p\,dx \leq Cr^{-p}\int_{B(\rho r)} \bar{u}^{(p-1)(1-\beta)}\,dx$$
$$\leq Cr^{n-p}(\inf_{B(\tau r)}\bar{u})^{(p-1)(1-\beta)}$$

and by (4.20) and (4.22) we have

(4.24) $$\int_{B(r)} f\bar{u}^{(p-1)(1-\beta)}\eta^p\,dx \leq Cr^{-\epsilon}\left(r^{n-p}(\inf_{B(\tau r)}\bar{u})^{p+\beta-1}\right)^{(p-1)(1-\beta)/pq}$$
$$\leq Cr^{n-p}(\inf_{B(\tau r)}\bar{u})^{(p-1)(1-\beta)}.$$

From (4.21) – (4.24) we obtain

$$\int_{B(r)} (b_1\eta + a_1|\nabla\eta|)|\nabla\bar{u}|^{p-1}\eta^{p-1}\,dx$$
$$\leq C\left(\int_{B(r)} |\nabla\bar{u}|^p \bar{u}^{\beta-1}\eta^p\,dx\right)^{(p-1)/p}\left(\int_{B(r)} (b_1^p\eta^p + |\nabla\eta|^p)\bar{u}^{(1-\beta)(p-1)}\,dx\right)^{1/p}$$
$$\leq C\left(r^{n-p}(\inf_{B(\tau r)}\bar{u})^{p+\beta-1}\right)^{(p-1)/p}\left(r^{n-p}(\inf_{B(\tau r)}\bar{u})^{(1-\beta)(p-1)}\right)^{1/p}$$
$$= Cr^{n-p}(\inf_{B(\tau r)}\bar{u})^{p-1}$$

which estimates the first integral on the left-hand side of (4.18). For the second integral, using the multiplier property of $\bar{a}^{p'}$ we obtain

$$\int_{B(r)} \bar{a}^{p'}\bar{u}^{p+\beta-1}\eta^p\,dx = \int_{B(r)} \bar{a}^{p'}(v\eta)^p\,dx$$
$$\leq C\int_{B(r)} \left(v^p|\nabla\eta|^p + |\nabla v|^p\eta^p\right)dx$$
$$= C\int_{B(r)} \bar{u}^{p+\beta-1}|\nabla\eta|^p\,dx + C'\int_{B(r)} \bar{u}^{\beta-1}|\nabla\bar{u}|^p\eta^p\,dx$$

and thus, utilizing (4.21)-(4.23),

$$\int_{B(r)} \bar{a}\bar{u}^{p-1}\eta^{p-1}|\nabla\eta|\,dx$$
$$\leq C\left(\int_{B(r)} \bar{a}^{p'}\bar{u}^{p+\beta-1}\eta^p\,dx\right)^{(p-1)/p}\left(\int_{B(r)} \bar{u}^{(1-\beta)(p-1)}|\nabla\eta|^p\,dx\right)^{1/p}$$
$$\leq C\left(r^{n-p}(\inf_{B(\tau r)}\bar{u})^{p+\beta-1}\right)^{(p-1)/p}\left(r^{n-p}(\inf_{B(\tau r)}\bar{u})^{(1-\beta)(p-1)}\right)^{1/p}$$
$$= Cr^{n-p}(\inf_{B(\tau r)}\bar{u})^{p-1}.$$

For the third integral we have by Hölder's inequality, (3.6) and (4.24)

$$\int_{B(r)} \bar{b}\bar{u}^{p-1}\eta^p \, dx$$
$$\leq \left(\int_{B(r)} \bar{b}\bar{u}^{(p-1)(1-\beta)}\eta^p \, dx\right)^{1/(1-\beta)} \left(\int_{B(r)} \bar{b}\eta^p \, dx\right)^{\beta/(\beta-1)}$$
$$\leq C\left(r^{n-p}\left(\inf_{B(\tau r)} \bar{u}\right)^{(p-1)(1-\beta)}\right)^{1/(1-\beta)} \left(r^{n-p}\right)^{\beta/(\beta-1)}$$
$$\leq C\, r^{n-p}\left(\inf_{B(\tau r)} \bar{u}\right)^{p-1}$$

which concludes the proof. $\square$

In addition to the notation of Theorem 4.4, we denote

(4.25) $$m_k(r) := \inf_{B(r)} u_k.$$

4.7 THEOREM. *Let $\sigma \in (0,1)$. Under the conditions of (H),*

$$\int_{B(\sigma r)} |\nabla u_k|^p \eta^p \, dx \leq C r^{n-p}[k - m_k(r) + \kappa(r)][m_k(r/2) - m_k(r) + \kappa(r)]^{p-1},$$

where
$$C = C(\sigma, \mathfrak{S}).$$

PROOF. In view of Lemma 3.4 we may assume that $b_0 = 0$. Also, we may assume that
$$m = \inf_{B(r)} u = m_k(r).$$
Let $\eta \in C_c^\infty(B(r))$, $0 \leq \eta \leq 1$ and $B(r/2) \subset \operatorname{spt} \eta \subset B(\sigma r)$. Let
$$\varphi := (k - u_k)\eta^p,$$
we see that φ is nonnegative, bounded, and vanishes identically on $\{u \geq k\}$. Hence, there exists $\varepsilon > 0$ such that $\varepsilon\varphi + u \leq l$. Thus, we may use $\varepsilon\varphi + u$ as a legitimate test function for (4.1), and consequently we may use φ as well. As in the proof of Theorem 4.4 we obtain

$$\int_{B(r)} |\nabla u_k|^p \eta^p \, dx$$
$$\leq \int_{B(r)} \bar{b}\bar{u}^{p-1}[\bar{u} + (k - u_k)]\eta^p \, dx$$
$$+ p\int_{B(r)} \bar{a}\bar{u}^{p-1}(k - u_k)\eta^{p-1}|\nabla\eta| \, dx$$
$$+ \int_{B(r)} |\nabla u_k|^{p-1}(k - u_k)\eta^{p-1}(pa_1|\nabla\eta| + b_1\eta) \, dx.$$

Since
$$k - u_k \leq \bar{u} + k - u_k = k - m_k(r) + \kappa(r),$$

by Lemma 4.6 we obtain

$$\int_{B(r)} |\nabla u_k|^p \eta^p \, dx$$
$$\leq C\left(k - m_k(r) + \kappa(r)\right) r^{n-p} (\inf_{B(r/2)} \overline{u})^{p-1}$$
$$= C r^{n-p} (k - m_k(r) + \kappa(r))[m_k(r/2) - m_k(r) + \kappa(r)]^{p-1}.$$

This proves the result. □

In the rest of the section we denote

(4.26) $$\boldsymbol{a}(r) = \left\|a_2^{1/(p-1)}\right\|_{B(r)} + \left\|a_3^{1/(p-1)}\right\|_{B(r)},$$

where the norms are in $M(W_0^{1,p}(B(r)), L^p(B(r)))$.

4.8 THEOREM. *Let $k \leq l \leq \infty$ with $b_0 = 0$ if $l = \infty$. Let u be a supersolution up to level l in $B := B(x_0, R)$, or u is a supersolution up to all levels if $l = \infty$. Then, (considering an appropriate representative of u), u_k has a Lebesgue point and is lower semicontinuous at each point of B provided $k < \infty$ and*

(4.27) $$\lim_{r \to 0} \boldsymbol{a}(r) = 0.$$

Furthermore, u_l is lower semicontinuous on B and p-finely continuous at x_0 provided

(4.28) $$\int_0 \boldsymbol{a}(r) \frac{dr}{r} < \infty.$$

If $u_l(x_0) < l$, then u is lower semicontinuous at x_0 and is p-finely continuous there.

PROOF. By Lemma 4.3, u is locally lower bounded. Since the change in structure due to the addition of a constant is not essential here, we may assume that $u \geq 0$ and thus also that $k \geq 0$. For $0 < r < R$, let

$$\kappa(r) = \boldsymbol{k}(r) + m_k(r) \, \boldsymbol{h}(r).$$

Suppose that $k < \infty$. Appealing to Theorem 4.5, we have

$$\fint_{B(r/2)} (u_k(x) - m_k(r))^\gamma \, dx \leq C(m_k(r/2) - m_k(r) + \kappa(r))^\gamma.$$

Since the right side tends to 0 with r, this implies

$$\lim_{r \to 0} \fint_{B(r/2)} |u_k(x) - m_k(0)|^\gamma \, dx = 0$$

where

$$m_k(0) := \lim_{r \to 0} m_k(r).$$

Thus, u_k has a Lebesgue point at x_0 and

$$u_l(x_0) = m_l(0) = \liminf_{x \to x_0} u(x),$$

which establishes the lower semicontinuity of u_k.

With the help of (3.7), (p. 162), we find that the hypothesis (4.28) implies

$$\int_0 \kappa(r)\frac{dr}{r} < \infty.$$

Moreover,

(4.29) $$\int_0^\rho \big(m_k(r/2) - m_k(r)\big)\frac{dr}{r} = \int_{\rho/2}^\rho (m_k(0) - m_k(r))\frac{dr}{r} < \infty$$

for any $\rho > 0$. Using Theorem 4.7, as in the proof of Theorem 2.121 we obtain that u_k is lower semicontinuous and p-finely continuous at x_0. Since this is true for any $k < l$, u_l is lower semicontinuous and p-finely continuous at x_0. Suppose now that $u_l(x_0) < l$. Since

$$p\text{-fine-}\limsup_{x \to x_0} u_l(x) = u_l(x_0),$$

there is a set E whose complement is p-thin at x_0, such that

$$\lim_{\substack{x \to x_0 \\ x \in E}} u_l(x) = u_l(x_0) < l.$$

Hence, $u = u_l < l$ on $B(r) \cap E$ for all small r, thus proving the p-fine continuity of u at x_0. $\square$

4.9 REMARK. If u is a subsolution of (3.1), then $-u$ is a supersolution of the equation $-\operatorname{div} \tilde{A} = \tilde{B}$ where

$$\tilde{A}(x,\tilde{\zeta},\tilde{\xi}) = A(x,-\tilde{\zeta},-\tilde{\xi}) \quad \text{and} \quad \tilde{B}(x,\tilde{\zeta},\tilde{\xi}) = B(x,-\tilde{\zeta},-\tilde{\xi}).$$

Furthermore, the structure for $\tilde{A}$ and $\tilde{B}$ is similar to that of (3.5). Consequently, we may apply (4.14) and Theorem 4.7 with u replaced by $\tilde{u}$ to find that our conclusions apply to subsolutions.

4.10 COROLLARY. *Suppose that (4.27) is valid. If u is either a bounded subsolution or supersolution of (3.1), then u has a Lebesgue point everywhere in its domain and is upper (lower) semicontinuous. In addition, if (4.28) is satisfied, then u is p-finely continuous everywhere.*

Another consequence of these results is an estimate of the modulus of continuity of u which establishes the Hölder continuity of u in the interior provided the coefficients are suitably well-behaved.

4.11 THEOREM. *Let u be a weak solution of (3.1) in Ω. Then u is continuous in Ω and has the following bound on its modulus of continuity: if $B(R) \subset\subset \Omega$ and $\beta \in (0,1)$, then there exist $C > 0$ and $\alpha \in (0,1)$ such that*

$$\operatorname*{osc}_{B(x_0,r)} u \leq C[\Big(\frac{r}{R}\Big)^\alpha \operatorname*{osc}_{B(x_0,R)} u + \kappa(R^{1-\beta}r^\beta)]$$

for all $0 < r < R$, where $C = C(\mathfrak{S}, \|u\|_{\infty,R})$ while α depends on β as well.

PROOF. First, let us notice that u is locally bounded by Theorem 3.9. The functions $\|u\|_{\infty;R} + u$, $\|u\|_{\infty;R} - u$, solve (3.1) with a similar structure. For the dependence of constants, notice that the norms of modified coefficients depend only on the norms of initial coefficients and $\|u\|_{\infty,R}$. Let $k(r)$ be a constant such that

$$\min\Big\{|\{x \in B(r/2)\colon u(x) \geq k(r)\}|, \, |\{x \in B(r/2)\colon u(x) \leq k(r)\}|\Big\} \geq 1/2|B(r/2)|.$$

Referring to (4.14), we have

(4.30) $$k(r) - m(r) \leq C\Big(\fint_{B(r)} (u(x) - m(r))^\gamma \, dx\Big)^{1/\gamma} \leq C\,[m(r/2) - m(r) + \kappa(r)],$$

where

$$m(r) := \inf_{B(r)} u.$$

Similarly, we obtain

(4.31) $$M(r) - k(r) \leq C[-M(r/2) + M(r) + \kappa(r)],$$

where

$$M(r) := \sup_{B(r)} u.$$

Adding (4.30) and (4.31) we obtain

$$\operatorname*{osc}_{B(x_0,r/2)} u \leq \frac{C-1}{C} \operatorname*{osc}_{B(x_0,r)} u + 2\kappa(r).$$

Consequently, our conclusion follows from the lemma below, which concludes this section. □

4.12 LEMMA. *Let f be a nondecreasing function on $(0, R]$ satisfying*

$$f(\tau r) < \mu f(r) + \varepsilon(r)$$

where ε is nondecreasing and $\tau, \mu \in (0, 1)$. Then, for any $\beta \in (0, 1)$, there exist $C = C(\mu)$ and $\alpha = \alpha(\mu, \tau, \beta)$ such that

$$f(r) \leq C[R^{-\alpha} f(R) r^\alpha + \varepsilon(R^{1-\beta} r^\beta)]$$

whenever $r \in (0, R]$.

PROOF. Let $b \leq R$. For any $r \leq b$, we have

$$f(\tau r) < \mu f(r) + \varepsilon(b)$$

since ε is nondecreasing. By iterating this inequality, we obtain

$$f(\tau^j b) \leq \mu^j f(b) + \varepsilon(b) \sum_{i=0}^{j-1} \mu^i$$

$$\leq \mu^j f(R) + \frac{\varepsilon(b)}{1 - \mu}.$$

For $r \leq b$, choose j such that

$$\tau^j b < r \leq \tau^{j-1} b.$$

Then
$$f(r) \le f(\tau^{j-1}b)$$
$$\le \mu^{j-1} f(R) + \frac{\varepsilon(b)}{1-\mu}$$
$$\le \frac{1}{\mu} \left(\frac{r}{b}\right)^{\ln\mu/\ln\tau} f(R) + \frac{\varepsilon(b)}{1-\mu}.$$

With
$$b := r^\beta R^{1-\beta},$$
we obtain
$$f(r) \le \frac{1}{\mu} \left(\frac{r^{1-\beta}}{R^{1-\beta}}\right)^{\ln\mu/\ln\tau} f(R) + \frac{\varepsilon(R^{1-\beta}r^\beta)}{1-\mu}. \qquad \square$$

4.2 Sufficiency of the Wiener condition for boundary regularity

Our previous results concerning removable sets for weak solutions suggest that if u is to continuously assume the value $f(x_0)$ at a point $x_0 \in \partial\Omega$, then at least the capacity of $B(x_0, r) \setminus \Omega$ must be positive for each $r > 0$. The Wiener criterion shows precisely how large (in the limit) the capacity of $B(x_0, r) \setminus \Omega$ must be in order for u to be continuous at x_0.

In this subsection we suppose that Ω is a p-finely open set and fix a point $x_0 \in \mathbf{R}^n \setminus \Omega$.

The results of the previous section can easily be adapted to obtain information on the behavior of solutions at the boundary. We begin with the following definition.

4.13 DEFINITION. Suppose u is defined on Ω and $L \in \mathbf{R}$. We say that $u \ge L$ weakly at x_0 if for each $l < L$ there exists $r > 0$ such that $(u-l)^- \eta \in W_0^{1,p}(\Omega \cap B(x_0, r))$ whenever $\eta \in C_c^\infty(B(x_0, r))$. We define $u \le L$ weakly in an analogous manner and then say that $u = L$ weakly if both $u \le L$ and $u \ge L$ hold simultaneously.

If u is a function on Ω and $x_0 \in \overline{\Omega} \setminus \Omega$, then we define
$$\operatorname*{w\,lim\,inf}_{x \to x_0} u(x) = \sup\Big\{\operatorname*{lim\,inf}_{\substack{x \to x_0 \\ x \in \Omega}} [u(x) + w(x)] : w \in W_0^{1,p}(\Omega)\Big\}.$$

Analogously we define $\operatorname*{w\,lim\,sup}_{x \to x_0} u(x)$. We write
$$L = \operatorname*{w\,lim}_{x \to x_0} u(x) \quad \text{if} \quad L = \operatorname*{w\,lim\,sup}_{x \to x_0} u(x) = \operatorname*{w\,lim\,inf}_{x \to x_0} u(x).$$

Obviously
$$\operatorname*{w\,lim\,inf}_{x \to x_0} u(x) \ge \operatorname*{lim\,inf}_{x \to x_0} [u(x) + (u-L)^-(x)] \ge L$$

whenever $u \ge L$ weakly at x_0. The converse is not valid in general, but the following lemma shows that it holds for weak supersolutions.

4.14 LEMMA. *Let v be a lower bounded weak supersolution of (3.1) in Ω, $x_0 \in \overline{\Omega} \setminus \Omega$ and $L \in \mathbf{R}$. Suppose that*

$$\operatorname*{w\,lim\,inf}_{x \to x_0} v(x) \geq L.$$

Then $v \geq L$ weakly at x_0. Furthermore, if

$$u = \begin{cases} v & \text{on } \Omega, \\ L & \text{outside } \Omega, \end{cases}$$

then for each $k < L$ there exists $r > 0$ such that u is a supersolution up to level k on $B(x_0, r)$.

PROOF. Choose $k < l < L$. Then there exists a function $w \in W_0^{1,p}(\Omega)$ such that

$$\liminf_{\substack{x \to x_0 \\ x \in \Omega}} [u(x) + w(x)] > l.$$

We may assume that

$$0 \leq w \leq M$$

for some constant M. There is a ball $B(x_0, R)$ such that

$$u + w > k \quad \text{on } B(x_0, R) \cap \Omega.$$

Write $u_k = \min\{u, k\}$. Consider $r \in (0, R)$. Let $\omega \in \mathcal{C}_c^\infty(B(x_0, R))$ be a cutoff function such that $\chi_{B(x_0,r)} \leq \omega \leq \chi_{B(x_0,R)}$. Set

$$\eta = \min\{w, M\omega\}.$$

There is a sequence η_j of functions from $W_c^{1,p}(B(x_0, R))$ such that $\eta_j \uparrow \eta$ and $\|\eta_j - \eta\|_{1,p} \to 0$. We use

$$\varphi_j = (k - u_k + \eta_j - w)^+$$

as test functions for (3.4). Similarly to the proof of Theorem 3.8 we obtain that there is a weak limit φ_∞ of $\{\varphi_j\}$ in $W_0^{1,p}(B(x_0, R))$ and that $k - u_k = \varphi_\infty$ on $B(x_0, r)$. It follows that $u \geq L$ weakly at x_0 and $u_k \in W^{1,p}(B(x_0, r))$. Fix $\varphi \in W_0^{1,p}(B(x_0, r))$ such that

$$0 \leq \varphi \leq k - u_k.$$

Then obviously $\varphi \in W_0^{1,p}(B(x_0, r) \cap \Omega)$. By Theorem 1.63, there is $f \in W^{1,p}(\mathbf{R}^n)$ such that $f = u_k$ on $B(x_0, r)$. By Remark 3.6, we can test (3.4) by φ and, taking into account that $u = u_k$ on $\{\varphi > 0\}$, we obtain

$$\int_\Omega \Big(\mathbf{A}(x, u_k, \nabla u_k) \cdot \nabla \varphi + B(x, u_k, \nabla u_k)\varphi \Big) \, dx \geq 0.$$

This proves that u is a supersolution up to level k on $B(x_0, r)$. □

4.15 THEOREM. *Let $x_0 \in \overline{\Omega} \setminus \Omega$. Suppose $u \in W^{1,p}_{\mathrm{loc}}(\Omega)$, $1 < p \leq n$, is a lower bounded supersolution of (3.1) in Ω. Assume condition (4.28). If*

$$\lambda := \liminf_{\substack{x \to x_0 \\ x \in \Omega}} u(x) < L := \operatorname{w}\liminf_{x \to x_0} u(x),$$

then the p-fine limit of u exists at x_0 and is equal to λ. If $\mathbf{R}^n \setminus \Omega$ is not p-thin at x_0, then

$$\lambda = L.$$

PROOF. Extend u by setting $u = L$ outside Ω. If $\lambda < k < L$, then by Lemma 4.14 there is $r > 0$ such that u is a supersolution up to level k on $B(x_0, r)$. By Theorem 4.8, $u_k := \min\{u, k\}$ is lower semicontinuous and p-finely continuous at x_0. Since $\{x \in B(x_0, r) : u_k(x) \geq k\}$ is p-thin at x_0, the set $\{x \in \mathbf{R}^n : u(x) \geq k\}$ is p-thin at x_0 as well. On the other hand, the set $\{x \in \mathbf{R}^n : u(x) < l\}$ is p-thin at x_0 for each $l < \lambda$ by the definition of λ. This proves that λ is the p-fine limit of u as $x \to x_0$.

Now, if $\mathbf{R}^n \setminus \Omega$ is not p-thin at x_0, we obtain that $\lambda = L$, as otherwise we are in the situation of the preceding part of the proof and $B(x_0, r) \setminus \Omega$ as a subset of $\{x \in B(x_0, r) : u_k(x) \geq k\}$ is p-thin at x_0. □

If u is a subsolution, then $-u$ is a supersolution of an equation of similar structure and consequently, the previous theorem has an obvious counterpart for subsolutions. Taken together, this yields the following result.

4.16 THEOREM. *Let $x_0 \in \overline{\Omega} \setminus \Omega$. Suppose $u \in W^{1,p}_{\mathrm{loc}}(\Omega)$, $1 < p \leq n$, is a bounded solution of (3.1) in Ω such that*

(4.32) $$\operatorname{w}\limsup_{x \to x_0} u(x) \leq L \leq \operatorname{w}\liminf_{x \to x_0} u(x).$$

Assume condition (4.28). If $\mathbf{R}^n \setminus \Omega$ is not p-thin at x_0, then

$$\lim_{\substack{x \to x_0 \\ x \in \Omega}} u(x) = L.$$

If $\mathbf{R}^n \setminus \Omega$ is p-thin at x_0, then there exists a p-fine limit of u at x_0.

4.17 REMARK. The condition (4.32) is slightly more general than that

$$\operatorname{w}\lim_{x \to x_0} u(x) = L.$$

It may happen that $B(x_0, r) \setminus \Omega$ is p-polar for some $r > 0$ and then the "weak limits at x_0" do not work well, namely

$$\operatorname{w}\liminf_{x \to x_0} u(x) = \infty, \qquad \operatorname{w}\limsup_{x \to x_0} u(x) = -\infty$$

for each bounded function u.

Assume that f is a function on $\overline{\Omega}$ which is continuous at all points of $\overline{\Omega} \setminus \Omega$. Then the condition (4.32) is satisfied at each $x_0 \in \overline{\Omega} \setminus \Omega$ with $L = f(x_0)$. Consequently, for the "solution of the Dirichlet problem with boundary data f" we have

4.18 COROLLARY. *Suppose* $u \in W^{1,p}_{loc}(\Omega)$, $1 < p \leq n$, *is a bounded solution of* (3.1) *in* Ω *such that* $u - f \in W^{1,p}_0(\Omega)$, *where* f *is as above, and* $x_0 \in \overline{\Omega} \setminus \Omega$. *If* $\mathbf{R}^n \setminus \Omega$ *is p-thin at* x_0, *then* u *has a p-fine limit at* x_0. *If* $\mathbf{R}^n \setminus \Omega$ *is not p-thin at* x_0, *then*

$$\lim_{\substack{x \to x_0 \\ x \in \Omega}} u(x) = f(x_0).$$

The preceding analysis can be used to obtain a modulus of continuity of u at the boundary. For this, let $f: \partial\Omega \to \mathbf{R}$ be continuous and assume u is a supersolution such that

$$w \liminf_{x \to y} u(x) \geq f(y)$$

for each $y \in \overline{\Omega} \setminus \Omega \cap B(x_0, r_0)$, where $x_0 \in \mathbf{R}^n \setminus \Omega$ (it is not important to assume that $x_0 \in \overline{\Omega}$) and $r_0 > 0$. Consider a ball $B = B(x_0, r)$ where $0 < r < r_0$ such that $B \cap \Omega \neq \emptyset$. We may assume that $u \geq 0$. By Lemma 4.14, if $l < f(y)$ for all $y \in B \cap \overline{\Omega} \setminus \Omega$, then there exists $r > 0$ such that $\eta(l - u)^+ \in W^{1,p}_0(\Omega)$ whenever $\eta \in \mathcal{C}^\infty_c(B(x_0, r))$ and that

$$u_l := \begin{cases} \inf(u, l) & \text{in } B(r) \cap \Omega \\ l & \text{in } B(r) \setminus \Omega \end{cases}$$

is a supersolution up to level l on B. We set

(4.33)
$$u_l := \begin{cases} \inf(u, l) & \text{in } B(r) \cap \Omega \\ l & \text{in } B(r) \setminus \Omega \end{cases}$$

$$m := \inf_{B(r)} u_l$$

$$\boldsymbol{\mu}_l(r) := \sup_{B(r)} (l - u_l)$$

$$\boldsymbol{\kappa}(r) := \boldsymbol{k}(r) + m\boldsymbol{h}(r) \quad \text{as in (4.4)}$$

$$\overline{u} := u_l - m + \boldsymbol{\kappa}(r)$$

where

$$l = l(r) := \inf_{B(x_0, r) \cap \partial\Omega} f - r.$$

Note that $\overline{u} \equiv \boldsymbol{\mu}_l(r) + \boldsymbol{\kappa}(r)$ on $B(x_0, r) \setminus \Omega$, and therefore we can test the capacity of $B(x_0, r/2) \setminus \Omega$ by

$$v := \frac{\overline{u}\eta}{\boldsymbol{\mu}_l(r) + \boldsymbol{\kappa}(r)}$$

where $\eta \in \mathcal{C}^\infty_c(B(x_0, r))$ is a cutoff function with $\eta = 1$ on $B(x_0, r/2)$. We use Lemma 2.5 to estimate that

$$\gamma_{p;r}(B(x_0, r/2) \setminus \Omega) \leq C\|v\eta\|_p^p.$$

By Theorem 4.5

$$\int_{B(r)} \overline{u}^p |\nabla\eta|^p \, dx \leq [\boldsymbol{\mu}_l(r) + \boldsymbol{\kappa}(r)] \int_{B(r)} \overline{u}^{p-1} |\nabla\eta|^p \, dx$$

$$\leq Cr^{n-p}[\boldsymbol{\mu}_l(r) + \boldsymbol{\kappa}(r)] \Big(\inf_{B(r/2)} \overline{u}\Big)^{p-1}$$

$$\leq Cr^{n-p}[\boldsymbol{\mu}_l(r) + \boldsymbol{\kappa}(r)][\boldsymbol{\mu}_l(r) - \boldsymbol{\mu}_l(r/2) + \boldsymbol{\kappa}(r)]^{p-1}.$$

Using Theorem 4.7 we obtain

(4.34) $$\int_{B(r)} |\nabla u_l|^p \eta^p \, dx \leq C r^{n-p} [\kappa(r) + \mu_l(r)][\mu_l(r) - \mu_l(r/2) + \kappa(r)]^{p-1}.$$

Hence
$$(\mu_l(r) + \kappa(r))^p \frac{\gamma_{p;r}[B(x_0, r/2) \setminus \Omega]}{r^{n-p}}$$
$$\leq C r^{p-n} \int_{B(x_0, r)} |\nabla(\eta \bar{u})|^p \, dx$$
$$\leq C[\mu_l(r) + \kappa(r)][\mu_l(r) - \mu_l(r/2) + \kappa(r)]^{p-1}.$$

We will assume
$$\underline{D}(x_0) := \inf_{0 < r < r_0} D(x_0, r/2) > 0$$
where

(4.35) $$D(r) = D(x_0, r) := \left(\frac{\gamma_{p;r}[B(x_0, r/2) \setminus \Omega]}{r^{n-p}} \right)^{1/(p-1)}$$

Then,
$$\mu_l(r) D(r) \leq C \left[\mu_l(r) - \mu_l(r/2) + \kappa(r) \right]$$
for all $r \in (0, r_0)$. That is, there exists $\gamma \in (0, 1)$ such that
$$\mu_l(r/2) \leq \gamma \mu_l(r) + \kappa(r)$$
for all $r \in (0, r_0)$. If $0 < \beta < 1$, Lemma 4.12 provides $\alpha > 0$ such that

(4.36) $$\mu_l(r) \leq C[r^\alpha + \kappa(r^\beta)]$$

for all $r \in (0, r_0)$. Hence, this can be used to obtain an estimate for the lower oscillation of u at x_0:

(4.37) $$\begin{aligned} \omega^-(r) &:= \left(f(x_0) - \inf_{B(x_0, r)} u \right)^+ \\ &= \mu_l(r) + (f(x_0) - l(r)) \\ &= \mu_l(r) + f(x_0) + r - \inf_{B(x_0, r) \cap \partial \Omega} f \\ &\leq f(x_0) - \inf_{B(x_0, r) \cap \partial \Omega} f + g(r) \end{aligned}$$

where
$$g(r) := r + C(r^\alpha + \kappa(r^\beta)).$$
If u is a subsolution, an analogous result holds for
$$\omega^+(r) := \left(\sup_{B(x_0, r)} u - f(x_0) \right)^+,$$
namely,

(4.38) $$\omega^+(r) \leq -f(x_0) + \sup_{B(x_0, r) \cap \partial \Omega} f + g(r),$$

see Remark 4.9. Consequently, if u is a solution, its oscillation $\omega(r)$ is bounded by
$$\omega(r) \leq \omega^+(r) + \omega^-(r).$$

In summary, we have the following result.

4.19 THEOREM. *Suppose u is a bounded solution of (3.1) in Ω and $f\colon \partial\Omega \to \mathbf{R}$ is continuous. If*

$$(4.39) \qquad \text{w}\limsup_{x \to y} u(x) \leq f(y) \leq \text{w}\liminf_{x \to y} u(x)$$

for each $y \in \overline{\Omega} \setminus \Omega \cap B(x_0, r_0)$, and

$$(4.40) \qquad \underline{D}(x_0) := \inf_{r \in (0, r_0)} \left(\frac{\gamma_{p;r}[B(x_0, r/2) \setminus \Omega]}{r^{n-p}} \right)^{1/(p-1)} > 0,$$

then, for any $\beta \in (0, 1)$,

$$\underset{B(x_0, r)}{\text{osc}}\, u \leq C[r^\alpha + \underset{B(x_0, r) \cap \partial\Omega}{\text{osc}} f + \kappa(r^\beta)]$$

where C depends on the same quantities as in Theorem 4.11 and α depends in addition on β and $\underline{D}(x_0)$.

Clearly, u is Hölder continuous if the same is true of κ. In view of (3.7) (p. 162) and (4.4) (p. 186), this holds if $\boldsymbol{a}(r)$ is a Hölder continuous function of r.

4.20 COROLLARY. *Assume (4.40) holds for all $x_0 \in \partial\Omega$ and $\boldsymbol{a}$ is a Hölder continuous function of r. If f is a Hölder continuous function on $\partial\Omega$ and u is a bounded solution of (3.1) such that (4.39) holds at each $y \in \overline{\Omega} \setminus \Omega \cap B(x_0, r_0)$, then u is Hölder continuous.*

With a slightly different assumption on the data, we also get the following estimate on the modulus of continuity.

4.21 LEMMA. *Let $k \in \mathbf{R}$ be a constant level. Suppose u is a bounded supersolution of (3.1) on Ω such that*

$$\text{w}\liminf_{x \to y} u(x) \geq k$$

for all $y \in \overline{\Omega} \setminus \Omega \cap B(x_0, r_0)$. Assume that $\boldsymbol{a}(r)$ is Hölder continuous. Then there is a constant C such that for all $r \in (0, r_0)$ and $t \in (0, r/2)$,

$$\underset{B(x_0, t) \cap \Omega}{\text{osc}} (k - u)^+ \leq C \exp\left(-\frac{1}{C} \int_{2t}^{r} D(x_0, s) \frac{ds}{s} \right) + \underset{B(x_0, t) \cap \partial\Omega}{\text{osc}} f,$$

where $D(x_0, \cdot)$ is as in (4.35).

PROOF. We may assume $u \geq 0$. The proof proceeds as in Theorem 4.19 and we obtain

$$[\mu_l(r) + \kappa(r)] D(r) \leq C[\mu_l(r) - \mu_l(r/2) + \kappa(r)]$$

or

$$\mu_l(r/2) \leq [\mu_l(r) + \kappa(r)](1 - A(r))$$

where $A(r) := C^{-1} D(r)$. Assuming that $\kappa(r) < Cr^\beta$ for some $0 < \beta < 1$, choose α so that $0 < \alpha < \beta$. Let

$$A_1 := \sup_{r \in (0, r_0)} \frac{A(r)}{1 - 2^{-\alpha}}$$

and define

$$B(r) = \frac{A(r)}{1 + A_1}.$$

4.2 SUFFICIENCY OF THE WIENER CONDITION FOR BOUNDARY REGULARITY

Note that $B(r) < 1 - 2^{-\alpha}$ for all r. With $b(r) := 1 - B(r) \geq 1 - A(r)$, we have
$$\mu_l(r/2) \leq [\mu_l(r) + Cr^\beta]b(r).$$
For each positive integer m, iteration of this inequality yields
$$\mu_l(2^{-m}r) \leq \mu_l(r) \prod_{j=0}^{m-1} b(2^{-j}r) + C \sum_{j=0}^{m-1} (2^{-j}r)^\beta \prod_{i=j}^{m-1} b(2^{-i}r)$$
$$\leq \mu_l(r) \prod_{j=0}^{m-1} b(2^{-j}r) + C \prod_{i=0}^{m-1} b(2^{-i}r) \sum_{j=0}^{m-1} \frac{(2^{-j}r)^\beta}{\prod_{i=0}^{j-1} b(2^{-i}r)}$$
$$= \mu_l(r) \prod_{j=0}^{m-1} b(2^{-j}r) + Cr^\beta \prod_{i=0}^{m-1} b(2^{-i}r) \sum_{j=0}^{m-1}\prod_{i=0}^{j-1} \frac{2^{-\beta}}{b(2^{-i}r)}.$$

Since $b(r) \geq 2^{-\alpha}$ for all r and $\alpha < \beta$, the last series converges. Hence,
$$\mu_l(r/2^m) \leq [\mu_l(r) + Cr^\beta] \prod_{i=0}^{m-1} (b(2^{-i}r))$$
$$= [\mu_l(r) + Cr^\beta] \prod_{i=0}^{m-1} (1 - B(2^{-i}r))$$
$$\leq [\mu_l(r) + Cr^\beta] \prod_{i=0}^{m-1} \exp(-B(2^{-i}r))$$
$$= [\mu_l(r) + Cr^\beta] \exp\left(-\sum_{i=0}^{m-1} B(2^{-i}r)\right),$$
for all positive integers m. When $y \leq s \leq 2y$,
$$B(s) \leq 2^{(n-p)/(p-1)} B(2y)$$
so that
$$\int_y^{2y} B(s) \frac{ds}{s} \leq 2^{(n-p)/(p-1)} B(2y).$$
Hence,
$$\int_{2^{-m}r}^{r} B(s) \frac{ds}{s} \leq 2^{(n-p)/(p-1)} \sum_{i=0}^{m-1} B(r/2^i)$$
thus implying
$$\mu_l(r/2^m) \leq [\mu_l(r) + Cr^\beta] \exp\left(-\frac{1}{C} \int_{2^{-m}r}^{r} B(s) \frac{ds}{s}\right)$$
$$\leq [\mu_l(r) + Cr^\beta] \exp\left(-\frac{1}{C} \int_{2^{-m}r}^{r} D(s) \frac{ds}{s}\right).$$
Then, as in (4.38), we obtain a bound for the oscillation. $\square$

Applying the preceding Lemma to both u and $-u$ we obtain the following theorem.

4.22 THEOREM. *Let f be a continuous function on $\partial\Omega$. Suppose u is a bounded solution of (3.1) such that (4.39) holds for all $y \in \overline{\Omega} \setminus \Omega \cap B(x_0, r_0)$. Assume that $\mathbf{a}(r)$ is Hölder continuous. Then there is a constant C such that for all $r \in (0, r_0)$,*

$$\underset{B(x_0,t) \cap \Omega}{\mathrm{osc}}\, u \leq C \exp\left(-\frac{1}{C}\int_{2t}^{r} D(x_0, s)\frac{ds}{s}\right) + \underset{B(x_0,t) \cap \partial\Omega}{\mathrm{osc}}\, f$$

whenever $t \leq r/2$.

4.2.1 The special case of harmonic functions.

Here we show how the techniques of the previous sections can be applied to establish the Wiener criterion for harmonic functions without appealing to the full strength of the theory, such as the Weak Harnack inequality. Indeed, the methods here are completely elementary. We use some properties of superharmonic function proved in Section 2.2 (p. 92), but not many of them. Namely, we need a connection between superharmonic functions and supersolutions of the Laplace equation and the mean value property. This development thus provides an alternative to the classical proof of Theorem 2.89 (p. 108), which is based on potential theory.

Throughout this section, we suppose $\Omega \subset \mathbf{R}^n$ ($n \geq 2$) is a bounded open set and $x_0 \in \partial\Omega$. As x_0 is fixed throughout this discussion, we will write $B(r) := B(x_0, r)$ for brevity. Given a radius $r > 0$, we will work with a cutoff function η such that

(4.41)
$$\begin{aligned}
&\eta \in \mathcal{C}_c^\infty B(r),\\
&0 \leq \eta \leq 1,\\
&\eta = 1 \text{ in } B(r/4),\\
&\eta = 0 \text{ in } B(r) \setminus B(r/2),\\
&|\nabla \eta| \leq C/r,\\
&|\nabla^2 \eta| \leq C/r^2.
\end{aligned}$$

We start with several simple observations. Let f be a continuous function on $\partial\Omega$. If u solves the Dirichlet problem with boundary data f and $f \geq l$ (l constant) on $\partial\Omega \cap B(r)$, then

$$u_l = \begin{cases} \min\{u, l\}, & x \in B(r) \cap \Omega, \\ l, & x \in B(r) \setminus \Omega \end{cases}$$

is a supersolution on $B(r)$. To prove this observation, it is enough to show that u_k is a supersolution for $k < l$ and then pass to the limit. The function u_k belongs to $W^{1,2}(B(r))$. Choose a nonnegative test function $\varphi \in W_c^{1,2}(B(r))$ and write

$$E := B(r) \cap \Omega \cap \{u < k\} = B(r) \cap \{u_k \neq k\},$$
$$\psi := \begin{cases} \min\{u + \varphi, k\} & \text{on } E, \\ u & \text{elsewhere in } \Omega. \end{cases}$$

Then, since u minimizes the Dirichlet integral in Ω and $u - \psi \in W_0^{1,2}(\Omega)$,

$$\int_\Omega |\nabla u|^2\, dx \leq \int_\Omega |\nabla \psi|^2\, dx$$

4.2 SUFFICIENCY OF THE WIENER CONDITION FOR BOUNDARY REGULARITY

and thus

$$\int_{B(r)} |\nabla u_k|^2 \, dx = \int_E |\nabla u|^2 \, dx \le \int_E |\nabla \psi|^2 \, dx$$
$$\le \int_{B(r)} |\nabla (u_k + \varphi)|^2 \, dx.$$

This shows that u_k is a superminimizer of the Dirichlet integral on $B(r)$ and thus a supersolution.

Another observation concerns the case when $g \in W^{1,2}(\Omega)$ is a supersolution on Ω and h is the solution of the Dirichlet problem in $\Omega' \subset \Omega$ such that $h - u \in W_0^{1,2}(\Omega')$. We want to show that

$$u := \begin{cases} h & \text{in } \Omega', \\ g & \text{in } \Omega \setminus \Omega', \end{cases}$$

is a supersolution in Ω. First notice that an elementary comparison principle yields that $u \le g$ in Ω. Now consider a test function $\varphi \in W_c^{1,2}(\Omega)$, $\varphi \ge 0$ and estimate

$$\int_\Omega |\nabla u|^2 \, dx \le \int_{\Omega \setminus \Omega'} |\nabla g|^2 \, dx + \int_{\Omega'} |\nabla h|^2 \, dx$$
$$\le \int_{\Omega \setminus \Omega'} |\nabla g|^2 \, dx + \int_{\Omega'} |\nabla \min\{u + \varphi, g\}|^2 \, dx$$
$$= \int_\Omega |\nabla g|^2 \, dx + \int_{\{u+\varphi<g\}} \left(|\nabla (u+\varphi)|^2 - |\nabla g|^2 \right) dx$$
$$\le \int_\Omega |\nabla \max\{u + \varphi, g\}|^2 \, dx + \int_{\{u+\varphi<g\}} \left(|\nabla (u+\varphi)|^2 - |\nabla g|^2 \right) dx$$
$$= \int_\Omega |\nabla (u + \varphi)|^2 \, dx$$

where we used first the minimizing property of h and then the superminimizing property of g.

4.23 LEMMA. *Let v be a superharmonic function on $B(r)$ and k a constant. Suppose that $0 \le v \le k$ on $B(r)$. Let η be as in (4.41). Then*

$$\int_{B(r)} |\nabla(\eta v)|^2 \, dx \le C r^{n-2} k \inf_{B(r/4)} v$$

where $C = C(n)$.

PROOF. Since v is superharmonic in $B(r)$ we have

$$\int_{B(r)} \nabla v \cdot \nabla \varphi \, dx \ge 0$$

whenever $\varphi \in W_0^{1,2}(B(r))$, $\varphi \geq 0$. Thus, with $\varphi := \eta^2[k-v]$, integration by parts yields

$$\int_{B(r)} \eta^2 |\nabla v|^2 \, dx \leq 2 \int_{B(r)} [k-v] \nabla v \cdot \eta \nabla \eta \, dx$$

$$= 2 \int_{B(r)} v \nabla v \cdot \eta \nabla \eta \, dx$$

$$- 2 \int_{B(r)} v[k-v] \operatorname{div}(\eta \nabla \eta) \, dx.$$

By Young's inequality and the bound on $|\nabla^2 \eta|$, we have

(4.42)
$$\int_{B(r)} \eta^2 |\nabla v|^2 \, dx \leq \frac{C}{r^2} \int_{B(r/2)} (v^2 + v[k-v]) \, dx$$

$$\leq \frac{Ck}{r^2} \int_{B(r/2)} v \, dx.$$

Since v is superharmonic in $B(r)$, it is super mean-valued and therefore,

$$\fint_{B(x_0, r/2)} v \, dx \leq C \fint_{B(x_1, 3r/4)} v \, dx \leq C v(x_1)$$

whenever $x_1 \in B(x_0, r/4)$. Thus, by (4.42),

$$\int_{B(r)} \eta^2 |\nabla v|^2 \, dx \leq C r^{n-2} k \inf_{B(r/4)} v.$$

Furthermore,

$$\int_{B(r)} v^2 |\nabla \eta|^2 \, dx \leq \frac{C}{r^2} \int_{B(r/2)} v^2 \, dx$$

$$\leq \frac{Ck}{r^2} \int_{B(r/2)} v \, dx$$

$$\leq C r^{n-2} k \inf_{B(r/4)} v.$$

This along with the previous inequality establishes the desired result. $\square$

With the help of the previous result, we now establish the Wiener criterion for harmonic functions.

4.24 Theorem. *A point $x_0 \in \partial \Omega$ is regular if and only if $\mathbf{R}^n \setminus \Omega$ is not 2-thin at x_0.*

Proof. First we will show that if x_0 is irregular, then

(4.43)
$$\int_0^R \frac{\gamma_{2;r}(B(r) \setminus \Omega)}{r^{n-2}} \frac{dr}{r} < \infty,$$

where R will be specified soon. Let f be a continuous function on $\partial \Omega$ and u the solution of the Dirichlet problem with boundary function f. Let us assume that

$$L := \limsup_{\substack{x \to x_0 \\ x \in \Omega}} u(x) > f(x_0)$$

4.2 SUFFICIENCY OF THE WIENER CONDITION FOR BOUNDARY REGULARITY

for some $x_0 \in \partial\Omega$. Choose $f(x_0) < l < L$ and select R such that $\eta(u-l)^+ \in W_0^{1,2}(B(r) \cap \Omega)$ for each $r \leq 4R$ and $\eta \in \mathcal{C}_c^\infty(B(r))$. Fix $r \in (0, 4R)$ and η as in (4.41). Let

$$v := \begin{cases} \boldsymbol{\mu}(r) - (u-l)^+ & \text{on } B(r) \cap \Omega, \\ \boldsymbol{\mu}(r) & \text{on } B(r) \setminus \Omega, \end{cases}$$

where

$$\boldsymbol{\mu}(r) := \sup_{B(r)} (u-l)^+.$$

As observed at the beginning of the subsection, v is a supersolution in $B(r)$. By Lemma 4.23, we have

$$\int_{B(r)} |\nabla(\eta v)|^2 \, dx \leq C r^{n-2} \boldsymbol{\mu}(r)[\boldsymbol{\mu}(r) - \boldsymbol{\mu}(r/4)].$$

Since

$$w := \frac{\eta v}{\boldsymbol{\mu}(r)} = 1 \quad \text{on} \quad B(r/4) \setminus \Omega$$

in the sense of Corollary 2.25 and

$$\liminf_{r \to 0} \boldsymbol{\mu}(r) \geq L - l > 0,$$

it follows that

$$\frac{\gamma_{2;r}[B(r/4) \setminus \Omega]}{r^{n-2}} \leq C r^{2-n} \int_{B(r)} |\nabla w|^2 \, dx$$

$$\leq C[\boldsymbol{\mu}(r) - \boldsymbol{\mu}(r/4)],$$

and consequently

$$\int_0^{4R} \frac{\gamma_{2;r}[B(r/4) \setminus \Omega]}{r^{n-2}} \frac{dr}{r} \leq C \int_0^{4R} [\boldsymbol{\mu}(r) - \boldsymbol{\mu}(r/4)] \frac{dr}{r} < \infty,$$

thus establishing (4.43).

The case

$$\liminf_{\substack{x \to x_0 \\ x \in \Omega}} u(x) < f(x_0)$$

is handled similarly.

As a next step we prove an estimate towards the proof of the converse. We assume (4.43) at x_0. Let $R_0 > 0$ and consider a function u which is superharmonic on $B(4R_0)$, harmonic on $B(4R_0) \cap \Omega$ and satisfies $0 \leq u \leq 1$. We claim there is a constant $C = C(n)$ such that

(4.44) $$\fint_{B(r)} u \, dx \leq C \left(\fint_{B(R)} u \, dx + \int_r^R \frac{\gamma_{2;r}[B(t) \setminus \Omega]}{t^{n-2}} \frac{dt}{t} \right)$$

for all $0 < r < R < R_0$. Set

$$F(t) := \fint_{B(t)} u \, dx$$

and obtain

$$F'(t) = \frac{d}{dt}\left(\fint_{B(1)} u(tx)\,dx\right)$$

$$= \frac{1}{t}\fint_{B(t)} \nabla u(x) \cdot x\,dx$$

$$= -\frac{1}{t}\fint_{B(t)} \nabla u(x) \cdot \nabla\left(\frac{t^2 - |x|^2}{2}\right)(1-\psi)\,dx$$

$$+ \frac{1}{t}\fint_{B(t)} \nabla u(x) \cdot x\psi\,dx$$

where ψ is chosen as an approximate capacity extremal for $B(t) \setminus \Omega$:

(4.45)
$$\begin{cases} \psi \equiv 1 \quad \text{in a neighborhood of } B(t) \setminus \Omega \\ \psi \in C_c^\infty(B(2t)) \\ \int_{B(2t)} |\nabla \psi|^2\,dx \leq C\gamma_{2;r}[B(t) \setminus \Omega]. \end{cases}$$

Since $\Delta u = 0$ in Ω and $1 - \psi \equiv 0$ near $B(t) \setminus \Omega$, we have

$$\int_{B(t)} \nabla u(x) \cdot \nabla\left(\frac{t^2 - |x|^2}{2}\right)(1-\psi(x))\,dx = \int_{B(t)} \nabla u(x) \cdot \nabla \psi(x)\left(\frac{t^2 - |x|^2}{2}\right)\,dx$$

$$\leq Ct^2 \|\nabla u\|_{2,B(t)} \|\nabla \psi\|_{2,B(2t)}.$$

This, combined with the estimate

$$\left|\int_{B(t)} \nabla u(x) \cdot x\psi\,dx\right| \leq t\|\nabla u\|_{2,B(t)} \|\psi\|_{2,B(2t)}$$

$$\leq Ct^2 \|\nabla u\|_{2,B(t)} \|\nabla \psi\|_{2,B(2t)}$$

yields, with the help of Young's inequality,

$$F'(t) \geq -\frac{C}{t^{n-1}} \|\nabla u\|_{2,B(t)} \|\nabla \psi\|_{2,B(2t)}$$

$$\geq -\frac{C\delta}{t^{n-1}} \int_{B(t)} |\nabla u|^2\,dx - \frac{C}{\delta\, t^{n-1}} \frac{\gamma_{2;r}[B(t) \setminus \Omega]}{t^{n-1}}$$

for every $\delta > 0$. Thus, integrating from r to R, we have

$$F(r) \leq F(R) + \frac{C}{\delta} \int_r^R \frac{\gamma_{2;r}[B(t) \setminus \Omega]}{t^{n-2}} \frac{dt}{t}$$

$$+ C\delta \int_r^R \left(\frac{1}{t^{n-2}} \int_{B(t)} |\nabla u|^2\,dx\right) \frac{dt}{t}.$$

Let

$$\lambda(s) := \inf_{B(s)} u$$

and use Lemma 4.23 with v replaced by $u - \lambda(s)$ to obtain

$$\frac{1}{s^{n-2}} \int_{B(s/4)} |\nabla u|^2\,dx \leq C[\lambda(s/4) - \lambda(s)]$$

4.2 SUFFICIENCY OF THE WIENER CONDITION FOR BOUNDARY REGULARITY

and
$$\int_r^R \left(\frac{1}{t^{n-2}} \int_{B(t)} |\nabla u|^2 \, dx \right) \le C \int_{4r}^{4R} \left(\frac{1}{s^{n-2}} \int_{B(s/4)} |\nabla u|^2 \, dx \right)$$
$$\le C \int_{4r}^{4R} [\lambda(s/4) - \lambda(s)] \frac{dt}{t} \le C\lambda(r)$$
$$\le C \fint_{B(r)} u \, dx.$$

With δ chosen small enough, this establishes (4.44).

We now show that (4.43) and (4.44) imply that x_0 is irregular. Set
$$g(x) = 1 - \frac{|x - x_0|^2}{(4R)^2}$$
for R to be selected below. Notice that g is superharmonic on $\mathbf{R}^n$ and $g = 0$ on $\partial B(4R)$. Write $\Omega' = \Omega \cap B(4R)$. Let h' be the solution of the Dirichlet problem in Ω' with the boundary function g and
$$u := \begin{cases} h' & \text{in } \Omega', \\ g & \text{in } \mathbf{R}^n \setminus \Omega'. \end{cases}$$

According to the observation at the beginning of the subsection, and elementary comparison principles, u is a supersolution in $\mathbf{R}^n$ and $0 \le u \le g \le 1$ in $B(4R)$. Let ψ be an approximate capacitary extremal for $B(t) \setminus \Omega$, as in (4.45), now with $t = 4R$. Let $\varphi := u(1-\psi)^2$ and note $\varphi \in W_0^{1,2}(\Omega \cap B(4R))$. Thus,
$$\int_{B(4R)} \nabla u \cdot \nabla \varphi \, dx = 0.$$

By Young's inequality, this leads to
$$\int_{B(4R)} |\nabla u|^2 (1-\psi)^2 \, dx \le C \int_{B(4R)} u^2 |\nabla \psi|^2 \le C \int_{B(4R)} |\nabla \psi|^2 \, dx$$
$$\le C \gamma_{2;r}[B(4R) \setminus \Omega].$$

With $v := u(1-\psi)$, Poincaré's inequality (Theorem 1.57, p.33) implies
$$\int_{B(4R)} u^2(1-\psi)^2 \, dx = \int_{B(4R)} v^2 \, dx \le CR^2 \int_{B(4R)} |\nabla v|^2 \, dx$$
$$\le CR^2 \int_{B(4R)} |\nabla u|^2 (1-\psi)^2 + CR^2 \int_{B(4R)} |\nabla \psi|^2 \, dx$$
$$\le CR^2 \gamma_{2;r}[B(4R) \setminus \Omega]$$

and
$$\int_{B(4R)} u^2 \psi^2 \, dx \le \int_{B(4R)} \psi^2 \, dx \le CR^2 \int_{B(4R)} |\nabla \psi|^2 \, dx$$
$$\le CR^2 \gamma_{2;r}[B(4R) \setminus \Omega].$$

Hence
$$\fint_{B(R)} u^2 \, dx \le 2 \fint_{B(R)} u^2(1-\psi)^2 \, dx + 2 \fint_{B(R)} u^2 \psi^2 \, dx$$
$$\le C \frac{\gamma_{2;r}[B(4R) \setminus \Omega]}{R^{n-2}}.$$

If R is chosen small enough, then (4.43) and (4.44) imply

$$\fint_{B(r)} u\,dx < 1/2$$

for all $0 < r < R$, and therefore

$$\liminf_{\substack{x \to x_0 \\ x \in \Omega'}} u(x) < 1 = g(x_0).$$

Now, let h be the solution of the Dirichlet problem with boundary function g in Ω. Then $h - u \in W_0^{1,2}(\Omega)$, and since u is a supersolution, this implies that $h \le u$ in Ω. Thus

$$\liminf_{\substack{x \to x_0 \\ x \in \Omega}} h(x) < 1 = g(x_0)$$

which shows that x_0 is not regular for Ω. $\square$

4.3 Necessity of the Wiener condition for boundary regularity

In this section we will investigate equations of the form

$$(4.46) \qquad -\operatorname{div} \boldsymbol{A}(x, u, \nabla u) + \boldsymbol{B}(x, u, \nabla u) = \mu,$$

where $\boldsymbol{A}$ and $\boldsymbol{B}$ are as in (3.5), p. 162, and $\mu \in (W_0^{1,p}(\Omega))^*$. The domain Ω is allowed to be p-finely open.

We still assume that $c_1 = 1$ and $b_2 = c_2$, but for easier orientation sometimes use distinguishing notation. We impose a more restrictive assumption on b_2 and b_1, namely that not only b_1^p and b_2 but also b_1^q and $b_2^{q/p}$ belong to $\mathcal{M}^{n/(p-\epsilon)}(\mathbf{R}^n)$, where $q > p$ is a fixed exponent such that $(n-p)q < np$ and ϵ is determined by q, namely

$$\epsilon = (n-p)\left(\frac{q}{p} - 1\right).$$

Also, let τ be defined by

$$\frac{1}{\tau} = \frac{p-1}{q} + \frac{1}{p}$$

and

$$\gamma = (p-1)\tau.$$

Notice that $\tau > 1$. For the dependence of the constant involving the radius, we use the scaling convention that

$$r^\epsilon \|b_1^p + b_0\|_{\mathcal{M}^{n/(p-\epsilon)}(B(r))} \le C, \qquad r^{q-p+\epsilon} \|(b_1^p + b_2)^{q/p}\|_{\mathcal{M}^{n/(p-\epsilon)}(B(r))} \le C,$$

which is compatible with Remark 3.2. We use $\boldsymbol{k} = \boldsymbol{k}(r)$ as in (3.7). We redefine

$$\boldsymbol{a}(r) = \|a_2^{1/(p-1)}\|_{M(p,p);B(r)},$$

write

$$\boldsymbol{b}(r) = \left(r^\epsilon \|b_1^p + b_2\|_{\mathcal{M}^{n/(p-\epsilon)}(B(r))}\right)^{1/(p-1)} + \left(r^\epsilon \|c_2\|_{\mathcal{M}^{n/(p-\epsilon)}(B(r))}\right)^{1/p}$$

and assume throughout the section that

$$(4.47) \qquad \int_0^{r_0} [\boldsymbol{k}(r) + \boldsymbol{a}(r)]\,\frac{dr}{r} < \infty.$$

4.3.1 Main estimate.

The main result of this subsection is Theorem 4.27 which, aside from its own intrinsic interest, is used later to establish the necessity of the Wiener condition.

4.25 LEMMA. *Let u be a subsolution of (4.46) in $\Omega \subset B(r)$ and $l \geq 0$. Suppose that either u is bounded above or $b_0 = 0$. Let Φ be a nonnegative bounded Borel measurable function on $(0, \infty)$ which vanishes on $(0, l)$ and λ be the L^1-norm of Φ. Let $\omega \in W_c^{1,p}(\Omega)$, $0 \leq \omega \leq 1$. Then*

(4.48)
$$\int_L \Phi(u)(c_1|\nabla u|^p - c_2 u^p - c_3)\omega^p\, dx$$
$$\leq \lambda \int_L (b_0|\nabla u|^p + b_1|\nabla u|^{p-1} + b_2 u^{p-1} + b_3)\omega^p\, dx$$
$$+ p\lambda \int_L (a_1|\nabla u|^{p-1} + a_2 u^{p-1} + a_3)\omega^{p-1}|\nabla \omega|\, dx$$
$$+ \lambda \mu(\{\omega > 0\}),$$

where $L = \{x \in \Omega: u(x) > l\}$.

PROOF. We write
$$\Psi(t) = \int_0^t \Phi(s)\, ds,$$
$$L = \Omega \cap \{u > l\}.$$

Using the test function
$$\varphi(x) = \Psi(u(x))\, \omega^p(x)$$
we obtain

(4.49)
$$\int_L \boldsymbol{A}(x, u, \nabla u) \cdot \nabla u\, \Phi(u)\, \omega^p\, dx$$
$$+ p \int_L \boldsymbol{A}(x, u, \nabla u) \cdot \Psi(u)\, \omega^{p-1} \nabla \omega\, dx$$
$$+ \int_L \boldsymbol{B}(x, u, \nabla u)\, \Psi(u)\, \omega^p\, dx$$
$$\leq \int_L \Psi(u)\, \omega^p\, d\mu.$$

If Ω is only p-finely open, we first use the test function $\Psi(u)([\omega - \varepsilon]^+)^p \in W_{cc}^{1,p}(\Omega)$ and at the end of the proof let $\varepsilon \downarrow 0$. Taking the structure into account, we have

(4.50)
$$\int_L \boldsymbol{A}(x, u, \nabla u) \cdot \nabla u\, \Phi(u)\, \omega^p\, dx$$
$$\geq \int_L \Phi(u)(|\nabla u|^p - c_2 u^p - c_3)\omega^p\, dx,$$

(4.51)
$$-\int_L \boldsymbol{A}(x, u, \nabla u) \cdot \Psi(u)\, \omega^{p-1} \nabla \omega\, dx$$
$$\leq \int_L \Psi(u)\, (a_1|\nabla u|^{p-1} + a_2 u^{p-1} + a_3)\omega^{p-1}|\nabla \omega|\, dx$$

and

(4.52)
$$-\int_L \boldsymbol{B}(x,u,\nabla u)\,\Psi(u)\,\omega^p\,dx$$
$$\leq \int_L \Psi(u)(b_0|\nabla u|^p + b_1|\nabla u|^{p-1} + b_2 u^{p-1} + b_2)\omega^p\,dx.$$

From (4.49) – (4.52) we obtain

(4.53)
$$\int_L \Phi(u)(c_1|\nabla u|^p - c_2 u^p - c_3)\,\omega^p\,dx$$
$$\leq \int_L \Psi(u)(b_0|\nabla u|^p + b_1|\nabla u|^{p-1} + b_2 u^{p-1} + b_3)\omega^p\,dx$$
$$+ \int_L \Psi(u)\,(a_1|\nabla u|^{p-1} + a_2 u^{p-1} + a_3)\omega^{p-1}|\nabla\omega|\,dx$$
$$+ \int_L \Psi(u)\omega^p\,d\mu.$$

Since $\omega \leq 1$ and $\Psi \leq \lambda$, we arrive at (4.48). □

4.26 LEMMA. *Let u be a subsolution of* (4.46) *in Ω. Suppose that either u is bounded above or $b_0 = 0$. Let $B = B(x_0, r)$ be an open ball in $\mathbf{R}^n$. Let $\eta, \varphi, \psi \in W^{1,p}(B)$ with $0 \leq \eta \leq 1$, $0 \leq \varphi \leq 1$, $0 \leq \psi \leq 1$, $\eta\psi \in W^{1,p}(B\cap\Omega)$, $(1-\varphi)(1-\psi) = 0$, $\nabla\eta \leq 5/r$ and*
$$\int_B |\nabla\omega|^p\,dx \leq Cr^{n-p}.$$

Suppose that $l \geq 0$. Write
$$\omega = \psi\eta,$$
$$\sigma = \omega\varphi,$$
$$M = \|u\|_\infty + \boldsymbol{k}(r),$$
$$L = B \cap \Omega \cap \{u > l\},$$
$$E = L \cap \{\varphi < 1\},$$
$$F = L \cap \{\varphi = 1\}.$$

(i) *There are constants $C, K > 0$ depending only on n, p, γ, the upper bound of $b_0 u$ and on the structure such that, if $\delta > 0$ and*
$$r^{-n}\int_E (1+v)^\gamma\,dx \leq K,$$

then

(4.54)
$$\int_L |\nabla w|^p \omega^p\,dx \leq Cr^{n-p}\left[\left(\frac{\boldsymbol{b}(r)(l+\delta) + \boldsymbol{k}(r)}{\delta}\right)^p\right.$$
$$\left.+ \left(\frac{\boldsymbol{b}(r)(l+\delta) + \boldsymbol{k}(r)}{\delta}\right)^{p-1} + \left(\frac{\boldsymbol{a}(r)l}{\delta}\right)^{p-1}\right]$$
$$+ Cr^{-p}\int_E (1+v)^\gamma\,dx$$
$$+ C\delta^{1-p}M^{p-1}\int_B |\nabla\sigma|^p\,dx + C\delta^{1-p}\mu(B),$$

where
$$w = w_\delta = (1+v)^{\gamma/q} - 1,$$
$$v = v_\delta = \frac{(u-l)^+}{\delta}.$$

(ii) There are constants C and $\kappa > 0$, depending only on n, p, γ, R_0, the upper bound of $b_0 u$ and on the structure, such that

(4.55)
$$\left(r^{-n}\int_L (u-l)^\gamma \omega^q \, dx\right)^{(p-1)/\gamma}$$
$$\leq C\bigl[\boldsymbol{b}(r)(l+\delta) + \boldsymbol{k}(r) + \boldsymbol{a}(r)l\bigr]^{p-1}$$
$$+ C r^{p-n} \mu(B(x_0, r))$$
$$+ CM^{p-1} r^{p-n} \int_B |\nabla\sigma|^p \, dx,$$

provided that

(4.56)
$$|E| \leq (2r)^n \kappa$$

and

(4.57)
$$\int_E (u-l)^\gamma \, dx \leq 2^{n+\gamma} \int_L (u-l)^\gamma \omega^q \, dx.$$

PROOF. By Lemma 3.4 we may assume that $b_0 = 0$.
(i) Notice that
$$\nabla w = \frac{\gamma}{q}(1+v)^{-\tau/p}\nabla v.$$
Since
$$w^q \leq (1+v)^\gamma \leq C(1+w^q)$$
$$w^p \leq C\min\{v^{p-\tau}, v^p\} \leq C\min\{(1+v)^\gamma, v^{p-1}\},$$
$$v^{p-1} \leq C(1+w^p),$$
(4.58) $\quad v^p(1+v)^{-\tau} \leq Cw^p,$
$$v^{p-1} \leq \delta^{1-p} u^{p-1} \leq \delta^{1-p} M^{p-1},$$
$$\omega = \eta \text{ on } E,$$
$$\omega = \sigma \text{ on } F,$$

it follows that
$$\int_L |\nabla(w\,\omega)|^p \, dx$$
(4.59) $\quad \leq C\left(\int_E (1+v)^\gamma |\nabla\eta|^p \, dx + M^{p-1}\delta^{1-p}\int_F |\nabla\sigma|^p \, dx\right)$
$$+ \delta^{-p}\int_L (1+v)^{-\tau}|\nabla u|^p\, \omega^p \, dx.$$

We use Lemma 4.25 with
$$\Phi(t) = \begin{cases} (1 + \frac{(t-l)^+}{\delta})^{-\tau}, & t > l, \\ 0, & t \leq l. \end{cases}$$

Note that the L^1-norm of Φ is bounded by $(\tau-1)^{-1}\delta$ and $u = l + \delta v$. We obtain

$$\int_L (1+v)^{-\tau} |\nabla u|^p \, \omega^p \, dx$$
$$\leq C \int_L \left[(c_2 u^p + c_3)(1+v)^{-\tau} + \delta(b_1 |\nabla u|^{p-1} + b_2 u^{p-1} + b_3) \right] \omega^p \, dx$$
$$+ C\delta \int_L (a_1 |\nabla u|^{p-1} + a_2 u^{p-1} + a_3) \omega^{p-1} |\nabla \omega| \, dx$$
$$+ C\delta \mu(\{\omega > 0\}).$$

Since $u = l + \delta v$, we have

(4.60)
$$\int_L (1+v)^{-\tau} |\nabla u|^p \, \omega^p \, dx$$
$$\leq C \int_L \left[c_2 l^p + \delta b_2 l^{p-1} + \delta b_3 + c_3 \right] \omega^p \, dx$$
$$+ C\delta \int_L [a_2 l^{p-1} + a_3] \omega^{p-1} |\nabla \omega| \, dx$$
$$+ C\delta \int_L (b_1 |\nabla u|^{p-1} \omega^p + a_1 |\nabla u|^{p-1} \omega^{p-1} |\nabla \omega|) \, dx$$
$$+ C \int_L \left[\delta^{p-1} a_2 v^{p-1} \omega^{p-1} |\nabla \omega| + \delta^p (b_2 v^p (1+v)^{-\tau} + b_2 v^{p-1}) \omega^p \right] dx$$
$$+ C\delta \mu(B).$$

If $\|a^{1/(p-1)}\| \in M(W_0^{1,p}(B), L^p(B))$, we have

(4.61)
$$\int_E a^{p-1} \omega^{p-1} |\nabla \omega| \, dx$$
$$\leq C \left(\int_B |\nabla \omega|^p \right)^{1/p} \left(\int_B a^{p'} \omega^p \, dx \right)^{1-1/p}$$
$$\leq C \|a^{1/(p-1)}\|_{M(p,p);B} \int_B |\nabla \omega|^p \, dx$$
$$\leq C r^{n-p} \|a^{1/(p-1)}\|_{M(p,p);B}$$

and for $b \in \mathcal{M}^{n/(p-\epsilon)}(B)$ we have

(4.62)
$$\int_B b \leq r^{n-p} \left(r^\epsilon \|b\|_{\mathcal{M}^{n/(p-\epsilon)}(B)} \right).$$

We abbreviate

$$\boldsymbol{d}(r) := r^{n-p} \left(\frac{\boldsymbol{b}(r)(l+\delta) + \boldsymbol{k}(r)}{\delta} \right)^p + \left(\frac{\boldsymbol{b}(r)(l+\delta) + \boldsymbol{k}(r)}{\delta} \right)^{p-1} + \left(\frac{\boldsymbol{a}(r) l}{\delta} \right)^{p-1}.$$

We split the integration domain as $L = E \cup F$. First, we estimate integration on E. Young's inequality together with the decomposition

$$1 = (1+v)^{-\tau/p'} (1+v)^{\gamma/p}$$

4.3 NECESSITY OF THE WIENER CONDITION FOR BOUNDARY REGULARITY

will be the main tools in subsequent estimates. For any $\varepsilon_1 > 0$ we have

$$
\begin{aligned}
\delta &\int_E \left(b_1|\nabla u|^{p-1}\omega^p + a_1|\nabla u|^{p-1}\omega^{p-1}|\nabla\omega|\right) dx \\
&\leq \varepsilon_1 \int_L (1+v)^{-\tau} |\nabla u|^p \,\omega^p \, dx \\
&\quad + C\varepsilon_1^{1-p}\delta^p \int_E (b_1^p \eta^p + a_1^p|\nabla\eta|^p)(1+v)^\gamma \, dx \\
&\leq C\varepsilon_1 \int_L (1+v)^{-\tau} |\nabla u|^p \,\omega^p \, dx \\
&\quad + C\varepsilon_1^{1-p}\delta^p \left(r^{n-p}\mathbf{b}(r)^{p-1} + \int_E b_1^p w^q \omega^p \, dx\right),
\end{aligned}
\tag{4.63}
$$

where one of the terms with b_1^p was treated as in (4.62). This is done also in the following. Since

$$
v^{p-1} \leq \tfrac{1}{p'} v^p (1+v)^{-\tau} + \tfrac{1}{p}(1+v)^\gamma,
\tag{4.64}
$$

we have

$$
\begin{aligned}
\int_E &\left[b_2 v^{p-1} + b_2 v^p (1+v)^{-\tau}\right] \omega^p \, dx \\
&\leq C \int_E b_2 (1+w^q) \omega^p \, dx \leq C\mathbf{b}(r)^p r^{n-p} + C \int_E b_2 w^q \omega^p \, dx.
\end{aligned}
$$

By Corollary 1.93,

$$
\begin{aligned}
\int_E &(b_1^p + b_2) w^q \omega^p \, dx \\
&\leq \left(r^{-n} \int_E (1+v)^\gamma\right)^{1-p/q} \left(\int_B r^{q-p+\epsilon}(b_1^p + b_2)^{q/p}(w\omega)^q\right)^{p/q} \\
&\leq C K^{1-p/q} \int_B |\nabla(w\omega)|^p \, dx.
\end{aligned}
\tag{4.65}
$$

Using the multiplier property of $a_2^{1/(p-1)}$ and (4.64),

$$
\begin{aligned}
\int_E &a_2 v^{p-1} \eta^{p-1} |\nabla\eta| \, dx \\
&\leq \varepsilon_1 \int_E a_2^{p'} (w\omega)^p \, dx + C\varepsilon_1^{1-p} \int_E (1+v)^\gamma |\nabla\eta|^p \, dx. \\
&\leq \varepsilon_1 \int_B |\nabla(w\omega)|^p \, dx + C\varepsilon_1^{1-p} \int_E (1+v)^\gamma |\nabla\eta|^p \, dx.
\end{aligned}
\tag{4.66}
$$

Getting together the estimates of integrals over E and the estimates of the remaining terms by (4.61) and (4.62) we arrive at

(4.67)
$$\begin{aligned}\int_E (1+v)^{-\tau} |\nabla u|^p \, \omega^p \, dx \\ \leq C\big[\varepsilon_1 + K^{1-p/q}\big] \int_B |\nabla(w\,\omega)|^p \, dx \\ + C(1+\varepsilon_1^{1-p})\Big(\delta^p r^{-p} \int_E (1+v)^\gamma \, dx \\ + \delta^p \boldsymbol{d}(r) + \delta\mu(B)\Big).\end{aligned}$$

Now, we will estimate the integration over F. Recall that $\omega = \sigma$ on F. We use (4.48) again with Φ being the characteristic function of the interval $[l, M]$ and with σ instead of ω. Then the L^1-norm of Φ is bounded by M and we get

$$\begin{aligned}\int_L |\nabla u|^p \, \sigma^p \, dx \\ \leq CM \int_L \big(a_1 |\nabla u|^{p-1} \sigma^{p-1} |\nabla\sigma| + b_1 |\nabla u|^{p-1} \sigma^p\big) \, dx \\ + C \int_L \big[(c_2 u^p + c_3) + M(b_2 u^{p-1} + b_3)\big]\sigma^p \\ + CM \int_L (a_2 u^{p-1} + a_3)\sigma^{p-1}|\nabla\sigma| \, dx + CM\mu(\{\sigma > 0\}).\end{aligned}$$

We obtain

(4.68)
$$\begin{aligned}\int_L |\nabla u|^p \, \sigma^p \, dx \\ \leq C \int_B [M^p b_2 + M b_3 + c_3]\sigma^p \, dx \\ + CM \int_B (a_2 M^{p-1} + a_3)\sigma^{p-1}|\nabla\sigma| \, dx \\ + CM \int_L \big(a_1 |\nabla u|^{p-1} \sigma^{p-1} |\nabla\sigma| + b_1 |\nabla u|^{p-1} \sigma^p\big) \, dx + CM \mu(B).\end{aligned}$$

Choose $\varepsilon_2 > 0$. A use of Young's inequality yields

(4.69)
$$\begin{aligned}|\nabla u|^{p-1} \sigma^{p-1}(b_1\sigma + a_1|\nabla\sigma|) \\ \leq \frac{\varepsilon_2}{M} |\nabla u|^p \sigma^p + \Big(\frac{\varepsilon_2}{M}\Big)^{1-p}(b_1^p \sigma^p + a_1^p |\nabla\sigma|^p).\end{aligned}$$

4.3 NECESSITY OF THE WIENER CONDITION FOR BOUNDARY REGULARITY

By the multiplier properties of coefficients we have

(4.70)
$$\int_B \left[a_2 \sigma^{p-1} |\nabla \sigma| + (b_1^p + b_2) \sigma^p \right] dx$$
$$\leq C \Big(\int_B a_2^{p'} \sigma^p \, dx \Big)^{1-1/p} \Big(\int_B |\nabla \sigma|^p \, dx \Big)^{1/p}$$
$$+ C \Big(r^{q-p+\epsilon} \int_B (b_1^p + b_2)^{q/p} \sigma^q \, dx \Big)^{p/q}$$
$$\leq C \int_B |\nabla \sigma|^p \, dx.$$

From (4.68), (4.69), (4.61), (4.62) and (4.70) we obtain

$$\delta^p \int_F \left[(b_2 v^p (1+v)^{-\tau} + b_2 v^{p-1}) \omega^p + a_2 v^{p-1} \omega |\nabla \omega| \right] dx$$
$$+ \delta \int_F (b_1 \omega + a_1 |\nabla \omega|) |\nabla u|^{p-1} \omega^{p-1} \, dx$$
$$\leq C \delta M^{p-1} \int_B (b_2 \sigma^p + a_2 \sigma^{p-1} |\nabla \sigma|) \, dx$$
$$+ C \frac{\delta \varepsilon_2}{M} \int_F |\nabla u|^p \sigma^p \, dx + C \delta \varepsilon_2^{1-p} M^{p-1} \int_F (b_1^p \sigma^p + a_1^p |\nabla \sigma|^p) \, dx$$
$$\leq C \varepsilon_2 \delta \int_L \left(a_1 |\nabla u|^{p-1} \sigma^{p-1} |\nabla \sigma| + b_1 |\nabla u|^{p-1} \sigma^p \right) dx$$
$$+ C(1 + \varepsilon_2^{1-p}) \Big(M^{p-1} \delta \int_B |\nabla \sigma|^p \, dx + \delta^p \mathbf{d}(r) \Big)$$
$$+ C \delta \mu B.$$

By an appropriate choice of ε_2 it follows

(4.71)
$$\delta^p \int_F \left[(b_2 v^p (1+v)^{-\tau} + b_2 v^{p-1}) \omega^p + a_2 v^{p-1} \omega |\nabla \omega| \right] dx$$
$$+ \delta \int_F (b_1 \omega + a_1 |\nabla \omega|) |\nabla u|^{p-1} \omega^{p-1} \, dx$$
$$\leq C M^{p-1} \delta \int_B |\nabla \sigma|^p \, dx + C \delta^p \mathbf{d}(r) + C \delta \mu B.$$

Now we are ready to get together the estimates of the right-hand side of (4.60). From (4.61), (4.62), (4.67) and (4.71) we deduce that

(4.72)
$$\int_L (1+v)^{-\tau} |\nabla u|^p \, \omega^p \, dx$$
$$\leq C \big[\varepsilon_1 + K^{1-p/q} \big] \int_B |\nabla (w \omega)|^p \, dx$$
$$+ C(1 + \varepsilon_1^{1-p}) \Big[\delta^p r^{-p} \int_E (1+v)^\gamma \, dx$$
$$+ C M^{p-1} \delta \int_B |\nabla \sigma|^p \, dx + C \delta^p \mathbf{d}(r) + \delta \mu(B) \Big].$$

Using (4.59) and an appropriate choice of ε_1 and K we infer that

$$\int_L |\nabla(w\,\omega)|^p \, dx$$
(4.73)
$$\leq C\delta^{-p} r^{-p} \int_E (1+v)^\gamma \, dx$$
$$+ C\boldsymbol{d}(r) + C\delta^{1-p} M^{p-1} \int_B |\nabla \sigma|^p \, dx + C\delta^{1-p} \mu(B).$$

This proves the (i).

(ii) Fix $\kappa > 0$ to be specified later. We use part (i) with the choice

$$\delta = \Big(\frac{1}{\kappa r^n} \int_L (u-l)^\gamma \omega^q \, dx\Big)^{1/\gamma},$$
$$K = 2^{n+2\gamma-1} \kappa.$$

The part (i) gives the first restriction to the choice of κ. Notice that

(4.74)
$$\kappa = r^{-n} \int_L v^\gamma \omega^q \, dx.$$

By (4.56) and (4.57),

(4.75)
$$\int_E (1+v)^\gamma \, dx \leq 2^{\gamma-1}\Big(|E| + \int_E v^\gamma \, dx\Big)$$
$$\leq 2^{\gamma+n-1} \kappa r^n + 2^{2\gamma+n-1} \int_L v^\gamma \omega^q \, dx$$
$$\leq 2^{2\gamma+n-1} \kappa r^n,$$

which justifies the choice of K. By (4.56) and (4.74),

$$2\kappa r^n = 2 \int_L v^\gamma \omega^q \, dx$$
$$\leq 2^{-n} \int_L \omega^q \, dx + \int_{L \cap \{v^\gamma \geq 2^{-n-1}\}} v^\gamma \omega^q \, dx$$
$$\leq 2^{-n}\Big(|E| + \int_F \sigma^q \, dx\Big) + \int_{L \cap \{v^\gamma \geq 2^{-n-1}\}} v^\gamma \omega^q \, dx$$
$$\leq \kappa r^n + \int_{L \cap \{v^\gamma \geq 2^{-n-1}\}} v^\gamma \omega^q \, dx + \int_B \sigma^q \, dx$$

and thus

$$\kappa r^n \leq \int_{B \cap \{v^\gamma \geq 2^{-n-1}\}} v^\gamma \omega^q \, dx + \int_B \sigma^q \, dx$$
$$\leq C\Big(\int_L w^q \omega^q \, dx + \int_B \sigma^q \, dx\Big).$$

We apply the Sobolev inequality to the functions $w\omega$ and σ and obtain

(4.76)
$$\Big(r^{-n} \int_{B \cap \Omega} w^q \omega^q \, dx\Big)^{p/q} \leq C r^{p-n} \int_{B \cap \Omega} |\nabla(w\,\omega)|^p \, dx$$

and

(4.77)
$$\left(r^{-n}\int_B \sigma^q\,dx\right)^{p/q} \leq C r^{p-n}\int_B |\nabla\sigma|^p\,dx.$$

Hence

(4.78)
$$\kappa^{p/q} \leq C r^{p-n}\left(\int_{B\cap\Omega}|\nabla(w\,\omega)|^p\,dx + \int_B |\nabla\sigma|^p\,dx\right).$$

We infer from (4.73) that there are constants C_1, C_2, C such that

(4.79)
$$\begin{aligned}\kappa^{p/q} \leq\ & C_1\kappa \\ & + C_2\Big[\Big(\frac{\boldsymbol{b}(r)(l+\delta)+\boldsymbol{k}(r)}{\delta}\Big)^p \\ & + \Big(\frac{\boldsymbol{b}(r)(l+\delta)+\boldsymbol{k}(r)}{\delta}\Big)^{p-1} + \Big(\frac{\boldsymbol{a}(r)l}{\delta}\Big)^{p-1} \\ & + M^{p-1}\int_B|\nabla\sigma|^p\,dx + \mu(B)\Big]\end{aligned}$$

holds for some constants C_1, C_2. If we specify κ to be so small that $C_3 := \kappa^{p/q} - C_1\kappa > 0$, we obtain

$$\begin{aligned}1 \leq\ & C\Big[\Big(\frac{\boldsymbol{b}(r)(l+\delta)+\boldsymbol{k}(r)}{\delta}\Big)^p \\ & + \Big(\frac{\boldsymbol{b}(r)(l+\delta)+\boldsymbol{k}(r)}{\delta}\Big)^{p-1} + \Big(\frac{\boldsymbol{a}(r)l}{\delta}\Big)^{p-1} \\ & + M^{p-1}\int_B|\nabla\sigma|^p\,dx + \mu(B)\Big].\end{aligned}$$

It follows that either
$$\frac{1}{2} \leq C\Big(\frac{\boldsymbol{b}(r)(l+\delta)+\boldsymbol{k}(r)}{\delta}\Big)^p$$
or

(4.80)
$$\begin{aligned}\frac{1}{2} \leq\ & C\Big[\Big(\frac{\boldsymbol{b}(r)(l+\delta)+\boldsymbol{k}(r)}{\delta}\Big)^{p-1} + \Big(\frac{\boldsymbol{a}(r)l}{\delta}\Big)^{p-1} \\ & + M^{p-1}\int_B|\nabla\sigma|^p\,dx + \mu(B)\Big].\end{aligned}$$

In any case we deduce that (4.80) holds (with a change of constant), which is equivalent to (4.55). □

4.27 Theorem. *Let u be a subsolution of (4.46) in Ω. Suppose that either u is bounded above or $b_0 = 0$. Then*

(4.81)
$$\begin{aligned}\text{p-fine-}\limsup_{x\to x_0} u(x) \leq\ & C\Bigg[\Big(r_0^{-n}\int_{B(x_0,r_0)\cap\Omega\cap\{u>0\}} u^\gamma\,dx\Big)^{1/\gamma} \\ & + \int_0^{r_0}\Big(\frac{\mu B(x_0,r)}{r^{n-p}}\Big)^{1/(p-1)}\frac{dr}{r} + \int_0^{r_0}\boldsymbol{k}(r)\frac{dr}{r} \\ & + (\boldsymbol{k}(r_0)+\|u\|_\infty)\int_0^{2r_0}\Big(\frac{\gamma_{p;r}(B(x_0,r)\setminus\Omega)}{r^{n-p}}\Big)^{1/(p-1)}\frac{dr}{r}\Bigg]\end{aligned}$$

for all $x_0 \in \bar{\Omega}$ and $0 < r_0 \leq R_0$. The constant C depends on n, p, γ, R_0, the upper bound of $b_0 u$ and on the structure, in particular on the integral (4.47).

PROOF. We denote $M = \boldsymbol{k}(r_0) + \|u\|_\infty$ and set $\kappa \in (0,1)$ to be the constant from Lemma 4.26. We set $r_j = 2^{-j} r_0$ and select cutoff functions η_j such that $0 \leq \eta_j \leq 1$, $\eta_j = 0$ outside $B(x_0, r_j)$, $\eta_j = 1$ on $B(x_0, r_{j+1})$ and $|\nabla \eta_j| \leq 5/r_j$. Further, we find functions $g_j \in W^{1,p}(\mathbf{R}^n)$ such that $0 \leq g_j \leq 1$, the interior of $\{g_j = 1\}$ contains $B(x_0, r_j) \setminus \Omega$ and

$$\int_{\mathbf{R}^n} (r_j^{-p} g_j^p + |\nabla g_j|^p) \, dx \leq C \gamma_{p;r_j}(B(x_0, r_j) \setminus \Omega). \tag{4.82}$$

We denote
$$\psi_j = \min(1, (2 - 3g_j)^+),$$
$$\varphi_j = \min(1, 3g_j + 3g_{j-1}), \quad j \geq 1,$$
$$B_j = B(x_0, r_j),$$
$$L_j = B_j \cap \Omega \cap \{u \geq l_j\},$$
$$E_j = L_j \cap \{\varphi_j < 1\},$$
$$F_j = L_j \cap \{\varphi_j = 1\},$$
$$\boldsymbol{a}_j = \boldsymbol{a}(r_j), \quad \boldsymbol{b}_j = \boldsymbol{a}(r_j), \quad \boldsymbol{k}_j = \boldsymbol{k}(r_j).$$

Then by (4.82),
$$\int_{B_j} (r_j^{-p} \varphi_j^p + |\nabla \varphi|_j^p) \, dx \leq C \boldsymbol{\gamma}_{p;r_j}(B_{j-1} \setminus \Omega),$$
$$\int_{B_j} |\nabla \psi|_j^p \, dx \leq C \boldsymbol{\gamma}_{p;r_j}(B_j \setminus \Omega). \tag{4.83}$$

We define recursively $l_0 = 0$,
$$l_{j+1} = l_j + \left(\frac{1}{\kappa r_j^n} \int_{L_j} (u - l_j)^\gamma \psi_j^q \eta_j^q \, dx \right)^{1/\gamma}, \qquad j = 0, 1, 2, \ldots.$$

We write
$$\delta_j = l_{j+1} - l_j.$$

We claim that for $j \geq 1$,

$$\delta_j \leq \frac{1}{2} \delta_{j-1} + C \left[(\boldsymbol{a}_j + \boldsymbol{b}_j) l_{j+1} + \boldsymbol{k}_j + \left(\frac{\mu B_j}{r_j^{n-p}} \right)^{1/(p-1)} \right. \tag{4.84}$$
$$\left. + M \left(\frac{\gamma_{p;r_j}(B_{j-1} \setminus \Omega)}{r_j^{n-p}} \right)^{1/(p-1)} \right].$$

This is trivial when $\delta_j \leq \frac{1}{2} \delta_{j-1}$, so assume that $\delta_{j-1} \leq 2\delta_j$. In this case, since $\psi_{j-1} \eta_{j-1} = 1$ on E_j, we have

$$|E_j| \leq \delta_{j-1}^{-\gamma} \int_{L_{j-1}} (u - l_{j-1})^\gamma \psi_{j-1} \eta_{j-1} \, dx \tag{4.85}$$
$$= \kappa r_{j-1}^n \leq 2^n \kappa r_j^n$$

4.3 NECESSITY OF THE WIENER CONDITION FOR BOUNDARY REGULARITY

and

(4.86)
$$\int_{E_j} (u-l_j)^\gamma \, dx$$
$$\leq \int_{L_{j-1}} (u-l_{j-1})^\gamma \psi_{j-1}^q \eta_{j-1}^q \, dx = \delta_{j-1}^\gamma \kappa r_{j-1}^n = 2^{n+\gamma} \delta_j^\gamma \kappa r_j^n$$
$$= 2^{n+\gamma} \int_{L_j} (u-l_j)^\gamma \psi_j^q \eta_j^q \, dx.$$

Thus (4.56) and (4.57) are verified and Lemma 4.26 yields

$$\delta_j \leq C \Bigg[(\boldsymbol{a}_j + \boldsymbol{b}_j) \, l_{j+1} + \boldsymbol{k}_j + \Big(\frac{\mu B_j}{r_j^{n-p}} \Big)^{1/(p-1)}$$
$$+ M \Big(\frac{\gamma_{p;r_j}(B_{j-1} \setminus \Omega)}{r_j^{n-p}} \Big)^{1/(p-1)} \Bigg]$$

which proves (4.84). Summing (4.84) for $j = 1, \ldots, k$ we get

$$\frac{1}{2} l_{k+1} = \frac{1}{2} (\delta_0 + \cdots + \delta_k) \leq \delta_k + \frac{1}{2}(\delta_0 + \cdots + \delta_{k-1})$$
$$\leq \delta_0 + C \Bigg[\sum_{j=1}^k (\boldsymbol{a}_j + \boldsymbol{b}_j) \, l_{j+1} + \boldsymbol{k}_j + \sum_{j=1}^k \Big(\frac{\mu B_j}{r_j^{n-p}} \Big)^{1/(p-1)}$$
$$+ M \sum_{j=1}^k \Big(\frac{\gamma_{p;r}(B_{j-1} \setminus \Omega)}{r_j^{n-p}} \Big)^{1/(p-1)} \Bigg]$$
$$\leq C \Bigg[\Big(r_0^{-n} \int_{E_0} u^\gamma \, dx \Big)^{1/\gamma}$$
$$+ l_{k+1} \sum_{j=1}^k \int_{r_j}^{r_{j-1}} \Big(l_{k+1} [\boldsymbol{a}(r) + \boldsymbol{b}(r)] + \boldsymbol{k}(r) \Big) \frac{dr}{r}$$
$$+ \sum_{j=1}^k \int_{r_j}^{r_{j-1}} \Big(\frac{\mu B(x_0, r)}{r^{n-p}} \Big)^{1/(p-1)} \frac{dr}{r}$$
$$+ M \sum_{j=1}^k \int_{r_j}^{r_{j-1}} \Big(\frac{\gamma_{p;r}(B(x_0, 2r) \setminus \Omega)}{r^{n-p}} \Big)^{1/(p-1)} \frac{dr}{r} \Bigg].$$

Hence

$$\lim_j l_j \leq C \Bigg[\Big(r_0^{-n} \int_{B(x_0,r_0) \cap \Omega \cap \{u>0\}} u^\gamma \, dx \Big)^{1/\gamma}$$
$$+ \lim_j l_j \int_0^{r_0} [\boldsymbol{a}(r) + \boldsymbol{b}(r)] \frac{dr}{r} + \int_0^{r_0} \boldsymbol{k}(r) \frac{dr}{r}$$
$$+ \int_0^{r_0} \Big(\frac{\mu B(x_0, r)}{r^{n-p}} \Big)^{1/(p-1)} \frac{dr}{r}$$
$$+ M \int_0^{2r_0} \Big(\frac{\gamma_{p;r}(B(x_0, r) \setminus \Omega)}{r^{n-p}} \Big)^{1/(p-1)} \frac{dr}{r} \Bigg].$$

Now we observe that
$$\int_0^{r_0} [a(r) + b(r)] \frac{dr}{r}$$
become small if r_0 is small. It follows that there is R_1 depending only on those quantities as C such that, if $r_0 \leq R_1$,

(4.87)
$$\begin{aligned}
\lim_j l_j \leq C \bigg[&\Big(r_0^{-n} \int_{B(x_0,r_0) \cap \Omega \cap \{u>0\}} u^\gamma \, dx\Big)^{1/\gamma} \\
&+ \int_0^{r_0} k(r) \frac{dr}{r} \\
&+ \int_0^{r_0} \Big(\frac{\mu B(x_0,r)}{r^{n-p}}\Big)^{1/(p-1)} \frac{dr}{r} \\
&+ M \int_0^{2r_0} \Big(\frac{\gamma_{p;r}(B(x_0,r) \setminus \Omega)}{r^{n-p}}\Big)^{1/(p-1)} \frac{dr}{r} \bigg].
\end{aligned}$$

If $R_1 < r_0 < R_0$, then (4.87) holds as well, because then
$$r_0/R_1 \leq R_0/R_1 \leq C.$$

It remains to prove that

(4.88)
$$p\text{-fine-lim sup}_{x \to z} u(x) \leq \lim_{j \to \infty} l_j.$$

We may assume that the right-hand side of (4.81) is finite, otherwise the assertion of the theorem is trivial. We can also assume again that r_0 is small. We choose $\varepsilon > 0$ and denote $l = \lim_j l_j$. Set
$$w_j = (2^{\gamma/q} - 1)^{-1}\Big(\Big(1 + \frac{(u - l - \varepsilon)^+}{\varepsilon}\Big)^{\gamma/q} - 1\Big).$$

Then $w_j \psi_j \eta_j \in W_0^{1,p}(B_j)$, $w_j \psi_j \eta_j \geq 1$ on $B_{j+1} \cap \Omega \cap \{u > l + 2\varepsilon\}$ in the sense of Corollary 2.25 and thus
$$\gamma_{p;r_j}(B_{j+1} \cap \Omega \cap \{u > l + 2\varepsilon\}) \leq C \int_{B_j \cap \Omega} |\nabla(w_j \psi_j \eta_j)|^p \, dx.$$

Denote
$$E'_j = B_j \cap \Omega \cap \{u > l + \varepsilon\} \cap \{\varphi_j < 1\}.$$
Using Lemma 4.26 (i) we obtain
$$\begin{aligned}
&\gamma_{p;r_j}(B_{j+1} \cap \Omega \cap \{u > l + 2\varepsilon\}) \\
&\leq C \int_{B_j \cap \Omega} |\nabla(w_j \psi_j \eta_j)|^p \, dx \\
&\leq C \Big[r_j^{n-p}\big(\varepsilon^{-p}[b_j(l+\varepsilon) + k_j]^p \\
&\quad + \varepsilon^{1-p}[b_j(l+\varepsilon) + a_j l + k_j]^{p-1}\big) \\
&\quad + r_j^{-p} \int_{E'_j} \Big(1 + \frac{u - l - \varepsilon}{\varepsilon}\Big)^\gamma dx + \varepsilon^{1-p} \mu(B_j) \\
&\quad + \varepsilon^{1-p}(k_0 + \|u\|_\infty)^{p-1} \int_{B_j} (r_j^{-p} \varphi_j^p + |\nabla \varphi_j|^p + |\nabla \psi_j|^p) \, dx \Big].
\end{aligned}$$

4.3 NECESSITY OF THE WIENER CONDITION FOR BOUNDARY REGULARITY

It follows

(4.89)
$$\sum_{j=1}^{\infty} \Big(\frac{\gamma_{p;r_j}(B_{j+1} \cap \Omega \cap \{u > l + 2\varepsilon\})}{r_j^{n-p}}\Big)^{1/(p-1)}$$
$$\leq C\Big[\sum_{j=1}^{\infty} \varepsilon^{-p'}\big[\boldsymbol{b}_j(l+\varepsilon) + \boldsymbol{k}_j\big]^{p'} + \sum_{j=1}^{\infty} \varepsilon^{-1}\big[\boldsymbol{b}_j(l+\varepsilon) + \boldsymbol{a}_j l + \boldsymbol{k}_j\big]$$
$$+ \sum_{j=1}^{\infty} \Big(r_j^{-n} \int_{E_j'} \Big(1 + \frac{u - l - \varepsilon}{\varepsilon}\Big)^{\gamma} dx\Big)^{1/(p-1)}$$
$$+ \varepsilon^{-1} \sum_{j=1}^{\infty} \Big(\frac{\mu(B_j)}{r_j^{n-p}}\Big)^{1/(p-1)}$$
$$+ \varepsilon^{-1}(\boldsymbol{k}_0 + \|u\|_\infty) \sum_{j=1}^{\infty} \Big(\frac{\gamma_{p;r_j}(B_j \setminus \Omega)}{r_j^{n-p}}\Big)^{1/(p-1)}\Big].$$

By our assumptions,

$$\sum_{j=1}^{\infty} \boldsymbol{a}_j < \infty, \quad \sum_{j=1}^{\infty} \boldsymbol{b}_j < \infty, \quad \sum_{j=1}^{\infty} \boldsymbol{k}_j < \infty, \quad \text{and} \quad \sum_{j=1}^{\infty} \delta_j < \infty.$$

If we raise each term of a convergent series to a power $\alpha > 1$, the resulting series converges again. Hence

(4.90)
$$\sum_{j=1}^{\infty} \boldsymbol{b}_j^{p'} < \infty, \quad \sum_{j=1}^{\infty} \boldsymbol{k}_j^{p'} < \infty, \quad \text{and} \quad \sum_{j=1}^{\infty} \delta_j^{\gamma/(p-1)} < \infty.$$

Using (4.83) and (4.90) we estimate

$$\sum_{j=1}^{\infty} \Big(r_j^{-n} \int_{E_j'} \Big(1 + \frac{u - l - \varepsilon}{\varepsilon}\Big)^{\gamma} dx\Big)^{1/(p-1)}$$
$$\leq C \sum_{j=1}^{\infty} \Big(r_j^{-n} \int_{E_j} \varepsilon^{-\gamma}(u - l_{j-1})^{\gamma} dx\Big)^{1/(p-1)}$$
$$\leq C \sum_{j=1}^{\infty} \Big(r_j^{-n} \int_{L_{j-1}} \varepsilon^{-\gamma}(u - l_{j-1})^{\gamma} \psi_{j-1}\varphi_{j-1} dx\Big)^{1/(p-1)}$$
$$\leq C \sum_{j=1}^{\infty} (\kappa \varepsilon^{-\gamma} \delta_{j-1}^{\gamma})^{1/(p-1)} < \infty.$$

If the right-hand side of (4.88) is finite, then the remaining sums on the right-hand side of (4.89) also converge, so that the set

$$\Omega \cap \{u > l + 2\varepsilon\}$$

is p-thin at x_0 for any $\varepsilon > 0$. This proves (4.88), which concludes the proof. □

4.3.2 Necessity of the Wiener condition.

4.28 THEOREM. *In addition to (3.5), suppose that for any $u_0 \in \mathcal{C}_c^1(\mathbf{R}^n)$ there exists $u \in W^{1,p}(\Omega)$ such that*

(4.91)
$$\begin{cases} -\operatorname{div} \boldsymbol{A}(x, u, \nabla u) + \boldsymbol{B}(x, u, \nabla u) = 0, \\ u - u_0 \in W_0^{1,p}(\Omega), \end{cases}$$

and

(4.92)
$$\|u\|_{1,p}^p \leq C(\|u\|_{1,p}^{p-1} + 1)\|u_0\|_{1,p},$$
$$\|u\|_\infty \leq C(\|u_0\|_\infty + 1)$$

with a constant C independent of u_0. Let $x_0 \in \overline{\Omega} \setminus \Omega$ and suppose that $\mathbf{R}^n \setminus \Omega$ is p-thin at x_0. Then x_0 is irregular for the equation

$$-\operatorname{div} \boldsymbol{A}(x, u, \nabla u) + \boldsymbol{B}(x, u, \nabla u) = 0.$$

PROOF. Choose $\varepsilon > 0, \rho \in (0, 1)$ to be specified later. The singleton $\{x_0\}$ has zero p-capacity. Hence, we find a $\mathcal{C}^1$-function u_0 on $\mathbf{R}^n$ supported in $B(x_0, 1)$ such that $u_0(x_0) = 1$ and $\|u_0\|_{1,p} < \varepsilon^p$. Let u be a continuous solution of (4.91), (4.92). Then

$$\left(\int_\Omega |u|^p\right)^{1/p} \leq C\varepsilon.$$

By Theorem 4.27,

$$p\text{-fine-}\limsup_{x \to x_0} u(x) \leq C_1 \left(\rho^{-n} \int_{B(x_0,\rho) \cap \Omega} u^\gamma \, dx\right)^{1/\gamma}$$
$$+ C_2(\|u\|_\infty + k(r_0)) \int_0^\rho \left(\frac{\gamma_{p;r}(B(x_0, r) \setminus \Omega)}{r^{n-p}}\right)^{1/(p-1)} \frac{dr}{r}$$
$$+ C_2 \int_0^\rho k(r) \frac{dr}{r}.$$

Hölder's inequality yields

$$\left(\rho^{-n} \int_{B(x_0,\rho) \cap \Omega} |u^\gamma| \, dx\right)^{1/\gamma} \leq C\rho^{-n/p} \left(\int_{B(x_0,\rho) \cap \Omega} |u|^p \, dx\right)^{1/p} \leq C_3 \rho^{-n/p} \varepsilon.$$

Since $\mathbf{R}^n \setminus \Omega$ is p-thin at x_0, we can find $\rho \in (0, 1)$ such that

$$C_2(\|u\|_\infty + k(r_0)) \int_0^\rho \left(\frac{\gamma_{p;r}(B(x_0, r) \setminus \Omega)}{r^{n-p}}\right)^{1/(p-1)} \frac{dr}{r}$$
$$+ C_2 \int_0^\rho k(r) \frac{dr}{r} < \frac{1}{3}.$$

Then we can specify the choice of ε so that

$$C_1 C_3 \rho^{-n/p} \varepsilon \leq \frac{1}{3}.$$

We obtain that

$$p\text{-fine-}\limsup_{x \to x_0} u(x) < 1 = u_0(x_0),$$

hence x_0 is not regular. $\square$

4.29 REMARK. The assumptions of Theorem 4.28, namely the existence of a solution satisfying the condition (4.92), are verified for example by the equation studied in Section 6.1 (Underlying Assumptions 6.1, p. 253), with $c_3 = 0$. See Theorem 6.12, Theorem 6.4 and the proof of Theorem 6.3, namely (6.12).

4.4 Equations with measure data

In this section, we study supersolutions of (3.1) up to all levels. We suppose that $b_0 = 0$. We show that, in a certain sense, they solve the equation

$$(4.93) \qquad -\operatorname{div} \boldsymbol{A}(x, u, Du) + \boldsymbol{B}(x, u, Du) = \mu,$$

with a nonnegative Radon measure μ. Then we estimate u in terms of the right-hand side μ. The dependence on μ leads to the Wolff nonlinear potential

$$\mathbf{W}_p^\mu(x) = \int_0^{r_0} \left(\frac{\mu(B(x,r))}{r^{n-p}}\right)^{1/(p-1)} \frac{dr}{r}.$$

We call attention to the similarity between the Wolff potential and the Wiener integral

$$\int_0^{r_0} \left(\frac{\gamma_{p;r}(E \cap B(x,r))}{r^{n-p}}\right)^{1/(p-1)} \frac{dr}{r}$$

occuring in the definition of thinness.

Recall that Du denotes the approximate derivative of a function u. If u is a nonnegative supersolution up to all levels and $B(r)$ are balls with a fixed center x_0, we write

$$m(r) = \inf_{B(r)} u,$$

$$\boldsymbol{\kappa}(r) = \boldsymbol{k}(r) + m(r)\boldsymbol{h}(r).$$

We consider the p-finely continuous representative of u (it exists and is unique by Theorem 4.8), so that in particular u is pointwise defined.

4.30 THEOREM. *Let u be a nonnegative supersolution of (3.1) up to all levels on Ω. Then*

$$\boldsymbol{A}(x, u, Du), \ \boldsymbol{B}(x, u, Du) \in L^1_{\text{loc}}(\Omega)$$

and

$$\int_\Omega \left(\boldsymbol{A}(x, u, Du) \cdot \nabla \varphi + \boldsymbol{B}(x, u, Du)\,\varphi\right) dx \geq 0$$

for each nonnegative $\varphi \in \mathcal{C}_c^\infty(\Omega)$.

PROOF. Without loss of generality we may assume that $\varphi \leq 1$. Let $k \in \{1, 2, \dots\}$. We use the test function

$$\psi := \begin{cases} \varphi & \text{on } \{u \leq k\}, \\ (2 - \frac{u}{k})\varphi & \text{on } \{k < u < 2k\}, \\ 0 & \text{on } \{u \geq 2k\} \end{cases}$$

to find that $\psi \in W_c^{1,p}(\Omega)$ and

$$0 \leq \psi \leq \begin{cases} 1 \leq 2k + 1 - u & \text{if } u \leq 2k, \\ 0 \leq (2k + 1 - u)^+ & \text{if } u \geq 2k. \end{cases}$$

Since u is a supersolution of (3.1) up to level $2k + 1$, we obtain

$$\int_\Omega \Big(\boldsymbol{A}(x, u, Du) \cdot \nabla \psi + \boldsymbol{B}(x, u, Du) \psi\Big) dx \geq 0.$$

This means that

(4.94)
$$\int_{\Omega \cap \{u<k\}} \Big(\boldsymbol{A}(x, u, Du) \cdot \nabla \varphi + \boldsymbol{B}(x, u, Du) \varphi\Big) dx$$
$$+ \int_{\Omega \cap \{k<u<2k\}} \Big(\boldsymbol{A}(x, u, Du) \cdot (2 - \frac{u}{k}) \nabla \varphi$$
$$+ \boldsymbol{B}(x, u, Du) (2 - \frac{u}{k}) \varphi\Big) dx$$
$$\geq \frac{1}{k} \int_{\Omega \cap \{k<u<2k\}} \boldsymbol{A}(x, u, Du) \cdot Du\, \varphi\, dx.$$

By Lemma 4.6 and the structure assumptions, the functions $\boldsymbol{A}(x, u, Du)$ and $\boldsymbol{B}(x, u, Du)$ are integrable on $\{\varphi > 0\}$. Hence by Lebesgue's dominated convergence theorem,

(4.95)
$$\lim_{k \to \infty} \Big(\int_{\Omega \cap \{u<k\}} \Big(\boldsymbol{A}(x, u, Du) \cdot \nabla \varphi + \boldsymbol{B}(x, u, Du) \varphi\Big) dx$$
$$+ \int_{\Omega \cap \{k<u<2k\}} \Big(\boldsymbol{A}(x, u, Du) \cdot (2 - \frac{u}{k}) \nabla \varphi$$
$$+ \boldsymbol{B}(x, u, Du) (2 - \frac{u}{k}) \varphi\Big) dx\Big)$$
$$= \int_\Omega \Big(\boldsymbol{A}(x, u, Du) \cdot \nabla \varphi + \boldsymbol{B}(x, u, Du) \varphi\Big) dx.$$

Due to structure assumptions, we have

(4.96)
$$- \int_{\Omega \cap \{k<u<2k\}} \boldsymbol{A}(x, u, Du) \cdot Du\, \varphi\, dx$$
$$\leq \int_{\Omega \cap \{k<u<2k\}} (\bar{b}\bar{u}^p - |Du|^p)\, \varphi\, dx$$
$$\leq C \int_{\operatorname{spt}\varphi \cap \{k<u<2k\}} b\bar{u}^p.$$

Let $\beta < 0$ be as in (4.19) and $\gamma = (p-1)(1-\beta)$. Using Theorem 4.5 and (4.20), we obtain

$$\int_{B \cap \{k<u<2k\}} \bar{b}\bar{u}^p\, dx \leq (k + \boldsymbol{\kappa})^{p-\gamma} \int_{B \cap \{k<u<2k\}} \bar{b}\bar{u}^\gamma\, dx$$
$$\leq C(k + \boldsymbol{\kappa})^{p-\gamma} r^n \inf_B (u + \boldsymbol{\kappa}(r))^\gamma$$

for each ball $B = B(z, r)$ with $B(z, 2r) \subset \Omega$. A simple covering argument shows that

$$\int_{\operatorname{spt}\varphi \cap \{k<u<2k\}} \bar{b}\bar{u}^p\, dx \leq C^*(1 + k^{p-\gamma})$$

with C^* independent of k. Hence

(4.97)
$$-\frac{1}{k}\int_{\Omega\cap\{k<u<2k\}} \boldsymbol{A}(x,u,Du)\cdot Du\,\varphi\,dx$$
$$\leq CC^* k^{p-1-\gamma} \to 0.$$

By (4.94), (4.95) and (4.97),
$$\int_\Omega \Big(\boldsymbol{A}(x,u,Du)\cdot\nabla\varphi + \boldsymbol{B}(x,u,Du)\,\varphi\Big)\,dx \geq 0.$$

This concludes the proof. □

4.31 COROLLARY. *Let u be a nonnegative supersolution of (3.1) up to all levels on Ω. Then there is a unique nonnegative Radon measure μ on Ω such that*

(4.98)
$$\int_\Omega \Big(\boldsymbol{A}(x,u,Du)\cdot\nabla\varphi + \boldsymbol{B}(x,u,Du)\,\varphi\Big)\,dx = \int_\Omega \varphi\,d\mu$$

for each $\varphi \in \mathcal{C}_c^\infty(\Omega)$.

PROOF. This follows from Theorem 4.30 using the Riesz representation theorem. □

4.32 REMARK. Now we can easily observe why supersolutions up to all levels and superharmonic functions are essentially the same if we work with the Laplace equation. Indeed, if u is a supersolution up to all levels, then by Theorem 4.30 it is a distributional solution of the Laplace equation and hence by Theorem 2.59 it has a superharmonic representative. Conversely, if u is a superharmonic function, the all truncations $\min\{u,k\}$ are superharmonic and thus u is a supersolution up to all levels.

4.33 DEFINITION. Let $\mathcal{S}$ be the class of all supersolutions of (3.1) up to all levels on Ω. We define an operator T on $\mathcal{S}$ with values in Radon measures on Ω, by
$$Tu = -\operatorname{div}\boldsymbol{A}(x,u,Du) + \boldsymbol{B}(x,u,Du),$$
where the divergence is understood in the sense of distributions.

4.34 THEOREM. *Let u be a nonnegative supersolution of (3.1) up to all levels on Ω and $\mu = Tu$. Suppose that $0 \leq m \leq u \leq l \leq \infty$, and that either $l < \infty$ or $b_0 = 0$. Let $B(r) = B(x_0, r) \subset \Omega$, $0 < r \leq r_0$. Then*
$$\mu(B(r/4)) \leq Cr^{n-p}\Big(\inf_{B(r/2)} u - m + \boldsymbol{\kappa}(r)\Big)^{p-1},$$
where C depends only on n, p, $b_0 l$ and the structure.

PROOF. In view of Lemma 3.4 we may assume that $b_0 = 0$. We denote $B = B(r)$, $\bar{u} = u - m + \boldsymbol{\kappa}(r)$. Let $\eta \in C^\infty(\mathbf{R}^n)$ be a nonnegative function supported in $B(r/2)$ such that $|\nabla\eta| < 5/r$ and $\eta \geq 1$ on $B(r/4)$. We use the test function $\varphi = \eta^p$ in (4.98) and obtain
$$p\int_B \boldsymbol{A}(x,u,Du)\cdot \eta^{p-1}\nabla\eta\,dx + \int_B \boldsymbol{B}(x,u,Du)\,\eta^p\,dx$$
$$\geq \int_B \eta^p\,d\mu.$$

Using the structure (3.5) and obvious estimates it follows

(4.99)
$$\int_B \eta^p \, d\mu \leq \int_B (a_1|\nabla \eta| + b_1\eta)|Du|^{p-1}\eta^{p-1}\, dx \\ + \int_B \left(\overline{a}u^{p-1}\eta^{p-1}|\nabla \eta| + \overline{b}u^{p-1}\eta^p\right) dx.$$

Letting $k \uparrow l$ in Lemma 4.6 we estimate the right-hand side in (4.99) by

$$Cr^{n-p}\left(\inf_{B(r/2)} u - m + \kappa(r)\right)^{p-1},$$

which establishes the theorem. □

We write
$$m(r) = \inf_{B(r)} u, \qquad m(0) = \lim_{r \to 0+} m(r).$$

4.35 Theorem. *Let u be a nonnegative supersolution of (3.1) up to all levels on Ω and $\mu = Tu$. Let $B(x_0, r_0) \subset \Omega$, $0 < r_0 \leq R_0$. Assume that*

(4.100)
$$\int_0^{r_0} a(r)\, \frac{dr}{r} < \infty.$$

Then
$$\mathbf{W}_p^\mu(x_0, r_0/4) \leq C\left(u(x_0) + \int_0^{r_0} k(r)\, \frac{dr}{r}\right),$$

where C depends only on n, p, R_0, the bound of $b_0\, u$ and the structure including the integral (4.100).

Proof. By integration of the estimate from Theorem 4.34 we obtain

$$\mathbf{W}_p^\mu(x_0, r_0/4) \leq C\left(m(0) - m(r_0) + \int_0^{r_0} \kappa(r)\, \frac{dr}{r}\right) \\ \leq C\left(m(0) - m(r_0) + \int_0^{r_0} [k(r) + m(0)h(r)]\, \frac{dr}{r}\right) \\ \leq Cm(0) + C\int_0^{r_0} k(r)\, \frac{dr}{r}.$$

Obviously $m(0) \leq u(x_0)$, which concludes the proof. □

Theorem 4.35 yields a lower estimate of u in terms of Wolff potentials. The corresponding upper estimate has been partially obtained in Theorem 4.27. We cannot apply this Theorem directly if u does not belong to $W^{1,p}_{\mathrm{loc}}(\Omega)$ and thus is not a (sub)solution of the problem 4.93 in the ordinary weak sense. However, it is possible to adapt the proof so that the situation of definition 4.33 will be covered.

Notice first that Lemma 4.25 was the only place where we utilized the definition of subsolution. Now, we are going to verify that we can obtain the conclusions of Lemma 4.25 assuming that u is a nonnegative supersolution of (3.1) up to all levels and $Tu = \mu$.

If we prove (4.48) under the additional assumption that there is $k \in (0, \infty)$ such that $\Phi = 0$ on (k, ∞), then we may easily pass to limit for $\Phi_k \to \Phi$, where

$$\Phi_k(s) = \begin{cases} \Phi(s), & s \leq k, \\ 0, & s > k, \end{cases}$$

and obtain (4.48) in the general case. Thus, we may assume that there is $k \in (0, \infty)$ such that

(4.101) $$\Phi = \Phi_k.$$

Recall that

$$\Psi(t) = \int_0^t \Phi(s)\, ds,$$
$$\varphi = \Psi(u)\, \omega^p.$$

We claim that

(4.102) $$\int_\Omega \bigl(\boldsymbol{A}(x, u, Du) \cdot \nabla \varphi + \boldsymbol{B}(x, u, Du)\, \varphi\bigr) \leq C\, \lambda \mu(\{\omega > 0\}).$$

To prove (4.102), we decompose

$$\varphi = \varphi_1 - \varphi_2,$$

where

$$\varphi_1 = \lambda\, \omega^p, \qquad \varphi_2 = (\lambda - \Psi(u))\, \omega^p.$$

Then φ_1 is smooth and thus, since $Tu = \mu$,

(4.103) $$\int_\Omega \bigl(\boldsymbol{A}(x, u, Du) \cdot \nabla \varphi_1 + \boldsymbol{B}(x, u, Du)\, \varphi_1\bigr)\, dx$$
$$\leq \int_\Omega \varphi_1\, d\mu \leq C\lambda \mu\{\omega > 0\}.$$

We have $\Psi(u) = \lambda$ on $\{u \geq k\}$, where k is as in (4.101). Since u is a supersolution of (3.1) up to level $k + \lambda + 1$, we have

(4.104) $$\int_\Omega \bigl(\boldsymbol{A}(x, u, Du) \cdot \nabla \varphi_2 + \boldsymbol{B}(x, u, Du)\, \varphi_2\bigr)\, dx \geq 0.$$

Subtracting (4.104) from (4.103) we get

$$\int_\Omega \bigl(\boldsymbol{A}(x, u, Du) \cdot \nabla \varphi + \boldsymbol{B}(x, u, Du)\, \varphi\bigr) \leq C\, \lambda \mu(\{\omega > 0\}).$$

We completed the proof of (4.102). This verifies that the methods of Subsection 4.3.1 are applicable and the following result holds.

4.36 THEOREM. *Let u be a nonnegative supersolution of (3.1) in Ω up to all levels and $\mu = Tu$. Let $B(x_0, r_0) \subset \Omega$, $0 < r_0 \leq R_0$. Suppose that either u is upper bounded or $b_0 = 0$. Then*

$$u(x_0) \leq C\Biggl[\biggl(r_0^{-n} \int_{B(x_0, r_0) \cap \Omega \cap \{u > 0\}} u^\gamma\, dx\biggr)^{1/\gamma}$$
$$+ \int_0^{r_0} \biggl(\frac{\mu B(x_0, r)}{r^{n-p}}\biggr)^{1/(p-1)} \frac{dr}{r} + \int_0^{r_0} \boldsymbol{k}(r)\, \frac{dr}{r}\Biggr].$$

The constant C depends on n, p, γ, R_0, the upper bound of $b_0 u$ and on the structure, in particular on the integral (4.47).

4.5 Historical notes

The result of [**GZ3**], which was announced in [**GZ2**], was predated by [**GZ1**] in which a sufficient condition for boundary regularity was established, but it was not the optimal Wiener condition. The results up to and including Theorem 4.22 were essentially established in [**GZ3**] with the notable exception that here the results are under the standing hypotheses of (H). This is an improvement of [**GZ3**] and plays an important role in developing the regularity theory for the double obstacle problem which is treated in the following chapter.

The treatment of the special case of harmonic functions in subsection 4.1.2 is due to Evans and Jensen [**EJ1**], [**EJ2**]. As stated earlier, their development is novel in the sense that the methods used in the proof are completely elementary, using only Sobolev's inequality and the mean value property of super harmonic functions. In particular, no potential-theoretic methods are employed.

The necessity of the Wiener condition as treated in Section 4.2 is adapted from [**KM4**]. Theorem 4.28 along with Theorem 4.16 thus shows that the Wiener condition is both necessary and sufficient for continuity at the boundary.

In a certain sense, this completes a story that was begun by Wiener in his celebrated paper [**Wie1**]. Since then, there have been many extensions of his result. Littman, Stampacchia, and Weinberger [**LSW**] proved that Wiener's criterion is also valid in the setting of linear uniformly elliptic second order equations with bounded, measurable coefficients. It was shown by Fabes, Jerrison and Kenig [**FJK**] that this result can be extended to degenerate equations in divergence form. Miller showed that the Wiener criterion does not hold for second order elliptic equations in non-divergence form [**Mi**]. However, both necessary and sufficient conditions were found by Bauman [**Bau**]. The main result proved in this paper is a Wiener test for a uniformly elliptic, second-order, linear partial differential equation in nondivergence form: $Lu = \sum_{i,j} a_{ij} D_{ij}^2 u + \sum_i b_i D_i u = 0$. The leading coefficients $a_{ij} = a_{ji}$ belong to L^∞ when $(n = 2)$ or continuous $(n \geq 3)$ on $\overline{\Omega}$. The capacity in this situation is related to the operator L and reduces to the classical capacity if the leading coefficients are Holder continuous. The estimate of the modulus of continuity for solutions in terms of the exponential of the capacity integral was obtained by Maz'ya [**Maz2**], [**Maz3**], [**Maz4**].

Maz'ya was the first to show that the Wiener condition was sufficient for boundary regularity for a class of nonlinear elliptic equations in divergence form including the p-Laplacian, [**Maz5**]. Also for these equations he obtained an estimate of the modulus of continuity similar to Theorem 4.22. Later, using different methods, it was shown in [**GZ3**] that the Wiener condition was sufficient for boundary regularity for a large class of equations of the form div $\boldsymbol{A}(x, u, \nabla u) - \boldsymbol{B}(x, u, \nabla u) = 0$. After this result, the question of the necessity of the Wiener condition was open for a long time. For a class of equations with p-Laplacian type growth, some necessary conditions were proved by Skrypnik [**Sk1**], [**Sk2**]. His necessary condition differed from the sufficiency condition in the power of the exponent in the Wiener expression. A proof of necessity of the sharp Wiener criterion which required $p > n-1$ was obtained by Lindqvist and Martio, [**LM**]. The question of necessity was resolved for all $p > 1$ by Kilpeläinen and Malý [**KM4**]. The extension to general structure (3.1) was done in [**Mal4**].

The Dirichlet problem on finely open sets was introduced as a part of *fine potential theory* which was developed for linear equations by Fuglede and summarized in his monograph [**Fu5**]. The fine potential theory also includes investigations of finely superharmonic functions on finely open sets. Among further contributions to linear fine potential theory we mention the monographs by Bliedtner and Hansen [**BH**] and Lukeš, Malý and Zajíček [**LMZ**]. For equation (2.69), the Dirichlet problem on p-finely open sets and further topics of p-fine potential theory were studied by Kilpeläinen and Malý [**KM3**]. The classes of p-finely superharmonic functions have been further investigated by Latvala [**Lat1**], [**Lat2**]. The equation with the structure (3.5) was treated in [**Mal2**], [**Mal3**].

The concept of "supersolution up to level l" is introduced in order to accommodate both regularity at the boundary and regularity of solutions of obstacle problems in a single framework. Theorem 4.5 and Lemma 4.6 are of central importance in proving regularity under the assumption of the Wiener condition.

The material in subsection 4.2.1 is due to Evans and Jensen, [**EJ1**] and [**EJ2**]. As stated previously, the interesting feature of this work is a completely elementary treatment of the Wiener criterion for boundary regularity of harmonic functions.

A linear counterpart of results in Section 4.4 consists of estimates of the Green function. These were given by Littman, Stampacchia and Weinberger [**LSW**]. A version of Corollary 4.31 for $\mathcal{A}$-superharmonic function has been proved by Kilpeläinen and Malý [**KM2**]. It is based on gradient integrability estimates, which for p-superharmonic functions were established by Lindqvist [**Lin1**]. Theorems 4.34 and 4.35 has been proved by Kilpeläinen and Malý [**KM2**] for the equation (2.69) and by Malý [**Mal2**] for supersolutions of (4.93). Uniform estimates of the solutions of (4.93) in terms of growth of $r \mapsto \mu(B(x,r))$ are due to Rakotoson and Ziemer [**RZ**] and Kilpeläinen [**Ki3**]. Lieberman [**Li3**] obtained Harnack type inequalities for the equation (4.93) with interesting consequences.

In our exposition of the problem with the right-hand side a measure we did not address the existence and uniqueness problems. The homogenous Dirichlet problem

$$-\operatorname{div} \boldsymbol{A}(x, \nabla u) = \mu$$

where $\boldsymbol{A}$ is e.g. as in (2.112), has a solution $u \in W_0^{1,p}(\Omega)$ only if μ has finite p-energy, i.e. $\mu \in \left(W_0^{1,p}(\Omega)\right)^*$. The Dirac measure is a typical example of a measure with infinite energy but even among absolutely continuous measures with L^1 density there exist measures with infinite energy. For very weak solutions, where the requirement $u \in W_0^{1,p}(\Omega)$ is left, the uniqueness does not hold even if $\mu = 0$, as shown by Serrin [**Se2**]. This is in contrast with Corollary 2.61 which shows that the distributional solutions of the Laplace equation are harmonic and thus nonuniqueness cannot occur.

The existence of a very weak solution for the problem whose right-hand side is a measure was established by Boccardo and Gallouët [**BG1**]. Further contributions, including the easing of some restrictions on the structure and p, are due to Boccardo and Gallouët [**BG2**], Bénilan, Boccardo, Gallouët, Gariepy, Pierre and Vazquez [**B+5**], Rakotoson [**R2**], [**R3**], and Kilpeläinen and Malý [**KM2**]. Questions related to Corollary 4.31 were studied by Boccardo and Murat [**BMu**]. (Very) weak minimizers of variational integrals were introduced by Iwaniec and Sbordone

[**IS**]. The problem for which sense of solution both existence and uniqueness are valid seems to be still open, although partial results are available. The case of discrete measures was solved by Kichenassamy [**Kic**]. The L^1 case is proved in the above mentioned paper [**B+5**] and independently by Lions and Murat [**LMu**]). Greco, Iwaniec and Sbordone [**GIS**] established the case $p = n$. Kilpeläinen and Xu [**KX**] studied the case of measures absolutely continuous with respect to the p-capacity.

CHAPTER 5

Variational Inequalities – Regularity

5.1 Differential operators with measurable coefficients

To motivate the investigations of the chapter, let us consider a simplified situation. Let Ω be a bounded open set of $\mathbf{R}^n$ and let ψ (the obstacle) be a function defined on Ω. For the moment we will not specify precisely the properties of ψ except to say that our treatment will admit obstacles that may not even be continuous. Let $\mathcal{K}$ be the family of all functions $v \in W_0^{1,p}(\Omega)$ such that $v \geq \psi$ p-quasi everywhere in Ω. When statements are made throughout this chapter concerning p-quasi everywhere pointwise behavior of functions in $W^{1,p}(\Omega)$, we will tacitly assume that p-quasicontinuous representatives are being used.

Suppose there is a function $u \in \mathcal{K}$ which minimizes the following variational problem:

$$\inf \left\{ \int_\Omega \mathbf{F}(x, v, \nabla v) \, dx \right\}$$

where the infimum is taken over all functions $v \in \mathcal{K}$. Assuming that $\mathbf{F} = \mathbf{F}(x, \zeta, \xi)$ has suitable regularity properties, it follows that the extremal u satisfies

$$\sum_{i=1}^{n} \int_\Omega \frac{\partial \mathbf{F}}{\partial \xi_i}(x, u, \nabla u) \, D_i \varphi \, dx + \int_\Omega \frac{\partial \mathbf{F}}{\partial \zeta}(x, u, \nabla u) \varphi \, dx \geq 0$$

for all $\varphi \in W_0^{1,p}(\Omega)$ with

$$\varphi(x) \geq \psi(x) - u(x)$$

for p-quasi all $x \in \Omega$. The primary example of this obstacle problem is given by the definition of p-capacity. That is, let $K \subset \Omega$ be a compact set and consider the problem

$$\inf \left\{ \int_\Omega |\nabla v|^p \, dx \right\}$$

where the infimum is taken over all $v \in W_0^{1,p}(\Omega)$ such that $v \geq 1$ p-quasi everywhere on K. The question of the existence of an extremal to this problem was addressed in Theorem 2.31 (p. 79). As in Lemma 2.30, it can be shown that the extremal u satisfies the variational inequality

$$\int_\Omega |\nabla u|^{p-2} \nabla u \cdot \nabla \varphi \, dx \geq 0$$

for all $\varphi \in W_0^{1,p}(\Omega)$ with $\varphi \geq \chi_K - u$ p-quasi everywhere in Ω. Thus, u is a weak supersolution of the p-Laplacian and therefore, by Corollary 4.10, we know that u is lower semicontinuous on Ω. Furthermore, u is locally Hölder continuous in $\Omega \setminus K$ and continuous at all points x_0 of ∂K at which K is not p-thin (Theorem 4.15).

Our objective in this section is to investigate regularity questions of this type for variational inequalities in the following general form.

5.1 DEFINITION. Let $\psi_1 \leq \psi_2$ be arbitrary functions on Ω and let

(5.1) $$\mathcal{K} := \{u \in W^{1,p}_{\text{loc}}(\Omega) : u \text{ is } p\text{-finely represented},$$
$$\psi_1 \leq u \leq \psi_2 \ p\text{-quasi everywhere}\}.$$

For the notion "p-finely represented" see the definition below. In this setting, a *test function* is a function $\varphi \in W^{1,p}_0(\Omega)$ such that $u + \varphi \in \mathcal{K}$ and the family of all test functions will be denoted by $\mathcal{T}$. A function u is called a solution to the *double obstacle problem* if

(5.2) $$\begin{cases} u \in \mathcal{K}, \\ \int_\Omega \boldsymbol{A}(x,u,\nabla u) \cdot \nabla\varphi + \boldsymbol{B}(x,u,\nabla u)\varphi\, dx \geq 0 \\ \qquad\qquad\qquad \text{for all } \varphi \in \mathcal{T}. \end{cases}$$

As special cases, we have single obstacle problems; namely, the *lower obstacle problem* when $\psi_2 \equiv +\infty$ and the *upper obstacle problem* when $\psi_1 \equiv -\infty$. The functions $\boldsymbol{A}$ and $\boldsymbol{B}$ are assumed to satisfy (3.5) and (3.6), p. 162.

We say that a function u is p-*finely represented* if u is defined everywhere in Ω and

$$p\text{-fine-}\liminf_{x \to z} u(x) \leq u(z) \leq p\text{-fine-}\limsup_{x \to z} u(x)$$

for each $z \in \Omega$. Each element of $W^{1,p}_{\text{loc}}(\Omega)$ can be p-finely represented. For example, if u is a p-quasicontinuous representative, then

(5.3) $$y \mapsto p\text{-fine-}\limsup_{x \to y} u(x)$$

is a p-fine representative.

5.1.1 Continuity in the presence of irregular obstacles.

Since we are concerned with pointwise regularity, we will consider a fixed point $x_0 \in \Omega$ throughout the remainder of this section. Also, we will assume throughout that ψ_2 is bounded below and ψ_1 bounded above.

5.2 DEFINITIONS. Let

(5.4) $$\overline{\psi}_1(r) = p\text{-}\sup_{B(x_0,r)} \psi_1 \qquad \underline{\psi}_1(r) = p\text{-}\inf_{B(x_0,r)} \psi_1$$
$$\overline{\psi}_1(x_0) = \lim_{r \to 0} \overline{\psi}_1(r) \qquad \underline{\psi}_1(x_0) = \lim_{r \to 0} \underline{\psi}_1(r).$$

Here p-sup and p-inf are the essential supremum and infimum in the sense of p-capacity. This means, for example, that p-sup of ψ_1 on $B(x_0,r)$ is the infimum of those numbers t for which $\{x \in B(x_0,r) : \psi_1(x) > t\}$ is p-polar. Similar notation will be used for ψ_2.

A point $x_0 \in \Omega$ is called a *Wiener point* relative to the obstacles ψ_1 and ψ_2 if

(5.5) $$p\text{-fine}\limsup_{x \to x_0} \psi_1(x) = \overline{\psi}_1(x_0),$$
$$p\text{-fine}\liminf_{x \to x_0} \psi_2(x) = \underline{\psi}_2(x_0)$$

and

(5.6) $$\overline{\psi}_1(x_0) \leq \underline{\psi}_2(x_0).$$

5.3 REMARK. If we consider a lower obstacle problem with the obstacle χ_E, $E \subset \Omega$, then (5.5) means exactly that E is not p-thin at x_0. Behavior of capacitary extremals illustrates that the condition is sharp in the sense that for some equations (like the p-Laplace equation) it is the best possible local condition. The inequality 5.6 is necessary even if we want only to insert a function continuous at x_0 between ψ_1 and ψ_2.

5.4 THEOREM. *Let u be a p-finely represented solution to the double obstacle problem as defined in (5.2). Assume the condition (4.28). If $x_0 \in \Omega$ is a Wiener point relative to ψ_1 and ψ_2, then*

$$\overline{\psi}_1(x_0) \leq u(x_0) \leq \underline{\psi}_2(x_0)$$

and u is continuous at x_0. In particular, if both ψ_1 and ψ_2 are continuous at x_0, then u is also continuous at x_0.

PROOF. Choose $k < \underline{\psi}_2(x_0)$. Then there exists $r > 0$ such that $B(x_0, r) \subset \Omega$ and $k < \psi_2(x)$ for p-quasi every $x \in B(x_0, r)$. Let $u_k := \min(u, k)$ and select $\varphi \in W_0^{1,p}(B(x_0, r))$ such that $0 \leq \varphi \leq k - u_k$. Then

(5.7) $$\begin{aligned} \psi_1 \leq u &\leq u + \varphi \quad p\text{-q.e. on} \quad \Omega, \\ u + \varphi = u_k + \varphi &\leq k < \psi_2 \quad p\text{-q.e. on} \quad \{u \leq k\} \cap B(x_0, r), \\ u + \varphi = u &\leq \psi_2 \quad p\text{-q.e. on} \quad \{u > k\} \cap B(x_0, r). \end{aligned}$$

Hence, using the fact that u is a solution to the double obstacle problem, we have

$$\int_\Omega \boldsymbol{A}(x, u, \nabla u) \cdot \nabla \varphi + \boldsymbol{B}(x, u, \nabla u) \varphi \, dx \geq 0.$$

Since $u = u_k$ on $\operatorname{spt} \varphi$ we also have

$$\int_\Omega \boldsymbol{A}(x, u_k, \nabla u_k) \cdot \nabla \varphi + \boldsymbol{B}(x, u_k, \nabla u_k) \varphi \, dx \geq 0.$$

We have thus verified that u_k is a supersolution of (3.1) up to level k. This allows us to conclude from Theorem 4.8 that u_k has a Lebesgue point at each point of Ω and u_k is both lower semicontinuous and p-finely continuous at x_0. This holds whenever $k < \underline{\psi}_2(x_0)$, but it also holds when $k = k_0 := \underline{\psi}_2(x_0)$ since $u_k \to u_{k_0}$ uniformly as $k \uparrow k_0$.

Now, consider k with $\overline{\psi}_1(x_0) \leq k \leq \underline{\psi}_2(x_0)$. Since u_k is p-finely continuous and x_0 satisfies (5.5), we have

$$u(x_0) \geq u_k(x_0) = p\text{-fine-lim}_{x \to x_0} u_k(x) \geq p\text{-fine-lim sup}_{x \to x_0} \psi_1(x) = \overline{\psi}_1(x_0)$$

which establishes the first half of the inequality in our theorem. With $w_k := \sup(u, k)$, it can be shown by an analogous argument that w_k is upper semicontinuous and p-finely continuous at x_0 thus leading to $u(x_0) \leq \underline{\psi}_2(x_0)$. In order to obtain the continuity of u at x_0, set $k = u(x_0)$. Then, since u_k is lower semicontinuous,

$$u(x_0) = k = u_k(x_0) \leq \liminf_{x \to x_0} u_k(x) \leq \liminf_{x \to x_0} u(x).$$

An analogous argument establishes

$$u(x_0) \geq \limsup_{x \to x_0} u(x).$$ □

5.1.2 The modulus of continuity.

In this section we will provide an estimate of the modulus of continuity of the solution discussed in Theorem 5.4. Thus, we assume x_0 is a Wiener point relative to the obstacles ψ_1 and ψ_2 and $B(r_0) \subset\subset \Omega$. Given $\lambda, r, t > 0$, we write

$$M_1(r) = \overline{\psi}_1(r) - \overline{\psi}_1(x_0),$$
$$M_2(r) = \underline{\psi}_2(x_0) - \underline{\psi}_2(r)$$
$$E_1^\lambda := \{\psi_1 > \overline{\psi}_1(x_0) - \lambda\},$$
$$E_2^\lambda := \{\psi_2 < \underline{\psi}_2(x_0) + \lambda\},$$
$$A_i^\lambda(s) := \left(\frac{\gamma_{p;r}[E_i^\lambda \cap B(x_0, s/2)]}{s^{n-p}}\right)^{1/(p-1)}, \quad i = 1, 2,$$
$$S_i(t,r) := \inf_{\lambda > 0}\left[\lambda + C\exp\left(-\frac{1}{C}\int_{2t}^r A_i^\lambda(s)\frac{ds}{s}\right)\right], \quad i = 1, 2,$$

where C is as in Theorem 4.22.

First, we observe that the solution u is locally bounded. Indeed, if M is the lower bound of ψ_2, then u is a supersolution up to level M and thus by Lemma 4.3, u is locally bounded below. In a similar way we obtain that u is locally bounded above.

5.5 THEOREM. *Assume that the hypotheses of Theorem 5.4 are in effect. Also, assume $\|a_2^{1/(p-1)}\|_{B(r)}$ and $\|a_3^{1/(p-1)}\|_{B(r)}$ are Hölder continuous functions of r. Then there are C and $\alpha > 0$, depending on $n, p, , r_0, \|u\|_{\infty; B(r_0)}$ and on the structure, such that the oscillation of of u at x_0 for $0 < r < r_0$ and all $t < r/2$ is estimated by*

$$\underset{B(x_0, t)}{\operatorname{osc}} u \leq C[S_1(t,r) + S_2(t,r) + (t/r)^\alpha] + M_1(r) + M_2(r).$$

PROOF. Referring to Theorem 5.4, we know that u is continuous at x_0 and

$$\overline{\psi}_1(x_0) \leq u(x_0) \leq \underline{\psi}_2(x_0).$$

Consider the functions

$$g_1(\tau) = \int_{2t}^r \left(\frac{\gamma_{p;r}[B(s) \cap \{u \geq \tau\}]}{s^{n-p}}\right)^{1/(p-1)} \frac{dr}{r},$$
$$g_2(\tau) = \int_{2t}^r \left(\frac{\gamma_{p;r}[B(s) \cap \{u \leq \tau\}]}{s^{n-p}}\right)^{1/(p-1)} \frac{dr}{r}.$$

Then g_1 is nonincreasing, g_2 is nondecreasing,

$$g_1 + g_2 \geq C^{-1}\int_{2t}^r \left(\frac{\gamma_{p;r}(B(s))}{s^{n-p}}\right)^{1/(p-1)} \frac{ds}{s} \geq C^{-1}\ln\frac{r}{t}$$

and

$$\lim_{\tau \to \infty} g_1(\tau) = \lim_{\tau \to -\infty} g_2(\tau) = 0.$$

It follows that there is a point τ_0 such that
$$\min\{\lim_{\tau\to\tau_0-} g_1(\tau), \lim_{\tau\to\tau_0+} g_2(\tau)\} \geq (2C)^{-1}\ln\frac{r}{t}.$$

Set
$$l = \begin{cases} \underline{\psi}_2(x_0) & \text{if } \tau_0 \geq \underline{\psi}_2(x_0), \\ \tau_0 & \text{if } \overline{\psi}_1(x_0) < \tau_0 < \underline{\psi}_2(x_0), \\ \overline{\psi}_1(x_0) & \text{if } \tau_0 \leq \overline{\psi}_1(x_0). \end{cases}$$

We will estimate
$$\overline{\operatorname{osc}}_{B(x_0,t)} u := \operatorname{osc}_{B(x_0,t)\cap\{u<l\}} u$$
by distinguishing two cases.

(i) Assume that $l > \overline{\psi}_1(x_0)$. Consider $k < \min\{l, \underline{\psi}_2(r)\}$. Then u is a supersolution of (3.1) up to level k and $g_1(k) \geq (2C)^{-1}\ln\frac{r}{t}$. If we denote by Ω' the p-fine interior of the set $\{u < k\}$, then
$$k \leq \operatorname{w}\liminf_{x\to y} u(x)$$
holds for each $y \in B(r) \cap \overline{\Omega}' \setminus \Omega'$. As in the proof of Theorem 4.22 we deduce that
$$\operatorname{osc}_{B(x_0,t)\cap\{u<k\}} u \leq C\exp\left(-\frac{g_1(k)}{C}\right) \leq C\exp\left(-\frac{\ln\frac{r}{t}}{C'}\right) = C\left(\frac{t}{r}\right)^{1/C'}.$$

(ii) Now we will assume that $l = \overline{\psi}_1(x_0)$. Choose $\lambda > 0$ and consider $k < \min\{l - \lambda, \underline{\psi}_2(r)\}$. Again, u is a supersolution of (3.1) up to level k, and if we denote by Ω' the p-fine interior of the set $\{u < k\}$, then
$$k \leq \operatorname{w}\liminf_{x\to y} u(x)$$
holds for each $y \in B(r) \cap \overline{\Omega}' \setminus \Omega'$ and
$$B(r) \setminus \Omega' \supset E_1^\lambda.$$

As in the proof of Theorem 4.22 we conclude that
$$\operatorname{osc}_{B(x_0,t)\cap\{u<k\}} u \leq C\exp\left(-\frac{1}{C}\int_{2t}^r A_1(s)\frac{ds}{s}\right).$$

In both cases (i) and (ii) we infer that
$$\overline{\operatorname{osc}}_{B(x_0,t)} u \leq \operatorname{osc}_{B(x_0,t)\cap\{u<k\}} u + (l-k)$$
$$\leq \exp\left(-\frac{1}{C}\int_{2t}^r A_1^\lambda(s)\frac{ds}{s}\right) + \lambda + C\left(\frac{t}{r}\right)^\alpha + M_2(r).$$

Similar estimates can be obtained for "upper oscillation". □

If ψ_1 and ψ_2 are continuous at x_0, then A_1^λ and A_2^λ can be taken as constant functions and
$$M_i(r) \leq \operatorname*{osc}_{B(x_0,r)} \psi_i \quad i = 1, 2.$$

This leads to the following.

5.6 COROLLARY. *In addition to the hypotheses of the previous theorem, assume that both ψ_1 and ψ_2 are continuous at x_0. Then*

$$\underset{B(x_0,r)}{\operatorname{osc}} u \leq C[\underset{B(x_0,r)}{\operatorname{osc}} \psi_1 + \underset{B(x_0,r)}{\operatorname{osc}} \psi_2 + r^\alpha].$$

5.2 Differential operators with differentiable structure

The object of this section is to investigate the smooth regularity of solutions to the double obstacle problem for a general class of equations with differentiable structure similar to (3.43), p. 180.

The method we employ is entirely self-contained and elementary; its essence is to reduce the regularity question of the obstacle problem to known existence and regularity results of an equation with a simple structure, namely, an equation of the form div $\boldsymbol{A}_0(Du) = 0$ where $|\boldsymbol{A}_0(\xi)| \sim |\xi|^{p-1}$. Here, both obstacles are assumed at the outset to have Hölder continuous gradients, thus requiring no subsequent approximation. This method can be used with virtually no change at the boundary.

The first step invokes a perturbation method of freezing the operator at a point $x_0 \in \Omega$. Thus, if $\boldsymbol{A}_0(\xi) := \boldsymbol{A}(x_0, u(x_0), \xi)$, we first show that the original solution u is related (or "linked") to a solution u_0 of the double obstacle problem involving the operator $\boldsymbol{A}_0$ by an inequality of the form

$$\int_{B(x_0,R)} |\nabla u - \nabla u_0|^p \, dx \leq CR^\sigma \int_{B(x_0,R)} [1 + |\nabla u|^2]^{p/2} \, dx$$

for some positive number σ. Here we use the $\mathcal{C}^{0,\alpha}$ regularity result of Corollary 5.6. We then show that u_0 is linked to a solution u_1 of a single obstacle problem by a similar inequality. Finally, we show that u_1 is linked to a solution v of the equation div $\boldsymbol{A}_0(\nabla v) = 0$, thus allowing us to conclude that u and v are linked by the inequality

$$\int_{B(x_0,R)} |\nabla u - \nabla v|^p \, dx \leq CR^\sigma \int_{B(x_0,R)} [1 + |\nabla u|^2]^{p/2} \, dx.$$

The "linking exponents" may differ from step to step but we will use the same symbol σ to simplify the notation. At the final stage, the minimum of linking exponents will suffice. Each of these intermediate functions, u_0, u_1 and v are assumed to have the same boundary values as u on $B(x_0, R)$ which is contained in the underlying domain Ω. Once this estimate is obtained, it is easy to conclude that $u \in \mathcal{C}^{1,\alpha}$ by employing the fact that the Campanato and Hölder spaces are isomorphic and the known results concerning the regularity of v. The information we need is that the solutions u_1, u_2 and v exist (Theorem 6.14) and v has Hölder continuous first derivatives (Theorem 3.19).

The setting for our development is the following. Let $\Omega \subset \mathbf{R}^n$ be an open set whose boundary, $\partial\Omega$, can be locally represented as a graph of a function with Hölder continuous derivatives. Let ψ_1, ψ_2 be obstacles defined on $\overline{\Omega}$ with

(i) $\psi_1, \psi_2 \in \mathcal{C}^{1,\alpha_1}(\overline{\Omega})$,
(ii) $\psi_1 \leq \psi_2$ on $\overline{\Omega}$,
(iii) $\psi_1 \leq f \leq \psi_2$ on $\partial\Omega$ where f is assumed to be defined on $\overline{\Omega}$ and has Hölder continuous derivatives. The function f is to be regarded as prescribed boundary data for the double obstacle problem described below.

5.2 DIFFERENTIAL OPERATORS WITH DIFFERENTIABLE STRUCTURE

We investigate the double obstacle problem for operators of the form

$$
(5.8) \qquad -\operatorname{div} \boldsymbol{A}(x, u, Du) + \boldsymbol{B}(x, u, Du)
$$

where $\boldsymbol{A}$ and $\boldsymbol{B}$ denote, respectively, vector and scalar valued functions defined on $\mathbf{R}^n \times \mathbf{R} \times \mathbf{R}^n$ satisfying the following structure conditions for fixed $p > 1$:

For each $(x, u) \in \mathbf{R}^n \times \mathbf{R}$, $\boldsymbol{A}(x, u, \cdot) \in \mathcal{C}^1(\mathbf{R}^n)$ for $p \geq 2$ and $\boldsymbol{A}(x, u, \cdot) \in \mathcal{C}^0(\mathbf{R}^n) \cap \mathcal{C}^1(\mathbf{R}^n - \{0\})$ for $1 < p < 2$,

$$
(5.9) \qquad \lambda |\xi|^{p-2} |h|^2 \leq \sum_{i,j=1}^n \frac{\partial \boldsymbol{A}^i}{\partial \xi_j}(x, u, \xi) h_i h_j,
$$

for all $u \in \mathbf{R}, h \in \mathbf{R}^n$, and all $x \in \mathbf{R}^n$,

$$
(5.10) \qquad \left|\frac{\partial \boldsymbol{A}^i}{\partial \xi_j}(x, u, \xi)\right| \leq \Lambda |\xi|^{p-2},
$$

$$
(5.11) \qquad |\boldsymbol{A}(x, u, \xi)| \leq \Lambda_1 (1 + |\xi|^2)^{(p-1)/2},
$$

$$
(5.12) \qquad |\boldsymbol{B}(x, u, \xi)| \leq \Lambda_1 (1 + |\xi|^2)^{p/2},
$$

and

$$
(5.13) \qquad |\boldsymbol{A}(x, u, \xi) - \boldsymbol{A}(x', u', \xi)| \leq \Lambda_1 (|x - x'|^{\alpha_2} + |u - u'|^{\alpha_2})(1 + |\xi|^2)^{(p-1)/2}.
$$

Here, $p > 1, \lambda, \Lambda$ and Λ_1 are positive constants, and $0 < \alpha_1, \alpha_2 < 1$. We consider the problem

$$
(5.14) \qquad \begin{cases} \int_\Omega (\boldsymbol{A}(x, u, \nabla u) \cdot \nabla \varphi + \boldsymbol{B}(x, u, \nabla u) \varphi) \, dx \geq 0 \text{ for all test functions } \varphi, \\ \psi_1 \leq u \leq \psi_2, \, u \in W^{1,p}_{\text{loc}}(\Omega) \\ \varphi \in W^{1,p}_c(\Omega), \, \psi_1 \leq u + \varphi \leq \psi_2 \text{ in } \Omega. \end{cases}
$$

We begin the analysis of the perturbation method of freezing the operator $\boldsymbol{A}$ at a point $x_0 \in \Omega$. Thus, we let

$$
\boldsymbol{A}_0(h) := \boldsymbol{A}(x_0, u(x_0), h)
$$

and choose $R > 0$ such that the closed ball $\overline{B}(x_0, R) \subset \Omega$. Let u_0 denote a weak solution of the problem

$$
(5.15) \qquad \begin{cases} \int_{B(x_0, R)} \boldsymbol{A}_0(\nabla u_0) \cdot \nabla \varphi \, dx \geq 0 \text{ for all test functions } \varphi, \\ \psi_1 \leq u_0 \leq \psi_2, \, u_0 - u \in W^{1,p}_0(B(x_0, R)) \\ \varphi \in W^{1,p}_0(B(x_0, R)), \, \psi_1 \leq u_0 + \varphi \leq \psi_2 \text{ in } B(x_0, R). \end{cases}
$$

Its existence follows from Theorem 6.14.

The following elementary result will be needed in the sequel.

5.7 LEMMA. *There is a positive constant λ_0 depending only on n, p, λ, and Λ such that*

$$[\boldsymbol{A}(x,u,\xi) - \boldsymbol{A}(x,u,\xi')] \cdot [\xi - \xi'] \geq \lambda_0 \begin{cases} (|\xi| + |\xi'|)^{p-2} \cdot |\xi - \xi'|^2 & \text{if } 1 < p < 2 \\ |\xi - \xi'|^p & \text{if } p \geq 2 \end{cases}$$

for all $\xi, \xi' \in \mathbf{R}^n$.

PROOF. We may suppose that $|\xi| \leq |\xi'|$. Hence,

$$1/4 \cdot |\xi - \xi'| \leq |\xi' + t \cdot (\xi - \xi')| \leq |\xi| + |\xi'|,$$

for all $t \in [0, 1/4]$. This, (5.9), and the inequality

$$[\boldsymbol{A}(x,u,\xi) - \boldsymbol{A}(x,u,\xi')] \cdot (\xi - \xi')$$
$$\geq \int_0^{1/4} \sum_{i,j=1}^n \frac{\partial \boldsymbol{A}_i}{\partial \xi_j}(x, u, \xi' + t(\xi - \xi')) \cdot (\xi_i - \xi'_i) \cdot (\xi_j - \xi'_j) \, dt$$

imply our desired result. $\square$

5.8 LEMMA. *If u_0 is a weak solution of (5.15), then for all $R < 1$,*

$$\int_{B(x_0, R)} |\nabla u - \nabla u_0|^p \, dx \leq C R^\sigma \int_{B(x_0, R)} [1 + |\nabla u|^2]^{p/2} \, dx.$$

The constant C depends on $C(n, p, \lambda, \Lambda, \Lambda_1)$, the exponent σ depends on α_2, p and on the exponent β from Corollary 5.6.

PROOF. We select a test function $\varphi := u - u_0$ and note that $\psi_1 \leq \varphi + u_0 \leq \psi_2$ and $\psi_1 \leq -\varphi + u \leq \psi_2$. Thus, we obtain

$$I \equiv \int_{B(x_0, R)} [\boldsymbol{A}_0(\nabla u) - \boldsymbol{A}_0(\nabla u_0)] \cdot [\nabla u - \nabla u_0] \, dx$$
$$\leq \int_{B(x_0, R)} [\boldsymbol{A}_0(\nabla u) - \boldsymbol{A}(x, u, \nabla u)] \cdot [\nabla u - \nabla u_0] \, dx$$
$$+ \int_{B(x_0, R)} \boldsymbol{A}(x, u, \nabla u) \cdot (\nabla u - \nabla u_0) \, dx$$
$$\leq \int_{B(x_0, R)} [\boldsymbol{A}_0(\nabla u) - \boldsymbol{A}(x, u, \nabla u)] \cdot [\nabla u - \nabla u_0] \, dx$$
$$- \int_{B(x_0, R)} \boldsymbol{B}(x, u, \nabla u)(u - u_0) \, dx.$$

Let J denote the right side of this inequality. From (5.12) and (5.13), and Young's inequality, we obtain

$$J \leq C \sup_{B(x_0, R)} (|x - x_0|^{\alpha_2} + |u(x) - u_0(x_0)|^{\alpha_2})$$
$$\cdot \left(\int_{B(x_0, R)} [1 + |\nabla u|^2]^{p/2} + \varepsilon \int_{B(x_0, R)} |\nabla u - \nabla u_0|^p \, dx \right)$$
$$+ C \sup_{B(x_0, R)} (|u(x) - u_0(x)|) \left(\int_{B(x_0, R)} |\nabla u|^p \, dx + R^n \right)$$

where ε is a small positive number to be determined below. Now $u = u_0$ on $\partial B(x_0, R)$ and Corollary 5.6 implies $\mathrm{osc}_{B(x_0,R)} u \leq CR^\beta$. This together with obvious minimum and maximum principles for u_0 yields that

$$\sup_{B(x_0,R)} |u - u_0| \leq CR^\beta.$$

Thus,

(5.16)
$$\begin{aligned} J &\leq C(R^{\alpha_2} + R^{\alpha_2\beta}) \left(\int_{B(x_0,R)} [1 + |\nabla u|^2]^{p/2} + \varepsilon \int_{B(x_0,R)} |\nabla u - \nabla u_0|^p \, dx \right) \\ &\quad + CR^\beta \left(\int_{B(x_0,R)} |\nabla u|^p \, dx + R^n \right) \\ &\leq CR^{\alpha_2\beta} \left(\int_{B(x_0,R)} [1 + |\nabla u|^2]^{p/2} \, dx + \varepsilon \int_{B(x_0,R)} |\nabla u - \nabla u_0|^p \, dx \right). \end{aligned}$$

Here we have used the assumption that $R < 1$. It follows from Lemma 5.7 that

$$CI \geq \begin{cases} \int_{B(x_0,R)} [|\nabla u|^2 + |\nabla u_0|^2]^{(p-2)/2} |\nabla u - \nabla u_0|^2 \, dx & \text{if } 1 < p < 2 \\ \int_{B(x_0,R)} |\nabla u - \nabla u_0|^p \, dx & \text{if } 2 \leq p. \end{cases}$$

We first consider the case $1 < p < 2$. In this case notice that

(5.17)
$$\begin{aligned} (|\nabla u|^2 + |\nabla u_0|^2)^{p/2} &= (|\nabla u|^2 + |\nabla u_0|^2)^{(p-2)/2}(|\nabla u|^2 + |\nabla u_0|^2) \\ &\leq (|\nabla u|^2 + |\nabla u_0|^2)^{(p-2)/2} \left(2|\nabla u - \nabla u_0|^2 + 3|\nabla u|^2 \right) \\ &\leq 2(|\nabla u|^2 + |\nabla u_0|^2)^{(p-2)/2}|\nabla u - \nabla u_0|^2 + 3|\nabla u|^p. \end{aligned}$$

Also,

(5.18)
$$\begin{aligned} |\nabla u - \nabla u_0|^p &= \left\{ (|\nabla u|^2 + |\nabla u_0|^2)^{(p-2)/2}|\nabla u - \nabla u_0|^2 \right\}^{p/2} \\ &\quad \cdot \left\{ (|\nabla u|^2 + |\nabla u_0|^2)^{p/2} \right\}^{(2-p)/2}. \end{aligned}$$

Apply Hölder's inequality to (5.18) and then appeal to (5.17), Lemma 5.7 and the fact that $I \leq J$, to obtain

(5.19)
$$\int_{B(x_0,R)} |\nabla u - \nabla u_0|^p \, dx \leq CJ^{p/2} \left(J + \int_{B(x_0,R)} |\nabla u|^p dx \right)^{1-p/2}.$$

From this and (5.16) we infer that

$$\int_{B(x_0,R)} |\nabla u - \nabla u_0|^p \, dx \leq CJ + C R^{\alpha_2 p \beta/2} \left(\int_{B(x_0,R)} [1 + |\nabla u|^2]^{p/2} \, dx \right.$$

$$+ \varepsilon \int_{B(x_0,R)} |\nabla u - \nabla u_0|^p \, dx \bigg)^{p/2}$$

$$\left(\int_{B(x_0,R)} |\nabla u|^p \, dx \right)^{1-p/2}$$

$$\leq CJ + CR^{\alpha_2 p \beta/2} \left(\int_{B(x_0,R)} [1 + |\nabla u|^2]^{p/2} \, dx \right.$$

$$+ \varepsilon \int_{B(x_0,R)} |\nabla u - \nabla u_0|^p \, dx \bigg).$$

By choosing ε sufficiently small, we have the desired estimate in case $1 < p < 2$:

$$\int_{B(x_0,R)} |\nabla u - \nabla u_0|^p \, dx \leq CR^\sigma \int_{B(x_0,R)} [1 + |\nabla u|^2]^{p/2} \, dx$$

where $\sigma = \alpha_2 p \beta / 2$.

When $p \geq 2$, we have from Lemma 5.7 and (5.16)

$$\int_{B(x_0,R)} |\nabla u - \nabla u_0|^p \, dx \leq CI \leq CJ$$

which yields the conclusion with $\sigma = \alpha_2 \beta$. $\square$

We now will show that the analysis of the double obstacle problem can be reduced to estimates involving a solution to the associated equation. To this end, consider the ball $B(x_0, R)$ as in (5.15) and denote by u_1 a weak solution of the variational inequality (a single obstacle problem) where $u_0 - u_1 \in W_0^{1,p}(B(x_0, R))$, $u_1 \geq \psi_1$ and

$$\int_{B(x_0,R)} \boldsymbol{A}_0(\nabla u_1) \cdot \nabla \varphi \, dx \geq 0$$

for all $\varphi + u_1 \geq \psi_1$, $\varphi \in W_0^{1,p} B(x_0, R)$. For brevity, we introduce the customary notation

$$f \wedge g := \min\{f, g\}$$

for functions f and g.

5.9 LEMMA. *There are constants $\sigma = \sigma(n, p, \alpha_2) > 0$ and*

$$C = C(n, p, \lambda, \Lambda, \Lambda_1, \|\boldsymbol{A}_0\|_{W^{1,\infty}(\Omega)}, \|\psi_1\|_{C^{1,\alpha_1}})$$

such that

(5.20) $$\int_{B(x_0,R)} |\nabla u_0 - \nabla u_1|^p \, dx \leq CR^\sigma \int_{B(x_0,R)} [1 + |\nabla u_0|^2]^{p/2} \, dx$$

whenever $R < 1$.

5.2 DIFFERENTIAL OPERATORS WITH DIFFERENTIABLE STRUCTURE

PROOF. Let $\varphi := u_0 - u_1$ and note that $\varphi \in W_0^{1,p}(B(x_0, R))$ and $\varphi + u_1 = u_0 \geq \psi_1$. Therefore,

$$\int_{B(x_0,R)} \boldsymbol{A}_0(\nabla u_1) \cdot \nabla \varphi \, dx = \int_{B(x_0,R)} \boldsymbol{A}_0(\nabla u_1) \cdot (\nabla u_0 - \nabla u_1) \, dx \geq 0.$$

Consequently, it follows that

$$
\begin{aligned}
(5.21) \quad I &:= \int_{B(x_0,R)} [\boldsymbol{A}_0(\nabla u_0) - \boldsymbol{A}_0(\nabla u_1)] \cdot [\nabla u_0 - \nabla u_1] \, dx \\
&= \int_{B(x_0,R)} \boldsymbol{A}_0(\nabla u_0) \cdot (\nabla u_0 - \nabla u_1) \, dx \\
&\quad - \int_{B(x_0,R)} \boldsymbol{A}_0(\nabla u_1) \cdot (\nabla u_0 - \nabla u_1) \, dx \\
&\leq \int_{B(x_0,R)} \boldsymbol{A}_0(\nabla u_0) \cdot (\nabla u_0 - \nabla u_1) \, dx \\
&= \int_{B(x_0,R)} \boldsymbol{A}_0(\nabla u_0) \cdot [\nabla u_0 - \nabla(u_1 \wedge \psi_2)] \\
&\quad + \int_{B(x_0,R)} \boldsymbol{A}_0(\nabla u_0) \cdot [\nabla(u_1 \wedge \psi_2) - \nabla u_1] \\
&= I_1 + I_2.
\end{aligned}
$$

Our first task is to show $I_1 \leq 0$. For this, let $\varphi = u_1 \wedge \psi_2 - u_0$ and observe that $\psi_2 \geq \varphi + u_0 = u_1 \wedge \psi_2 \geq \psi_1$. Moreover, if $u_1 \geq \psi_2$, then $0 \leq \varphi = \psi_2 - u_0 \leq u_1 - u_0 = 0$ on $\partial B(x_0, R)$. On the other hand, if $u_1 \leq \psi_2$, then $\varphi = u_1 - u_0 = 0$ on $\partial B(x_0, R)$. Hence, $\varphi \in W_0^{1,p}(B(x_0, R))$. Therefore, $I_1 \leq 0$ and

$$
\begin{aligned}
(5.22) \quad I &\leq \int_{B(x_0,R)} \boldsymbol{A}_0(\nabla u_0) \cdot [\nabla(u_1 \wedge \psi_2) - \nabla u_1] \, dx \\
&\leq \left(\int_{B(x_0,R)} |\boldsymbol{A}_0(\nabla u_0)|^{p'} \, dx \right)^{1/p'} \left(\int_{B(x_0,R)} |\nabla(u_1 - \psi_2)^+|^p \, dx \right)^{1/p}
\end{aligned}
$$

since $u_1 - (u_1 \wedge \psi_2) = (u_1 - \psi_2)^+$.

Now consider

$$
\begin{aligned}
(5.23) \quad J &:= \int_{B(x_0,R)} [\boldsymbol{A}_0(\nabla u_1) - \boldsymbol{A}_0(\nabla \psi_2)] \cdot [\nabla(u_1 - \psi_2)^+] \, dx \\
&= \int_{B(x_0,R)} \boldsymbol{A}_0(\nabla u_1) \cdot [\nabla(u_1 - \psi_2)^+] \, dx \\
&\quad - \int_{B(x_0,R)} \boldsymbol{A}_0(\nabla \psi_2) \cdot [\nabla(u_1 - \psi_2)^+] \, dx \\
&= J_1 - J_2.
\end{aligned}
$$

Notice that $J_1 \leq 0$ for if a test function φ is defined by $\varphi := -(u_1 - \psi_2)^+$, then $u_1 + \varphi \geq \psi_1$. Also, the fact that $\psi_2 \geq u_0 = u_1$ on $\partial B(x_0, R)$ shows that $\varphi \in W_0^{1,p}(B(x_0, R))$. Since $\boldsymbol{A}_0 \in \mathcal{C}^1$, we have that $\boldsymbol{A}_0$ is locally Lipschitz with Lipschitz constant K, say. Hence, by Hölder's inequality,

$$J \leq -\int_{B(x_0,R)} \boldsymbol{A}_0(\nabla\psi_2) \cdot [\nabla(u_1 - \psi_2)^+] \, dx$$

$$= -\int_{B(x_0,R)} [\boldsymbol{A}_0(\nabla\psi_2) - \boldsymbol{A}_0(\nabla\psi_2(x_0))] \cdot [\nabla(u_1 - \psi_2)^+] \, dx$$

(5.24)
$$\leq CK \Big(\int_{B(x_0,R)} |\nabla\psi_2 - \nabla\psi_2(x_0)|^{p'} \, dx\Big)^{1/p'}$$
$$\Big(\int_{B(x_0,R)} |\nabla(u_1 - \psi_2)^+|^p \, dx\Big)^{1/p}$$
$$\leq CK \, \|\psi_2\|_{\mathcal{C}^{1,\alpha_1}}^{1/p'} \, R^{n/p'+\alpha_1} \Big(\int_{B(x_0,R)} |\nabla(u_1-\psi_2)^+|^p \, dx\Big)^{1/p}.$$

We claim that

(5.25)
$$\int_{B(x_0,R)} |\nabla(u_1-\psi_2)^+|^p \, dx \leq CR^{n+\varepsilon_1},$$
$$\text{where } \varepsilon_1 = \begin{cases} \alpha_1 p' & \text{if } p \geq 2, \\ \alpha_1 p & \text{if } 1 < p < 2. \end{cases}$$

For this first consider the case $p \geq 2$. Then, from Lemma 5.7, we have

$$J = \int_{B(x_0,R) \cap \{u_1 \geq \psi_2\}} [\boldsymbol{A}_0(\nabla u_1) - \boldsymbol{A}_0(\nabla\psi_2)] \cdot [\nabla u_1 - \nabla\psi_2] \, dx$$
$$\geq \lambda_0 \int_{B(x_0,R)} |\nabla(u_1-\psi_2)^+|^p \, dx.$$

Consequently, (5.25) follows from (5.23) and (5.24).

In the case $1 < p < 2$, we repeat the arguments leading to (5.19) and obtain

(5.26)
$$\int_{B(x_0,R)} |\nabla(u_1-\psi_2)^+|^p \, dx \leq CJ^{p/2} \Big(J + \int_{B(x_0,R)} |\nabla\psi_2|^p \, dx\Big)^{1-p/2}$$
$$\leq CJ + CR^{n(1-p/2)} J^{p/2}$$

where C now depends on $\|\psi_2\|_{\mathcal{C}^{1,\alpha_1}}$. Then, for arbitrary $\varepsilon > 0$,

$$R^{n(1-\frac{p}{2})} J^{p/2} \leq CK^{p/2} R^{n(1-\frac{p}{2}) + \frac{n+\alpha_1 p'}{p'} \cdot \frac{p}{2}} \Big(\int_{B(x_0,R)} |\nabla(u_1-\psi_2)^+|^p \, dx\Big)^{1/2}$$
$$\leq \frac{1}{4\varepsilon} CK^p R^{n+\alpha_1 p} + \varepsilon \int_{B(x_0,R)} |\nabla(u_1-\psi_2)^+|^p \, dx,$$

by (5.24) and Young's inequality. Thus, from (5.26) and (5.24) we obtain

$$\int_{B(x_0,R)} |\nabla(u_1-\psi_2)^+|^p \, dx \leq CR^{n+\alpha_1 p} + CR^{n+\alpha_1 p'}$$
$$+ 2\varepsilon \int_{B(x_0,R)} |\nabla(u_1-\psi_2)^+|^p \, dx$$
$$\leq CR^{n+\alpha_1 p}$$
$$+ 2\varepsilon \int_{B(x_0,R)} |\nabla(u_1-\psi_2)^+|^p \, dx,$$

since $R < 1$. A choice of $\varepsilon = 1/4$ establishes our claim, (5.25), in the case $1 < p < 2$.

By (5.22), (5.25) and (5.11) we have

$$I \leq CR^{(n+\varepsilon_1)/p} \left(\int_{B(x_0,R)} |A_0(\nabla u_0)|^{p'} \, dx \right)^{1/p'}$$
$$\leq CR^{\frac{n+\varepsilon_1-\varepsilon_1/p'}{p}} R^{\frac{\varepsilon_1}{pp'}} \left(\int_{B(x_0,R)} |\nabla u_0|^p \, dx + R^n \right)^{1/p'}.$$

By Young's inequality, it follows

(5.27)
$$I \leq CR^{\varepsilon_1/p} \left(\int_{B(x_0,R)} |\nabla u_0|^p \, dx + R^n \right) + CR^{n+\varepsilon_1-\varepsilon_1/p'}$$
$$\leq CR^{\varepsilon_1/p} \int_{B(x_0,R)} (|\nabla u_0|^2 + 1)^{p/2} \, dx.$$

We now are in a position to establish (5.20). In case $p \geq 2$, it follows from (5.21) and Lemma 5.7 that

$$\int_{B(x_0,R)} |\nabla u_0 - \nabla u_1|^p \, dx \leq CI.$$

Thus, (5.20) holds with $\sigma = \varepsilon_1/p$.

Now for the case $1 < p < 2$. Using (5.18) and Lemma 5.7, as in (5.19) we obtain

(5.28)
$$\int_{B(x_0,R)} |\nabla u_0 - \nabla u_1|^p \, dx \leq CI^{p/2} \left(I + \int_{B(x_0,R)} |\nabla u_0|^p \, dx \right)^{1-p/2}.$$

Hence by (5.27),

$$\int_{B(x_0,R)} |\nabla u_0 - \nabla u_1|^p \, dx$$

$$\leq CI + CI^{p/2} \Big(\int_{B(x_0,R)} |\nabla u_0|^p \, dx\Big)^{1-p/2}$$

$$\leq CR^{\varepsilon_1/p} \int_{B(x_0,R)} (|\nabla u_0|^2 + 1)^{p/2} \, dx$$

$$+ C\Big(R^{\varepsilon_1/p} \int_{B(x_0,R)} (|\nabla u_0|^2 + 1)^{p/2} \, dx\Big)^{p/2} \Big(\int_{B(x_0,R)} |\nabla u_0|^p \, dx\Big)^{1-p/2}$$

$$\leq CR^{\sigma} \int_{B(x_0,R)} (|\nabla u_0|^2 + 1)^{p/2} \, dx,$$

where

$$\sigma = \min\Big\{\frac{\varepsilon_1}{p}, \frac{\varepsilon_1}{2}\Big\} = \frac{\varepsilon_1}{2}.$$

This concludes the proof. □

The previous lemma states that if u_0 is a weak solution to the double obstacle problem relative to the operator $\boldsymbol{A}_0$, then there is a function u_1 such that $u_0 - u_1 \in W_0^{1,p}(B(x_0,R))$ and u_1 is a solution to a single obstacle problem (the obstacle being ψ_2) which is "linked" to u_0 by the inequality

(5.29) $$\int_{B(x_0,R)} |\nabla u_0 - \nabla u_1|^p \, dx \leq CR^{\sigma} \int_{B(x_0,R)} [1 + |\nabla u_0|^2]^{p/2} \, dx$$

for some positive number σ. In particular, if we take $\psi_1 \equiv -\infty$, the result states that a solution of a single obstacle problem is linked to a solution of the equation div $\boldsymbol{A}_0(Dv) = 0$. Of course, in obtaining this conclusion, the direction of the inequalities is immaterial in the formulation of the variational inequality. If we now apply these observations to u_1 which is a solution to a single obstacle problem, we then can conclude that there is a function v such that $u_1 - v \in W_0^{1,p}(B(x_0,R))$, div $\boldsymbol{A}_0(Dv) = 0$ and

(5.30) $$\int_{B(x_0,R)} |\nabla u_1 - \nabla v|^p \, dx \leq CR^{\sigma} \int_{B(x_0,R)} [1 + |\nabla u_1|^2]^{p/2} \, dx.$$

Lemma 5.9 states that u and u_0 are linked, (5.29) states that u_0 and u_1 are linked and (5.30) states that u_1 and v are linked. These links thus form a chain between u and v which allow us to conclude that u and v are linked.

5.10 THEOREM. *There is a function v such that $u - v \in W_0^{1,p}(B(x_0,R))$, div $\boldsymbol{A}_0(Dv) = 0$, and*

$$\int_{B(x_0,R)} |\nabla u - \nabla v|^p \, dx \leq CR^{\sigma} \int_{B(x_0,R)} [1 + |\nabla u|^2]^{p/2} \, dx$$

for all $R < 1$ where $C = C(n, p, \lambda, \Lambda, \Lambda_1, \|\boldsymbol{A}_0\|_{W^{1,\infty}(\Omega)}, \|\psi_1\|_{C^{1,\alpha_1}}, \|\psi_2\|_{C^{1,\alpha_1}})$ and $\sigma = \sigma(\alpha_1, \alpha_2, \beta)$ is a positive number.

PROOF. We show first that u and u_1 are linked. We may assume that the linking exponent are equal. By Lemma 5.9 and (5.29) we have

$$\int_{B(x_0,R)} |\nabla u - \nabla u_1|^p \, dx$$
$$\leq 2^{p-1} \Big[\int_{B(x_0,R)} |\nabla u - \nabla u_0|^p \, dx$$
$$+ \int_{B(x_0,R)} |\nabla u_0 - \nabla u_1|^p \, dx \Big]$$
$$\leq CR^\sigma \int_{B(x_0,R)} |\nabla u|^p \, dx + CR^\sigma \int_{B(x_0,R)} |\nabla u_0|^p \, dx + CR^{n+\sigma}.$$

Since

$$\int_{B(x_0,R)} \big(|\nabla u|^p + |\nabla u_0|^p\big) \, dx$$
$$\leq C \int_{B(x_0,R)} \big(|\nabla u_0 - \nabla u|^p\big) \, dx$$
$$+ C \int_{B(x_0,R)} |\nabla u|^p \, dx,$$

we obtain

$$\int_{B(x_0,R)} |\nabla u - \nabla u_1|^p \, dx \leq CR^\sigma \int_{B(x_0,R)} \big(|\nabla u|^2 + 1\big)^{p/2} \, dx$$

which shows that u and u_1 are linked. We know that u_1 and v are linked by (5.30) and thus, by a similar argument, we can show that u and v are linked. □

We will need the following Lemma which is an immediate consequence of (3.44) and (3.45). We will use the notation

$$(\nabla v)_{x_0,\rho} := \fint_{B(x_0,\rho)} \nabla v \, dx.$$

5.11 LEMMA. *Let $B(x_0, R) \subset \Omega$ and suppose $v \in W^{1,p}(B(x_0, R))$ is a weak solution of* $\operatorname{div} \boldsymbol{A}_0(\nabla v) = 0$ *in $B(x_0, R)$. Then*

(5.31) $$\int_{B(x_0,\rho)} |\nabla v|^p \, dx \leq C \left(\frac{\rho}{R}\right)^n \int_{B(x_0,R)} |\nabla v|^p \, dx$$

(5.32) $$\int_{B(x_0,\rho)} |\nabla v - (\nabla v)_{x_0,\rho}|^p \, dx \leq C \left(\frac{\rho}{R}\right)^{n+\alpha_3 p} \int_{B(x_0,R)} |\nabla v|^p \, dx$$

for all $0 \leq \rho \leq R$. Here $C = C(n, p, \Lambda/\lambda)$ and $0 \leq \alpha_3 = \alpha_3(n, p, \Lambda/\lambda) \leq 1$.

PROOF. From (3.45) and (3.44) it follows that for each $y \in B(x_0, \rho)$, with $\rho \leq R/4$,

$$|\nabla v(y) - (\nabla v)_{x_0,\rho}| \leq C \operatorname{osc}_{B(x_0,\rho)} |\nabla v|$$
$$\leq C \sup_{B(x_0,R/2)} |\nabla v| \left(\frac{\rho}{R}\right)^{\alpha_3}$$
$$\leq C \left(\fint_{B(x_0,R)} |\nabla v|^p \, dx\right)^{1/p} \left(\frac{\rho}{R}\right)^{\alpha_3}.$$

Thus, for $\rho \leq R/4$,

$$\int_{B(x_0,\rho)} |\nabla v - (\nabla v)_{x_0,\rho}|^p \, dx \leq C \left(\frac{\rho}{R}\right)^{n+\alpha_3 p} \int_{B(x_0,R)} |\nabla v|^p \, dx.$$

On the other hand, for $R/4 \leq \rho \leq R$, we have $\rho/R \geq 1/4$ and

$$\int_{B(x_0,\rho)} |\nabla v - (\nabla v)_{x_0,\rho}|^p \, dx \leq C \int_{B(x_0,\rho)} |\nabla v|^p \, dx$$
$$\leq C \left(\frac{\rho}{R}\right)^{n+\alpha_3 p} \int_{B(x_0,R)} |\nabla v|^p \, dx.$$

The first inequality is established with the help of Jensen's inequality. This proves (5.32).

For the proof of (5.31), note that (3.44) implies

$$\fint_{B(x_0,\rho)} |\nabla v|^p \, dx \leq C \fint_{B(x_0,R)} |\nabla v|^p \, dx$$

for all $0 \leq \rho \leq R/2$. As this inequality is trivial for $R/2 \leq \rho \leq R$, (5.31) follows. □

We will also need the following iteration lemma.

5.12 LEMMA. *For certain constants A, B, α, and β with $\beta < \alpha$, suppose φ is a nonnegative nondecreasing function satisfying*

$$\varphi(\rho) \leq A \left[\left(\frac{\rho}{R}\right)^\alpha + \delta\right] \varphi(R) + BR^\beta$$

for $\rho \leq R \leq R_0$. Then there exist $\delta_0 = \delta_0(A, \alpha, \beta)$ and $C = C(A, \alpha, \beta)$ such that if $\delta < \delta_0$, then

$$\varphi(\rho) \leq C \left(\frac{\rho}{R}\right)^\beta [\varphi(R) + BR^\beta]$$

for $0 \leq \rho \leq R \leq R_0$.

PROOF. For $\tau \in (0,1)$ and $0 < R < R_0$, we have

$$\varphi(\tau R) \leq A[\tau^\alpha + \delta]\varphi(R) + BR^\beta$$
$$= A\tau^\alpha[1 + \delta\tau^{-\alpha}]\varphi(R) + BR^\beta.$$

With γ chosen to satisfy $\alpha > \gamma > \beta$, let $\tau < 1$ be such that

$$2A\tau^\alpha \leq \tau^\gamma$$

and then let δ_0 be such that $\delta_0 < \tau^\alpha$. Then, for all $0 < R < R_0$,
$$\varphi(\tau R) \leq \tau^\gamma \varphi(R) + BR^\beta.$$
Hence, for all positive integers i,
$$\varphi(\tau^{i+1}R) \leq \tau^\gamma \varphi(\tau^i R) + B\tau^{i\beta} R^\beta$$
$$\leq \tau^{(i+1)\gamma} \varphi(R) + B\tau^{i\beta} R^\beta \sum_{j=0}^{i} \tau^{j(\gamma-\beta)}$$
$$\leq \tau^{(i+1)\gamma} \varphi(R) + B\tau^{i\beta} R^\beta \frac{1}{1-\tau^{\gamma-\beta}}.$$
Since
$$\frac{1}{1-\tau^{\gamma-\beta}} \leq C\tau^\beta$$
it follows that
$$\varphi(\tau^{i+1}R) \leq C\tau^{(i+1)\beta}[\varphi(R) + BR^\beta].$$
For given ρ and R, choose i such that
$$\tau^{i+1}R < \rho \leq \tau^i R.$$
Then
$$\varphi(\rho) \leq \varphi(\tau^i R) \leq C\tau^{-\beta}\tau^{(i+1)\beta}[\varphi(R) + BR^\beta],$$
which implies our desired result. $\square$

In the following, we let
$$\Phi(R) := \int_{B(x_0,R)} |\nabla u|^p \, dx.$$

5.13 LEMMA. *For each $x_0 \in \Omega$ and for any $\tau > 0$, there are constants $C = C(\tau)$ and $\overline{R} = \overline{R}(\tau) < 1$ that also depend on the data as in Theorem 5.10 such that*
$$\Phi(\rho) \leq C\left(\frac{\rho}{R}\right)^{n-\tau}[\Phi(R) + R^{n-\tau}]$$
for all $0 \leq \rho \leq R \leq \overline{R}$.

PROOF. Using (5.31) we estimate
$$\Phi(\rho) \leq 2^{p-1} \int_{B(x_0,\rho)} |\nabla v|^p \, dx + 2^{p-1} \int_{B(x_0,\rho)} |\nabla u - \nabla v|^p \, dx$$
$$\leq C\left(\frac{\rho}{R}\right)^n \int_{B(x_0,R)} |\nabla v|^p \, dx + C\int_{B(x_0,\rho)} |\nabla u - \nabla v|^p \, dx$$
$$\leq C\left(\frac{\rho}{R}\right)^n \left[\int_{B(x_0,R)} |\nabla u|^p \, dx + \int_{B(x_0,R)} |\nabla u - \nabla v|^p \, dx\right]$$
$$+ \int_{B(x_0,R)} |\nabla u - \nabla v|^p \, dx.$$

By Theorem 5.10, it follows

$$\Phi(\rho) \leq C\left[\left(\frac{\rho}{R}\right)^n \int_{B(x_0,R)} |\nabla u|^p \, dx \right.$$
$$\left. + R^\sigma \int_{B(x_0,R)} |\nabla u|^p \, dx + R^{n+\sigma}\right]$$
$$\leq C\left(\frac{\rho}{R}\right)^{n-\tau}\left[\int_{B(x_0,R)} |\nabla u|^p \, dx + R^{n-\tau}\right],$$

where the last inequality follows from Lemma 5.12 provided R is sufficiently small, i.e., for $R \leq \overline{R}$. □

5.14 THEOREM. *Let u be a solution of the double obstacle problem (5.14). Then, $u \in \mathcal{C}^{1,\alpha}_{\text{loc}}(\Omega)$ for some*

$$0 < \alpha = \alpha(n, p, \Lambda/\lambda, \alpha_1, \alpha_2, \alpha_3, \beta) < 1.$$

PROOF. Let $d_0 < 1$ be a small positive number and select $x_0 \in \Omega$ such that $\text{dist}(x_0, \partial\Omega) < d_0$. We first apply Lemma 5.13 where τ is chosen so that $0 < \tau < \sigma$. A further condition will be placed on τ below. Thus, there exists $0 < R_0 = R_0(n, p, \lambda, \Lambda, \tau) < d_0$ such that

(5.33) $$\Phi(R) \leq C\left(\frac{R}{R_0}\right)^{n-\tau}[\Phi(R_0) + R_0^{n-\tau}]$$

for all $0 \leq R \leq R_0$. Let $0 < \rho < R$. By (5.32) and Theorem 5.10,

$$\int_{B(x_0,\rho)} |\nabla u - (\nabla u)_{x_0,\rho}|^p \, dx$$
$$\leq C\Big[\int_{B(x_0,\rho)} |\nabla u - \nabla v|^p \, dx + \int_{B(x_0,\rho)} |\nabla v - (\nabla v)_{x_0,\rho}|^p \, dx$$
$$+ \int_{B(x_0,\rho)} |(\nabla v)_{x_0,\rho} - (\nabla u)_{x_0,\rho}|^p \, dx\Big]$$
$$\leq C\left(\frac{\rho}{R}\right)^{n+\alpha_3 p}\int_{B(x_0,R)} |\nabla v|^p \, dx + C\int_{B(x_0,\rho)} |\nabla u - \nabla v|^p \, dx$$
$$\leq C\left(\frac{\rho}{R}\right)^{n+\alpha_3 p}\int_{B(x_0,R)} |\nabla u|^p \, dx + C\int_{B(x_0,\rho)} |\nabla u - \nabla v|^p \, dx$$
$$\leq C\left(\frac{\rho}{R}\right)^{n+\alpha_3 p}\int_{B(x_0,R)} |\nabla u|^p \, dx$$
$$+ CR^\sigma \int_{B(x_0,R)} |\nabla u|^p \, dx + CR^{n+\sigma}.$$

Hence, by (5.33), we obtain

$$\int_{B(x_0,\rho)} |\nabla u - (\nabla u)_{x_0,\rho}|^p \, dx$$

$$\leq C \left(\frac{\rho}{R}\right)^{n+\alpha_3 p} \left(\frac{R}{R_0}\right)^{n-\tau} \left[\int_{B(x_0,R_0)} |\nabla u|^p + CR_0^{n-\tau}\right]$$

$$+ CR^\sigma \left[\left(\frac{R}{R_0}\right)^{n-\tau} \left(\int_{B(x_0,R_0)} |\nabla u|^p \, dx + R_0^{n-\tau}\right)\right]$$

$$+ CR^{n+\sigma}.$$

Now we choose ρ and R so that they are related by $R = \rho^\vartheta$ where $\vartheta < 1$. We choose τ so small that

$$0 < \frac{n}{\sigma - \tau} < \frac{\alpha_3 p}{\tau}$$

and insert a number $\tilde{\vartheta}$ between these positive numbers. Setting $\vartheta = \tilde{\vartheta}/(1+\tilde{\vartheta})$ we obtain

(i) $n + \alpha_3 p - \vartheta(\tau + \alpha_3 p) > n$

and

(ii) $\vartheta(n + \sigma_{(u,v)} - \tau) > n$.

With these choices of ϑ and τ, it is an easy matter to verify that there exists $\delta > 0$ such that

$$\int_{B(x_0,\rho)} |\nabla u - (\nabla u)_{x_0,\rho}|^p \, dx$$

$$\leq C\rho^{n+\delta} R_0^{\tau-n} \left[\int_{B(x_0,R_0)} |\nabla u|^p \, dx + R_0^{n-\tau}\right]$$

$$+ C\rho^{n+\delta} \left[R_0^{\tau-n} \left(\int_{B(x_0,R_0)} |\nabla u|^p \, dx + R_0^{n-\tau}\right)\right] + C\rho^{n+\delta},$$

for all sufficiently small $\rho > 0$, independent of the point x_0. In particular, $\rho < d_0$. Since $d_0 > 0$ is arbitrary, the Campanato theorem (Theorem 1.54,31) implies that $u \in \mathcal{C}^{1,\alpha}_{\text{loc}}(\Omega)$, for some $0 < \alpha < 1$. $\square$

5.3 Historical notes

One of the basic regularity results for the obstacle problem states that the the solution of the lower obstacle problem is locally Hölder continuous provided that obstacle is Hölder continuous. This was done in the case of linear equations by Beirao da Veiga [**BeV**]. A standard reference to both existence and regularity theory for obstacle problems is the book by Kinderlehrer and Stampacchia [**KS**] which also provides a valuable source for historical information.

For a large class of variational inequalities, it is known that the solution remains continuous provided that the (discontinuous) obstacles satisfy an appropriate Wiener-type condition. For linear equations this was investigated by Frehse and Mosco [**FM1**], [**FM2**] and the necessity part is due to Mosco [**Mosc**]. The situation involving linear degenerate equations was considered by Biroli and Mosco [**BMo**]. Michael and Ziemer [**MZ2**] proved sufficiency of the Wiener condition for the equation (3.1). Heinonen and Kilpeläinen [**HK3**] studied lower obstacle problems for the equation (2.69). They obtained necessity for $p > n - 1$ and established various relations to nonlinear potential theory. The Wiener criterion for the double obstacle problems was established by Dal Maso, Mosco and Vivaldi [**DMV**] in the linear case.

The material in Section 5.1 was adapted from [**Mal2**] and [**KZ**], which, in turn, had its origins in [**MZ2**]. Others who contributed to this theory included Frehse and Mosco, [**FM1**] and [**FM2**].

In the framework of elliptic equations based on Hörmander fields, necessity and sufficiency of the Wiener condition was recently obtained by Gianazza, Marchi and Villani [**GMV**].

The treatment in Section 5.2 is due to [**MuZ**] and its forerunner Mu [**Mu**]. Other authors have also have shown $C^{1,\alpha}$ regularity under varying hypotheses on the structure of the equation, cf. Lindqvist, [**Lin2**], Fuchs [**Fuc**], Norando [**Nor**], Choe and Lewis [**CL**], and Lieberman [**Li2**].

A survey article on L^p potential theory techniques related to nonlinear partial differential equations was written by Adams [**Ad2**]. Most of the article is concerned with the p-Laplacian and some closely related nonlinear equations. The topics covered include the Wiener test, removable singularities and the obstacle problem.

CHAPTER 6

Existence Theory

In this chapter we investigate the existence of solutions to variational inequalities and the Dirichlet problem. In Section 6.1 we prove existence of weak solutions of variational inequalities in which the obstacles are assumed to be only bounded functions. These solutions possess some further regularity properties which will not stated explicitly in the existence theorems. For this we refer to the regularity theory in Chapters 3, 4 and 5. For example, the solutions are Hölder continuous in the complement of the coincidence set, continuous at Wiener points of the obstacles, and tend to boundary values at regular points of $\partial\Omega$. One of the main components of this treatment is the use of pseudomonotone operators.

Section 6.2 is a brief section in which we consider the Dirichlet problem within the framework of a differentiable structure, which however, is not a special case of the structure considered in 6.1. Indeed, the growth of B leads to a lack of coercivity. The method here utilizes an approximation process in which the approximating problems fall within the classical existence theory.

6.1 Existence of solutions to variational inequalities

6.1 UNDERLYING ASSUMPTIONS. Let $\Omega \subset \mathbf{R}^n$ be an arbitrary p-finely open set with $|\Omega| < \infty$. Also, let $\psi_1 \leq \psi_2$ be an arbitrary pair of functions defined at all points of Ω. We will investigate the existence of solutions u to the obstacle problem

$$\int_\Omega \boldsymbol{A}(x, u, \nabla u) \cdot \nabla\varphi + \boldsymbol{B}(x, u, \nabla u)\varphi\, dx \geq 0$$

for all φ with $\psi_1(x) \leq u(x) + \varphi(x) \leq \psi_2(x)$ for p-quasi every $x \in \Omega$. We also require u to satisfy a Dirichlet condition with a boundary function f. In what follows, Ω will be as above and the properties of f will be specified according to the situation.

Furthermore, we impose the following assumptions:

(i) Both $\boldsymbol{A}$ and $\boldsymbol{B}$ are Carathéodory functions on $\mathbf{R}^n \times \mathbf{R} \times \mathbf{R}^n$; i.e., for almost all $x \in \mathbf{R}^n$, $\boldsymbol{A}(x, \cdot, \cdot)$ and $\boldsymbol{B}(x, \cdot, \cdot)$ are continuous on $\mathbf{R} \times \mathbf{R}^n$, while for all $\zeta \in \mathbf{R}$ and $\xi \in \mathbf{R}^n$, $\boldsymbol{A}(\cdot, \zeta, \xi)$ and $\boldsymbol{B}(\cdot, \zeta, \xi)$ are measurable on $\mathbf{R}^n$.

(ii) $\boldsymbol{A}$ and $\boldsymbol{B}$ satisfy a structure similar to (3.5), p. 162. The first two conditions of (3.5) are assumed to hold as well as

(6.1) $$[\boldsymbol{A}(x,\zeta,\xi) - \boldsymbol{A}(x,\zeta,\xi')] \cdot (\xi - \xi') > 0$$

and

(6.2) $$\boldsymbol{A}(x,\zeta,\xi) \cdot \xi + \zeta \boldsymbol{B}(x,\zeta,\xi) \geq c_1|\xi|^p - c_2|\zeta|^p - c_3,$$

for all $x \in \Omega$, $\zeta \in \mathbf{R}$ and $\xi, \xi' \in \mathbf{R}^n$.

(iii) We suppose that the coefficients are in the function spaces described in (3.6). Also, we suppose that $b_0 = 0$, $a_3^{p'} + b_1^p + b_2 + b_3 + c_3 \in L^1(\Omega)$ and $a_2^{1/(p-1)}$ is a *compact* multiplier. By this we mean that the weak convergence $v_j \to v$ in $W^{1,p}(\mathbf{R}^n)$ implies the strong convergence $a_2^{1/(p-1)} v_j \to a_2^{1/(p-1)} v$ in $L^p(\Omega)$. See Remark 6.2 for more comments on this requirement.

(iv) Finally, we assume that the Poincaré type inequality

$$(6.3) \qquad \int_\Omega c_2 |v|^p \, dx \leq \mu \int_\Omega c_1 |\nabla v|^p \, dx, \quad \text{for each } v \in W_0^{1,p}(\Omega),$$

holds with $\mu < 1$.

Notice that, by Young's inequality,

$$(6.4) \qquad \begin{aligned} \mathbf{A}(x,\zeta,\xi) \cdot \xi &\geq \Big[\mathbf{A}(x,\zeta,\xi) \cdot \xi + \zeta \mathbf{B}(x,\zeta,\xi)\Big] - \zeta \mathbf{B}(x,\zeta,\xi) \\ &\geq c_1 |\xi|^p - c_2 |\zeta|^p - c_3 - b_1 \zeta |\xi|^{p-1} - b_2 |\zeta|^p - b_3 |\zeta| \\ &\geq c_1' |\xi|^p - c_2' |\zeta|^p - c_3' \end{aligned}$$

where

$$(6.5) \qquad \begin{aligned} c_1' &= c_1/p, \\ c_2' &= c_2 + b_2 + c_1^{1-p} b_1^p / p + b_3 / p \\ c_3' &= c_3 - b_3 / p' \end{aligned}$$

which shows that the third condition of (3.5) is satisfied. From here on we simplify the notation by assuming that $c_1 = 1$.

Let $u \in W^{1,p}(\mathbf{R}^n)$ and denote

$$\mathbf{A}_\Omega := \begin{cases} \mathbf{A}(x,u,\nabla u) & \text{on } \Omega, \\ 0 & \text{outside } \Omega, \end{cases}$$

$$\mathbf{B}_\Omega := \begin{cases} \mathbf{B}(x,u,\nabla u) & \text{on } \Omega, \\ 0 & \text{outside } \Omega. \end{cases}$$

Similar to the proof of Proposition 3.5 it follows that $\mathbf{A}_\Omega \in L^{p'}(\mathbf{R}^n)$ and $\mathbf{B}_\Omega \in (W^{1,p}(\mathbf{R}^n))^*$ with

$$(6.6) \qquad \begin{aligned} \|\mathbf{A}_\Omega\| &\leq C(\|u\|_{1,p}^{p-1} + 1), \\ \|\mathbf{B}_\Omega\| &\leq C(\|u\|_{1,p}^{p-1} + 1), \end{aligned}$$

where the constant C depends only on the structure and Ω and the norms are relative to $\mathbf{R}^n$. We let

$$\mathbf{M}_\Omega \colon W^{1,p}(\mathbf{R}^n) \to W^{1,p}(\mathbf{R}^n)^*$$

denote the mapping

$$\langle \mathbf{M}_\Omega(v), w \rangle := \int_\Omega \mathbf{A}(x,v,\nabla v) \cdot \nabla w \, dx + \int_\Omega \mathbf{B}(x,v,\nabla v) w \, dx$$

for all $v, w \in W^{1,p}(\mathbf{R}^n)$.

6.2 REMARK. If a multiplier $g \in M(W^{1,p}(\mathbf{R}^n), L^p(\mathbf{R}^n))$ belongs to the closure of $L^\infty(\mathbf{R}^n)$ in this space, then it is compact to $L^p(\Omega)$. Indeed, we may decompose g as a sum $g_1 + g_2$, where g_1 is multiplier-norm small and g_2 is bounded. Then we apply the Rellich-Kondrachov imbedding theorem 1.61 to the sequence $g_2 v_j$ and the rest is easy.

If $p < n$, any $g \in L^n(\mathbf{R}^n)$ is such a compact multiplier. Corollary 1.95 shows that $|g|^p \in \mathcal{M}^{n/(p-\varepsilon)}$ implies the compact multiplier property of g as well.

Also, condition (iv) above requires special attention. The condition $\mu < 1$ would follow if $\|c_2\|$ were small, as can be seen with the help of (3.11), p. 163. Furthermore, if c_1 and c_2 are known *a priori*, then (6.3) would result from an assumption on the smallness of $|\Omega|$.

If the growth in ζ is slightly better, (iv) holds irrespective of the domain. For example, if $q < p$ and the condition

$$(6.7) \qquad \boldsymbol{A}(x,\zeta,\xi) \cdot \xi + \zeta \boldsymbol{B}(x,\zeta,\xi) \geq c_1 |\xi|^p - c_2 |\zeta|^q - c_3,$$

is satisfied, then for any Ω with $|\Omega| < \infty$ there is $\varepsilon = \varepsilon(\Omega)$ such that εc_2 is small enough. Then our results will be valid because Young's inequality yields

$$\boldsymbol{A}(x,\zeta,\xi) \cdot \xi + \zeta \boldsymbol{B}(x,\zeta,\xi) \geq c_1 |\xi|^p - \varepsilon c_2 |\zeta|^p - C(\varepsilon) - c_3.$$

6.3 THEOREM. *Let $f, v \in W^{1,p}(\mathbf{R}^n)$, $v = f$ outside Ω, and*

$$(6.8) \qquad v - f \in W_0^{1,p}(\Omega).$$

Then there is a constant C depending on μ and the structure such that

$$(6.9) \qquad \|v - f\|_{1,p;\Omega}^p \leq C \Big[\langle M_\Omega u, u - f \rangle + \|f\|_{1,p;\mathbf{R}^n}^p + 1 \Big].$$

PROOF. We have

$$\langle M_\Omega v, v - f \rangle = \int_\Omega \boldsymbol{A}(x, v, \nabla v) \cdot (\nabla v - \nabla f) dx + \int_\Omega \boldsymbol{B}(x, v, \nabla v)(v - f) \, dx$$

and by the structure conditions and (6.6),

$$(6.10) \qquad \int_\Omega (|\nabla v|^p - c_2 |v|^p - c_3) \, dx \leq \langle M_\Omega v, v - f \rangle + C(\|v\|_{1,p}^{p-1} + 1) \|f\|_{1,p}.$$

Choose $\varepsilon \in (0,1)$. Recalling the inequality

$$(6.11) \qquad |a + b|^p \leq (1+\varepsilon)^{p-1} |a|^p + (1 + 1/\varepsilon)^{p-1} |b|^p$$

which holds for any a, b in $\mathbf{R}$ (or in $\mathbf{R}^n$) and $\varepsilon > 0$, (Lemma 1.1), and using (6.3), we have

$$\int_\Omega |\nabla v - \nabla f|^p \, dx \leq (1+\varepsilon)^{p-1} \int_\Omega |\nabla v|^p \, dx + C\varepsilon^{1-p} \int_\Omega |\nabla f|^p \, dx$$

$$\leq (1+\varepsilon)^{p-1} \int_\Omega (c_2|v|^p + c_3) \, dx + 2\langle \mathbf{M}_\Omega v, v - f \rangle$$

$$+ \varepsilon \|v\|_{1,p}^p + C\varepsilon^{1-p}\|f\|_{1,p}^p + C(\|v\|_{1,p}^{p-1} + 1)\|f\|_{1,p}$$

$$\leq (1+\varepsilon)^{2p-2} \int_\Omega c_2|v - f|^p \, dx + 2\langle \mathbf{M}_\Omega v, v - f \rangle$$

$$+ C\varepsilon \|v - f\|_{1,p}^p + C\varepsilon^{1-p}(\|v\|_{1,p}^{p-1} + 1)\|f\|_{1,p} + C_3$$

$$\leq [(1+\varepsilon)^{2p-2}\mu + C\varepsilon] \int_\Omega |\nabla v - \nabla f|^p \, dx + 2\langle \mathbf{M}_\Omega v, v - f \rangle$$

$$+ C\varepsilon^{1-p}(\|v\|_{1,p}^{p-1} + 1)\|f\|_{1,p} + C_3,$$

where

$$C_3 = \varepsilon^{1-p} \int_\Omega c_3 \, dx.$$

Choosing ε small enough, a cancellation is possible and we obtain

(6.12) $$\int_\Omega |\nabla v - \nabla f|^p \, dx \leq C\langle \mathbf{M}_\Omega v, v - f \rangle + C(\|v\|_{1,p}^{p-1} + 1)\|f\|_{1,p} + C_3.$$

Using Poincaré's and Young's inequality we conclude the proof. □

With the estimate of the previous theorem in hand, we can easily establish a maximum principle.

6.4 THEOREM. *Let $L > 0$ be a constant, $u \in W^{1,p}_{\text{loc}}(\Omega)$,*

$$(u - L)^+ \in W^{1,p}_0(\Omega)$$

and

$$\int_\Omega (\mathbf{A}(x, u, \nabla u) \cdot \varphi + \mathbf{B}(x, u, \nabla u)\varphi) \, dx \leq 0$$

for all $\varphi \in W^{1,p}_{cc}(\Omega)$ such that

$$0 \leq \varphi \leq (u - L)^+.$$

Then there exists a constant C depending on μ and the structure, such that

(6.13) $$\sup_\Omega [u(x) - L]^+ \leq (C+1)|L|.$$

PROOF. We may assume that $L = 0$ because the difference in structure is not essential. We refer to the proof of Theorem 3.11 to find that

$$\sup_\Omega u^+ \leq C \left[\left(\fint_\Omega (u^+)^p \, dx \right)^{1/p} + k \right]$$

where C and k depend on the same quantities as in Theorem 3.11. Our result now follows from the previous theorem applied to $v = u^+$ and $f = 0$. □

6.1 EXISTENCE OF SOLUTIONS TO VARIATIONAL INEQUALITIES

6.5 THEOREM. *Let $\{u_k\}$ be a sequence in $W^{1,p}(\mathbf{R}^n)$ which converges weakly to u in $W^{1,p}(\mathbf{R}^n)$ and strongly to u in $L^p(\Omega)$. Let $\eta \in W^{1,p}(\Omega)$ be such that $0 \leq \eta \leq 1$ and*

$$(6.14) \qquad (u_k - u)\,\eta \in W_0^{1,p}(\Omega)$$

for all k. Suppose that either η is Lipschitz or $\{u_k\}$ is bounded in $L^\infty(\mathbf{R}^n)$. Suppose that

$$(6.15) \qquad \limsup_{k \to \infty} \langle \eta^{p-1} \boldsymbol{M}_\Omega(u_k), (u_k - u)\,\eta \rangle \leq 0.$$

Then

$$(6.16) \qquad \int_\Omega \eta^p(x) |\nabla u_k(x) - \nabla u(x)|^p dx \to 0$$

as $k \to \infty$. Also

$$(6.17) \qquad \eta^{p-1} \boldsymbol{M}_\Omega(u_k) \to \eta^{p-1} \boldsymbol{M}_\Omega(u)$$

weakly in $W_0^{1,p}(\Omega)^*$ as $k \to \infty$ and

$$(6.18) \qquad \langle \eta^{p-1} \boldsymbol{M}_\Omega(u_k), \eta u_k \rangle \to \langle \eta^{p-1} \boldsymbol{M}_\Omega(u), \eta u \rangle$$

as $k \to \infty$.

PROOF. After passing to a subsequence, we may assume that

$$(6.19) \qquad u_k \to u \quad \text{pointwise almost everywhere.}$$

Corresponding to this subsequence, define

$$(6.20) \qquad \begin{aligned} h_k(x) := &[\boldsymbol{A}(x, u_k(x), \nabla u_k(x)) - \boldsymbol{A}(x, u_k(x), \nabla u(x))] \\ & \cdot [\nabla u_k(x) - \nabla u(x)] \end{aligned}$$

for $x \in \Omega$ and all k. Then each h_k is measurable on Ω and it follows from (6.1) that

$$(6.21) \qquad h_k(x) \geq 0$$

for almost all $x \in \Omega$ and all k.

The remainder of the proof will be carried out after the following three lemmas have been established.

6.6 LEMMA. *We have*

$$(6.22) \qquad \int_\Omega h_k(x)\eta^p \, dx \to 0$$

as $k \to \infty$.

PROOF. Since $u_k \to u$ weakly in $W^{1,p}(\mathbf{R}^n)$, it follows that

(6.23) $$\int_\Omega \boldsymbol{A}(x,u,\nabla u) \cdot (\nabla u_k - \nabla u)\eta^p\, dx \to 0$$

as $k \to \infty$. Also, from the continuity assumption on $\boldsymbol{A}(x,\cdot,\nabla u(x))$, for almost all $x \in \Omega$ we obtain

(6.24) $$\boldsymbol{A}(x,u_k(x),\nabla u(x)) \to \boldsymbol{A}(x,u(x),\nabla u(x))$$

as $k \to \infty$.

Utilizing the structure, we have

(6.25) $$\begin{aligned}|\boldsymbol{A}(x,u_k,\nabla u) - \boldsymbol{A}(x,u,\nabla u)| \\ \leq 2a_1|\nabla u|^{p-1} + a_2|u_k|^{p-1} + a_2|u|^{p-1} + 2a_3 \\ \leq 2a_1|\nabla u|^{p-1} + 2^{p-1}a_2|u_k - u|^{p-1} + (2^{p-1}+1)a_2|u|^{p-1} + 2a_3 \\ \leq \begin{cases} 2\Big[2a_1|\nabla u|^{p-1} + (2^{p-1}+1)a_2|u|^{p-1} + 2a_3\Big] & \text{on } E_k, \\ 2^p a_2|u_k - u|^{p-1} & \text{on } \Omega \setminus E_k, \end{cases}\end{aligned}$$

where

$$E_k := \{2a_1|\nabla u|^{p-1} + (2^{p-1}+1)a_2|u|^{p-1} + 2a_3 \geq 2^{p-1}a_2|u_k - u|^{p-1}\}.$$

Hence

(6.26) $$\begin{aligned}|\boldsymbol{A}(x,u_k,\nabla u) - \boldsymbol{A}(x,u,\nabla u)|^{p'}\eta^p \\ \leq |\boldsymbol{A}(x,u_k,\nabla u) - \boldsymbol{A}(x,u,\nabla u)|^{p'}\eta^p \chi_{E_k} \\ + 2^{pp'}a_2^{p'}|u_k - u|^p \eta^p.\end{aligned}$$

Using the compact multiplier property of $a_2^{1/(p-1)}$, we obtain

(6.27) $$\int_\Omega a_2^{p'}|u_k - u|^p \eta^p\, dx \to 0.$$

Reference to (6.24) and (6.25) along with an application of Lebesgue's dominated convergence theorem yields

(6.28) $$\int_\Omega |\boldsymbol{A}(x,u_k,\nabla u) - \boldsymbol{A}(x,u,\nabla u)|^{p'} \eta^p \chi_{E_k}\, dx \to 0$$

as $k \to \infty$. By (6.26), (6.27) and (6.28),

(6.29) $$\int_\Omega |\boldsymbol{A}(x,u_k,\nabla u) - \boldsymbol{A}(x,u,\nabla u)|^{p'} \eta^p\, dx \to 0.$$

Also, we have

$$\left|\int_\Omega [\boldsymbol{A}(x,u_k,\nabla u) - \boldsymbol{A}(x,u,\nabla u)] \cdot (\nabla u_k - \nabla u)\eta^p\, dx\right|$$
$$\leq \left[\int_\Omega |\boldsymbol{A}(x,u_k,\nabla u) - \boldsymbol{A}(x,u,\nabla u)|^{p'} \eta^p\, dx\right]^{1/p'} \|\nabla(u_k - u)\eta\|_{p;\Omega}.$$

Hence by (6.29),

(6.30) $$\int_\Omega [\boldsymbol{A}(x, u_k, \nabla u) - \boldsymbol{A}(x, u, \nabla u)] \cdot (\nabla u_k - \nabla u) \eta^p \, dx \to 0$$

as $k \to \infty$, so that by (6.23),

(6.31) $$\int_\Omega \boldsymbol{A}(x, u_k, \nabla u) \cdot (\nabla u_k - \nabla u) \eta^p \, dx \to 0$$

as $k \to \infty$. We estimate

(6.32) $$\int_\Omega \boldsymbol{B}(x, u_k, \nabla u_k)(u_k - u) \eta^p \, dx$$
$$\leq C \Big(\int_\Omega (|\nabla u_k|^p + b_2 |u_k|^p + b_3) \eta^p \, dx\Big)^{1/p'}$$
$$\Big(\int_\Omega (b_1^p + b_2 + b_3)|u_k - u|^p) \eta^p \, dx\Big)^{1/p}$$

The first integral on the right is bounded with respect to k and the second one tends to zero by Corollary 1.95. By (6.15),

(6.33) $$\limsup_{k \to \infty} \Big[\int_\Omega \boldsymbol{A}(x, u_k, \nabla u_k) \cdot ((\nabla u_k - \nabla u)\eta^p + (u_k - u)\eta^{p-1} \nabla \eta) \, dx$$
$$+ \int_\Omega \boldsymbol{B}(x, u_k, \nabla u_k)(u_k - u) \eta^p \, dx\Big] \leq 0.$$

It follows from the assumptions, structure conditions and (6.32) that the 2nd and 3rd terms on the left of (6.33) approach zero as $k \to \infty$. Thus

(6.34) $$\limsup_{k \to \infty} \int_\Omega \boldsymbol{A}(x, u_k, \nabla u_k) \cdot (\nabla u_k - \nabla u) \eta^p \, dx \leq 0.$$

The required result now follows from (6.31), (6.34) and (6.21). □

6.7 LEMMA. *There exists a subsequence of* (6.19) *(still denoted by the full sequence) such that*
$$\lim_{k \to \infty} \eta(x) \nabla u_k(x) = \eta(x) \nabla u(x)$$
for almost all $x \in \Omega$.

PROOF. It follows from Lemma 6.6 that there is a subsequence of $\{h_k \eta^p\}$ which converges pointwise to zero almost everywhere on Ω. We will assume that $\{u_k\}$ corresponds to this subsequence. Due to structure assumptions, we have

$$|\nabla u_k|^p \eta^p \leq C\Big(h_k + c_2|u_k|^p + c_3 + (a_1|\nabla u_k|^{p-1} + a_2|u_k|^{p-1} + a_3)|\nabla u|$$
$$+ (a_1|\nabla u|^{p-1} + a_2|u_k|^{p-1} + a_3)|\nabla u_k - \nabla u|\Big) \eta^p.$$

Using Young's inequality and cancellation we obtain

(6.35) $\quad |\nabla u_k|^p \eta^p \leq C\big[h_k + (c_2 + a_2^{p'})|u_k|^p + (a_1^{p'} + a_1^p + 1)|\nabla u|^p + a_3^{p'} + c_3\big]\eta^p.$

There exists a subset Z of Ω such that

(6.36) $$|\Omega - Z| = 0,$$

(6.37) $$|u(x)| < \infty, \quad |\nabla u(x)| < \infty$$

for all $x \in Z$, inequality (6.35) holds at x, all structure coefficients are finite at x,

(6.38) $$h_k(x)\eta^p(x) \to 0 \quad \text{and} \quad u_k(x) \to u(x)$$

as $k \to \infty$ and

(6.39) $$\boldsymbol{A}(x, \cdot, \cdot)$$

is continuous on $\mathbf{R} \times \mathbf{R}^n$.

Consider $x \in Z$ with $\eta(z) > 0$ and suppose $\{\nabla u_k(x)\}$ does not converge to $\nabla u(x)$. By (6.35), (6.37) and (6.38), there is a subsequence (again denoted by the full sequence) such that $\{\nabla u_k(x)\}$ converges to $\xi \in \mathbf{R}^n$, where $\xi \neq \nabla u(x)$. Using (6.38), we see that

$$[\boldsymbol{A}(x, u_k(x), \nabla u_k(x)) - \boldsymbol{A}(x, u_k(x), \nabla u(x))] \cdot [\nabla u_k(x) - \nabla u(x)]\eta^p(x) \to 0$$

as $k \to \infty$. Since $u_k(x) \to u(x)$ as $k \to \infty$ and $\boldsymbol{A}(x, \cdot, \cdot)$ is continuous on $\mathbf{R} \times \mathbf{R}^n$,

$$0 = [\boldsymbol{A}(x, u(x), \xi) - \boldsymbol{A}(x, u(x), \nabla u(x))] \cdot [\xi - \nabla u(x)] > 0,$$

a contradiction. $\square$

6.8 LEMMA. *There is a subsequence of the one established in the previous lemma such that*

$$\lim_{k \to \infty} \int_\Omega \eta^p(x) |\nabla u_k(x) - \nabla u(x)|^p \, dx = 0.$$

PROOF. We proceed as in the proof of (6.29). From (6.35) we obtain pointwise the inequality

(6.40) $$\begin{aligned}
\eta^p |\nabla u_k - \nabla u|^p &\leq 2^{p-1} \eta^p \Big(|\nabla u_k^p| + |\nabla u|^p \Big) \\
&\leq C\Big(h_k + (c_2 + a_2^{p'})|u_k|^p + (a_1^{p'} + a_1^p + 2)|\nabla u|^p + a_3^{p'} + c_3 \Big)\eta^p \\
&\leq C_1\Big(h_k + (c_2 + a_2^{p'})|u_k - u|^p + |\nabla u|^p \\
&\qquad + (c_2 + a_2^{p'})|u|^p + a_3^{p'} + c_3 \Big)\eta^p \\
&\leq \begin{cases} 2C_1 \Big(|\nabla u|^p + (c_2 + a_2^{p'})|u|^p + a_3^{p'} + c_3 \Big)\eta^p & \text{on } E_k \\ 2C_1 \big[h_k + (c_2 + a_2^{p'})|u_k - u|^p \big]\eta^p & \text{on } \Omega \setminus E_k \end{cases}
\end{aligned}$$

where now

$$E_k = \Big\{ |\nabla u|^p + (c_2 + a_2^{p'})|u|^p + a_3^{p'} + c_3 \geq h_k + (c_2 + a_2^{p'})|u_k - u|^p \Big\}$$

and C_1 depends also on a_1. We have

(6.41) $$\begin{aligned}
\eta^p |\nabla u_k - \nabla u|^p &\leq \eta^p |\nabla u_k - \nabla u|^p \chi_{E_k} \\
&\quad + 2C_1 \big[h_k + (c_2 + a_2^{p'})|u_k - u|^p \big]\eta^p.
\end{aligned}$$

As in (6.27) we obtain that
$$\int_\Omega (c_2 + a_2^{p'})|u_k - u|^p \eta^p \, dx \to 0.$$
By Lemma 6.6,
$$\int_\Omega h_k \eta^p \, dx \to 0.$$
From the previous lemma we know that $\nabla u_k \to \nabla u$ pointwise almost everywhere for a subsequence and thus, by Lebesgue's dominated convergence theorem,
$$\int_\Omega |\nabla u_k - \nabla u|^p \eta^p \chi_{E_k} \, dx \to 0.$$
Hence, the right-hand side of (6.41) tends to zero in L^1 and the lemma is thus established. □

PROOF OF THEOREM 6.5 (continued). Corresponding to the subsequence of Lemma 6.8, let
$$g_k(x) := \boldsymbol{A}(x, u_k(x), \nabla u_k(x))\eta^{p-1}(x)$$
and
$$g(x) := \boldsymbol{A}(x, u(x), \nabla u(x))\eta^{p-1}(x)$$
for $x \in \Omega$. Then, by (6.19) and Lemma 6.7, $g_k \to g$ pointwise almost everywhere on Ω. Furthermore, since $\{g_k\}$ is a bounded sequence in $L^{p'}(\Omega)$, it follows (for a subsequence) that $g_k \to g$ weakly in $L^{p'}(\Omega, \mathbf{R}^n)$; that is, for each $w \in W_0^{1,p}(\Omega)$,

$$\lim_{k \to \infty} \int_\Omega \boldsymbol{A}(x, u_k(x), \nabla u_k(x))\eta^{p-1} \cdot \nabla w \, dx$$
(6.42)
$$= \int_\Omega \boldsymbol{A}(x, u(x), \nabla u(x))\eta^{p-1} \cdot \nabla w \, dx.$$

Similarly,

(6.43) $\quad \lim_{k \to \infty} \int_\Omega \boldsymbol{B}(x, u_k(x), \nabla u_k(x))\eta^{p-1} w \, dx = \int_\Omega \boldsymbol{B}(x, u(x), \nabla u(x))\eta^{p-1} w \, dx.$

By (6.42) and (6.43)
$$\langle \boldsymbol{M}_\Omega(u_k)\eta^{p-1}, w \rangle \to \langle \boldsymbol{M}_\Omega(u)\eta^{p-1}, w \rangle \quad \text{as} \quad k \to \infty,$$
and consequently, we have established (6.17).

Finally, we prove (6.18). We have
$$\left| \langle \eta^{p-1} \boldsymbol{M}_\Omega(u_k), \eta(u_k - u) \rangle \right| \leq \left(\left\| \boldsymbol{A}(\cdot, u_k(\cdot), \nabla u_k(\cdot),)\eta^{p-1} \right\|_{p';\Omega} \right.$$
$$\left. + \left\| \boldsymbol{B}(\cdot, u_k(\cdot), \nabla u_k(\cdot),)\eta^{p-1} \right\|_{p';\Omega} \right) \|\eta(u_k - u)\|_{1,p;\Omega}.$$

From the strong convergence of $u_k \to u$ in L^p and Lemma 6.8, it follows that
$$\|\eta(u_k - u)\|_{1,p;\Omega} \to 0,$$
and therefore
$$\langle \eta^{p-1} \boldsymbol{M}_\Omega(u_k), \eta(u_k - u) \rangle \to 0.$$

This proves (6.18) since, by (6.17),

$$\langle \eta^{p-1} M_\Omega(u_k), \eta u \rangle \to \langle \eta^{p-1} M_\Omega(u), \eta u \rangle.$$

This completes the proof of Theorem 6.5. □

6.1.1 Pseudomonotone operators.

Along with the results of the previous section, the main ingredient needed for the existence of solutions is the theory of pseudomonotone operators.

6.9 DEFINITION. Let V be a real reflexive Banach space and let V^* denote its dual. A bounded mapping $P: V \to V^*$ is said to be *pseudomonotone* if for each sequence $\{u_k\}$ converging weakly to u with

$$\limsup_{k \to \infty} \langle P(u_k), u_k - u \rangle \leq 0,$$

we have

$$\liminf_{k \to \infty} \langle P(u_k), u_k - w \rangle \geq \langle P(u), u - w \rangle$$

for all $w \in V$.

The keystone of the theory of pseudomonotone operators is the following existence theorem which is found in [**Li**], Theorem 8.2.

6.10 THEOREM. *Let K be a closed convex subset of V. Suppose $P: K \to V'$ is a pseudomonotone operator satisfying the coercive condition that there exists $v_0 \in K$ such that*

$$\frac{\langle P(v), v - v_0 \rangle}{\|v\|} \to +\infty$$

when $\|v\| \to \infty$, $v \in K$. Then, for each $f \in V'$, there exists $u \in K$ such that

$$\langle P(u), u - v \rangle \leq \langle f, v - u \rangle$$

for all $v \in K$.

6.11 THEOREM. *Let $f \in W^{1,p}(\mathbf{R}^n)$. Then the mapping*

$$P: W_0^{1,p}(\Omega) \to W^{1,p}(\Omega)^*$$

defined by

(6.44) $$\langle P(v), w \rangle = \langle M_\Omega(f + v), w \rangle$$

for $v, w \in W_0^{1,p}(\Omega)$, is pseudomonotone.

PROOF. Consider a sequence $\{v_k\}$ in $W_0^{1,p}(\Omega)$ that converges weakly to $v \in W_0^{1,p}(\Omega)$ and denote $u_k = f + v_k$, $u = f + v$. Suppose

$$\limsup_{k \to \infty} \langle M_\Omega(u_k), u_k - u \rangle \leq 0.$$

We must show that

(6.45) $$\liminf_{k \to \infty} \langle M_\Omega(u_k), u_k - w \rangle \geq \langle M_\Omega(u), u - w \rangle$$

for all $w \in W_0^{1,p}(\Omega)$. Note that it is sufficient to establish (6.45) for a subsequence of $\{v_k\}$. Since $v_k \to v$ weakly, there is a subsequence such that $v_k \to v$ strongly in $L^p(\Omega)$. By Theorem 6.5 with $\eta = 1$,

$$M_\Omega(u_k) \to M_\Omega(u) \quad \text{and} \quad \langle M_\Omega(u_k), u_k \rangle \to \langle M_\Omega(u), u \rangle$$

as $k \to \infty$. Hence,

$$M_\Omega(u_k)(w) \to M_\Omega(u)(w)$$

for $w \in W_0^{1,p}(\Omega)$ and (6.45) thus follows. $\square$

6.12 THEOREM. *Let $f \in W^{1,p}(\mathbf{R}^n)$ and X be a closed convex subset of $W^{1,p}(\Omega)$ such that*

$$f \in X \subset f + W_0^{1,p}(\Omega).$$

Then there exists $u \in X$ such that

$$\langle M_\Omega(u), w - u \rangle \geq 0$$

for all $w \in X$.

PROOF. The operator P defined on $W_0^{1,p}(\Omega)$ by (6.44) is pseudomonotone by Theorem 6.11 and it coercivity follows from Theorem 6.3. Hence the existence follows from Theorem 6.10. $\square$

6.13 REMARK. If $f \in W^{1,p}(\Omega)$, then there is a unique minimizer of

$$\int_\Omega |\nabla u|^p \, dx$$

in $f + W_0^{1,p}(\Omega)$, where Ω is of a finite measure but otherwise arbitrary. However, for a general equation where the coefficients are in the spaces (3.6), the price we pay for allowing such generality of the coefficients is at the expense of requiring f to be extendable to a Sobolev function on $\mathbf{R}^n$. This is guaranteed if either the function f itself offers a transparent method of extension, or if Ω is an extension domain, see Remark 1.60.

6.1.2 Variational problems – existence of bounded solutions.

In this subsection we solve a double obstacle problem with continuous Dirichlet data. A new difficulty arises if the boundary function is not a trace of Sobolev function.

We keep the convention that all locally Sobolev function are represented by their p-quasicontinuous versions.

We extend the notion of weak boundary values to a "point at infinity": namely a function u on Ω is said to vanish weakly at infinity, if for each $l > 0$ there is a function $\eta \in C_c^\infty(\mathbf{R}^n)$ such that $(u - l)^+(1 - \eta)$ and $(-u - l)^+(1 - \eta)$ belong to $W_0^{1,p}(\Omega)$.

6.14 THEOREM. *Let $f \in C_0(\mathbf{R}^n)$. Let $\psi_1 \leq \psi_2$ be arbitrary functions defined at all points of Ω. Suppose that ψ_1 is bounded above and ψ_2 is lower bounded. Let Y be the set of all functions $v \in W_{\text{loc}}^{1,p}(\Omega)$ such that $\psi_1 \leq v \leq \psi_2$ p-q.e., $v = f$*

weakly at each point of $\overline{\Omega}\setminus\Omega$ and, if Ω is unbounded, v vanishes weakly at infinity. If Y is nonempty, then there exists $u \in Y$ such that

$$\int_\Omega (\boldsymbol{A}(x,u,\nabla u)\cdot \nabla\varphi + \boldsymbol{B}(x,u,\nabla u)\varphi)\,dx \geq 0 \tag{6.46}$$

for all $\varphi \in W^{1,p}_{cc}(\Omega)$ with $u+\varphi \in Y$.

First, we need an L^∞ estimate for solutions of the problem, which will be applied to approximate solutions.

6.15 LEMMA. *Let $u \in W^{1,p}_{\text{loc}}(\Omega)$ solve the problem described in Theorem 6.14. Then there exists a constant C independent of ψ_1, ψ_2 and f such that*

$$u(x) \leq C(1 + \sup_{x\in\partial\Omega} f(x) + \sup_{x\in\Omega} \psi_1(x)). \tag{6.47}$$

PROOF. Let L be a number with

$$L > \sup_{x\in\partial\Omega} f(x) + \sup_{x\in\Omega} \psi_1(x).$$

Then, $v := (u-L)^+$ belongs to $W^{1,p}_0(\Omega)$ and

$$\int_\Omega (\boldsymbol{A}(x,v,\nabla v)\cdot \nabla\varphi + \boldsymbol{B}(x,v,\nabla v)\varphi)\,dx \leq 0$$

for all test functions φ satisfying

$$0 \leq \varphi \leq (u-L)^+.$$

Appealing to Theorem 6.4, (6.47) is established. $\square$

PROOF OF THEOREM 6.14. Since most of readers are perhaps interested in open sets, we will give the proof in the case when Ω is open. In Remark 6.16 we indicate how the proof should be modified to cover the case of p-finely open domains.

(i) Assume initially that f itself belongs to $W^{1,p}(\Omega)\cap Y$. We define X as the set of all $v \in Y$ such that $v-f \in W^{1,p}_0(\Omega)$. Notice that X is closed. Indeed, if $\{v_k\}$ is a sequence in X converging strongly to v, then clearly $v-f \in W^{1,p}_0(\Omega)$. Furthermore, from the proof of Theorem 2.24, p. 75, it follows that there is a subsequence that converges pointwise to v p-q.e., which shows that $\psi_1 \leq v \leq \psi_2$ p-q.e. Therefore X satisfies the assumptions of Theorem (6.12) and thus there exists a solution $u \in X$ of our problem.

(ii) We now consider the general case when f is assumed to be in $\mathcal{C}_0(\mathbf{R}^n)$ and there exist a function $g \in Y$. We may assume that g is bounded. We will consider a sequence of approximating problems. Let $h_k \in \mathcal{C}^\infty_c(\mathbf{R}^n)$, $\|h_k - f\|_\infty \leq 2^{-k-1}$. Set

$$f_k(x) = \begin{cases} g(x)+2^{-k} & \text{if } h_k(x) > g(x)+2^{-k}, \\ g(x)-2^{-k} & \text{if } h_k(x) < g(x)-2^{-k}, \\ h_k(x) & \text{otherwise.} \end{cases}$$

Due to the weak boundary values of f, the functions f_k belong to $h_k + W^{1,p}_0(\Omega)$. Also we notice that $\psi_1 - 2^{-k} \leq f_k \leq \psi_2 + 2^{-k}$. We consider the problem with

$$X_k = \{v \in f_k + W^{1,p}_0(\Omega) : \psi_1 - 2^{-k} \leq v \leq \psi_2 + 2^{-k}\}$$

and define u_k as its solution which exists according to part (i). By Lemma 6.15 and its symmetric counterpart,
$$\sup_\Omega |u_k| \leq M$$
with M independent of k. Now, consider $\Omega' \subset\subset \Omega'' \subset\subset \Omega$. Let $\eta \in W_c^{1,p}(\Omega)$, $-1 \leq \eta \leq 1$, $\eta = 1$ on Ω' and $\eta = -1$ on $\partial\Omega''$. Set
$$\tilde{g} = \begin{cases} \min\{M\eta, g\} & \text{on } \Omega'', \\ M\eta & \text{on } \mathbf{R}^n \setminus \Omega'', \end{cases}$$
$$v_k = \begin{cases} \min\{\tilde{g}, u_k\} & \text{on } \Omega'', \\ M\eta & \text{on } \mathbf{R}^n \setminus \Omega''. \end{cases}$$
$$\varphi_k = \tilde{g} - v_k.$$

Since $u_k + \varphi_k \in X_k$, we have
$$\int_\Omega (\boldsymbol{A}(x, u_k, \nabla u_k) \cdot \nabla \varphi_k + \boldsymbol{B}(x, u_k, \nabla u_k)\varphi_k) \geq 0,$$
which can be read as
$$\int_\Omega (\boldsymbol{A}(x, v_k, \nabla v_k) \cdot \nabla \varphi_k + \boldsymbol{B}(x, v_k, \nabla v_k)\varphi_k) \geq 0.$$

By Theorem 6.3 we have
$$(6.48) \qquad \|\varphi_k\|_{1,p;\Omega} \leq C + C\|\tilde{g}\|_{1,p} \leq C(\|g\|_{1,p;\operatorname{spt}\eta} + \|\eta\|_{1,p} + 1),$$
so that
$$\|(u_k - g)^-\|_{1,p;\Omega'} \leq C.$$
Similarly we have
$$\|(u_k - g)^+\|_{1,p;\Omega'} \leq C.$$

Hence, the sequence $\{u_k \, \llcorner \, \Omega'\}$ is bounded in $W^{1,p}(\Omega')$. From reflexivity we infer that, after passing to a subsequence, $\{u_k \, \llcorner \, \Omega'\}$ is weakly convergent in $W^{1,p}(\Omega')$. Moreover, using Mazur's lemma, we find functions $\tilde{u}_k$ such that $\tilde{u}_k \, \llcorner \, \Omega' \to u \, \llcorner \, \Omega'$ strongly in $W^{1,p}(\Omega')$ and each $\tilde{u}_k$ is a convex combination of u_j's with $j \geq k$. The proof of Theorem 2.24 shows that $\tilde{u}_k$ converge p-q.e. on Ω' to the p-quasicontinuous representative of u. A diagonal procedure leads to a subsequence, labeled again by u_k, and a limit function u such that $u_k \to u$ $W^{1,p}$-weakly in each open subset of Ω and pointwise p-q.e. on Ω.

Now we will investigate the properties of u. By the p-q.e. convergence $u_k \to u$ we have
$$\psi_1 \leq u \leq \psi_2 \quad p\text{-q.e. in } \Omega.$$

Let us consider again $\Omega' \subset\subset \Omega$. Let $\eta \in W_c^{1,p}(\Omega)$ such that $0 \leq \eta \leq 1$ and $\eta = 1$ on Ω'. We may modify u_k and u outside the support of η so that the change leads to convergence in $W^{1,p}(\mathbf{R}^n)$. This justifies the use of Theorem 6.5. We obtain
$$\int_\Omega \eta^p |\nabla u_k - \nabla u|^p \, dx \to 0,$$
$$M_\Omega u_k \eta^{p-1} \to M_\Omega v \eta^{p-1}$$

weakly in $W_0^{1,p}(\Omega)^*$ and
$$\langle \eta^{p-1} M_\Omega(u_k), \eta u_k \rangle \to \langle \eta^{p-1} M_\Omega(u), \eta u \rangle.$$

Let $\varphi \in W_0^{1,p}(\Omega)$ be a bounded test function with support in Ω' such that $u+\varphi \in Y$. With $\varphi_k := \varphi + u - u_k$, we have
$$\langle M_\Omega u_k \eta^{p-1}, \varphi_k \eta \rangle = \langle M_\Omega u_k, \varphi_k \eta^p \rangle \geq 0.$$

Letting $k \to \infty$ we obtain
$$\langle M_\Omega u, \varphi \rangle \geq 0.$$

It remains only to show that the required boundary behavior is satisfied by u. Let $x_0 \in \partial\Omega$. We will show $u \leq f(x_0)$ weakly, the proof of the opposite inequality being similar. If $l > f(x_0)$, then $l > f_k$ on $B(x_0, r) \cap \partial\Omega$ for some $r > 0$ and all k sufficiently large. Let $\eta \in \mathcal{C}_c^\infty(B(x_0, r))$ be such that $\chi_{B(x_0, r/2)} \leq \eta \leq \chi_{B(x_0, r)}$. Then
$$v_k := \Big(\min\{u_k - l, -M - l + 2M\eta\}\Big)^+ \in W_0^{1,p}(\Omega \cap B(x_0, r)).$$

Similar to the proof of (6.48) we use Theorem 6.3 to show that
$$\|v_k\|_{1,p;\Omega \cap B(x_0)} \leq C.$$

If v is the weak limit of a subsequence of $\{v_k\}$, then clearly
$$v = \Big(\min\{u - l, -M - l + 2M\eta\}\Big)^+$$

on $\Omega \cap B(x_0, r)$. This shows that $u \leq l$ weakly at x_0. In a similar way we could establish that u vanishes weakly at infinity. $\square$

6.16 REMARK. If we prove Theorem 6.14 on p-finely open sets, it is more difficult to realize localization procedures. If $x_0 \in \Omega$, then by Theorem 2.137 there exists a p-finely open set $\Omega' \subset\subset \Omega$ such that $x_0 \in \Omega'$. The "double inclusion" for p-finely open sets means that there is a function $\eta \in W_c^{1,p}(\Omega)$ such that $\eta = 1$ on Ω'.

Consider a level $l > 0$ and denote by Ω_l the p-fine interior of $\{\eta > l\}$. By Theorem 2.144 and Theorem 2.161, $\eta - l \, \llcorner \, \Omega_l \in W_0^{1,p}(\Omega_l)$. This provides a useful scale of sets. Notice that $\Omega_l \subset\subset \Omega$ and there is $\eta_l \in W_c^{1,p}(\Omega)$ such that $\chi_{\Omega'} \leq \eta_l \leq \chi_{\Omega_l}$. Using Theorem 2.146 we construct a sequence $\{\Omega_j\}$ of p-finely open sets such that $\Omega_j \subset\subset \Omega$ and $\{\Omega_j\}$ covers Ω up to a p-polar set. This is useful to define the limit of the sequence of approximate solutions globally. Taking into account these techniques and the proof of Theorem 3.8, it is possible to follow the lines of the proof of Theorem 6.14 with minor modifications.

6.1.3 Variational problems leading to unbounded solutions.

In this subsection we suppose that Ω is a bounded open set and consider the problem with a lower obstacle, which however may be bounded above. We do not impose the assumption that the set of competitors is nonempty and so the existence problem then requires a condition that will ensure this.

6.17 THEOREM. *Let f be a continuous function on $\partial\Omega$ and $\psi\colon \Omega \to \mathbf{R}$ be an arbitrary function with the property that*

(6.49)
$$\limsup_{\substack{x \to x_0 \\ x \in \Omega}} \psi(x) \leq f(x_0)$$

for all $x_0 \in \partial\Omega$. Also, assume

(6.50)
$$\int_0^\infty \mathbf{C}_p[A(t)] t^{p-1}\, dt < \infty$$

where
$$A(t) := \{x \in \Omega : \psi(x) > t\}.$$
Then there exists $u \in W^{1,p}_{\mathrm{loc}}(\Omega)$ such that

  (i) $u(x) \geq \psi(x)$ *for p-quasi all $x \in \Omega$,*

  (ii) $\langle M_{\Omega'}(u), \varphi \rangle \geq 0$ *for all $\Omega' \subset\subset \Omega$ and $\varphi \in W^{1,p}_0(\Omega')$ satisfying $\varphi(x) \geq \psi(x) - u(x)$ for p-quasi all $x \in \Omega'$,*

  (iii) *u is bounded below and is also bounded above if ψ is bounded above,*

  (iv) $u(x_0) = f(x_0)$ *weakly for all $x_0 \in \partial\Omega$.*

First, we need the following lemma.

6.18 LEMMA. *Suppose that f and ψ satisfy the assumptions of Theorem 6.17. Let λ be a positive number such that $\lambda > f(x)$ for all $x \in \partial\Omega$. Then there exists a lower semicontinuous function $g \in W^{1,p}(\Omega)$ such that $g = \lambda + 2$ on $U_\varepsilon := \Omega \cap \{x : \mathrm{dist}\,(x, \partial\Omega) < \varepsilon\}$ for some $\varepsilon > 0$ and $g \geq \sup(\lambda + 2, \psi)$ on Ω.*

PROOF. Denote
$$\Omega_\varepsilon := \{x \in \Omega : \mathrm{dist}\,(x, \partial\Omega) > \varepsilon\}.$$
In view of (6.49) we have $\psi \leq \lambda$ on $\Omega \setminus \Omega_{2\varepsilon}$ for some $\varepsilon > 0$. Furthermore, (6.50) implies
$$\sum_{k=1}^\infty \mathbf{C}_p[A(\lambda + 2^k)](\lambda + 2^{k-1})^{p-1}(2^k - 2^{k-1}) < \infty$$
and therefore

(6.51)
$$\sum_{k=1}^\infty 2^{p(k+1)} \mathbf{C}_p[A(\lambda + 2^k)] < \infty.$$

Note that $A(\lambda + 2^k) \subset \Omega_{2\varepsilon}$ for all integers $k \geq 1$. Let φ_k denote the capacitary extremal of $\mathbf{C}_p(A(\lambda + 2^k))$. From Corollary 4.10, p. 195, we conclude that φ_k is lower semicontinuous. Choose $\eta \in \mathcal{C}_c^\infty(\Omega)$ so that $\mathrm{spt}\,\eta \subset \Omega_\varepsilon$ and $\eta = 1$ on the closure of $\Omega_{2\varepsilon}$. Then, with $\omega_k := \eta \varphi_k$, there is a constant C independent of k such that
$$\|\omega_k\|_{1,p} \leq C \|\nabla \varphi_k\|_{1,p} \leq C\mathbf{C}_p\bigl(A(\lambda + 2^k)\bigr)^{1/p}.$$
Let

(6.52)
$$g_m(x) := \lambda + 2 + \sup_{1 \leq k \leq m} 2^{k+1} \omega_k(x) \quad \text{and} \quad g(x) := \lim_{m \to \infty} g_m(x).$$

Observe that
$$\mathrm{spt}\, g_m \subset \Omega_\varepsilon$$

and
$$g_m(x) \geq \psi(x)$$
if $\lambda + 2^{k+1} \geq \psi(x)$, $1 \leq k \leq m$. Therefore,
$$g(x) \geq \psi(x) \text{ for all } x \in \Omega.$$
Furthermore,
$$\|g_m\|_{1,p}^p \leq (\lambda + 2)|\Omega| + \sum_{k=1}^m 2^{p(k+1)} \|g_m\|_{1,p}^p$$
and thus, by (6.51),
$$\limsup_{m \to \infty} \|g_m\|_{1,p}^p < \infty.$$
This implies that there is a subsequence of $\{g_m\}$ that converges weakly to some element $g^* \in W^{1,p}(\Omega)$. Since $g_m \to g^*$ strongly in L^p, a further subsequence converges pointwise a.e. to g^*. From (6.52) we conclude that $g^* = g$ a.e., thus establishing our result. □

PROOF OF THEOREM 6.17. Assume initially that f is a Lipschitz function on $\mathbf{R}^n$ and that there exists $\delta > 0$ such that

(6.53) $$\limsup_{\substack{x \to x_0 \\ x \in \Omega}} \psi(x) \leq f(x_0) - \delta$$

for each $x_0 \in \partial\Omega$. Let X denote the convex set consisting of all $v \in W^{1,p}(\Omega)$ such that $v - f \in W_0^{1,p}(\Omega)$ and $v(x) \geq \psi(x)$ for p-quasi all $x \in \Omega$. We will show that X is nonempty. Let g be the function of Lemma 6.18. It follows from (6.53) that there exists an open set $\Omega' \subset\subset \Omega$ such that
$$f(x) > \psi(x) + \frac{1}{2}\delta$$
for all $x \in \Omega \setminus \Omega'$. Let h be a compactly supported Lipschitz function on Ω with $0 \leq h \leq 1$ and $h \geq \chi_{\Omega'}$. Then, with
$$\varphi(x) := h(x)g(x) + (1 - h(x))f(x), \quad x \in \Omega,$$
we have $\varphi \in W^{1,p}(\Omega)$, $\varphi(x) \geq \psi(x)$ for p-quasi all $x \in \Omega$, and
$$\varphi - f = h(g - f) \in W_0^{1,p}(\Omega),$$
thus proving that $\varphi \in X$. As in the proof of Theorem 6.14 we see that X is closed. Thus appeal to Theorem 6.12 to conclude that there exists $u \in X$ such that
$$\langle M_\Omega(u), v - u \rangle \geq 0$$
for all $v \in X$, thus establishing parts (i) and (ii) of our theorem under the assumption that f is Lipschitz.

We now consider the general case when f is assumed to be a continuous function on $\partial\Omega$. Let $\{f_k\}$ be a decreasing sequence of Lipschitz functions on $\mathbf{R}^n$ converging uniformly to f on $\partial\Omega$ and having the property that
$$f_k \geq f + \frac{1}{k} \text{ on } \partial\Omega.$$

From the preceding step we know there exists a solution u_k corresponding to f_k. As in the proof of Theorem 6.14 we verify that there is a limit function u of the sequence $\{u_k\}$ which possesses the required properties. □

6.2 The Dirichlet problem for equations with differentiable structure

The main objective of this section is to investigate the Dirichlet problem for equations in divergence form having a differentiable structure. We will focus our attention on prescribing continuous data on the boundary of an arbitrary open set Ω. The classical Dirichlet problem is concerned with finding a solution corresponding to prescribed smooth data on the boundary of a smooth domain. Thus, if these results are particularized to finding solutions corresponding to only prescribed continuous boundary data, the assumption that Ω has a smooth boundary is unnecessarily strong. The main thrust of the results is to show, under suitable hypotheses, the existence of solutions corresponding to prescribed continuous data on the boundary of an arbitrary bounded domain that continuously assume their boundary values at p-quasi all points of $\partial \Omega$.

We will consider equations of type (3.42) satisfying the structure assumptions of (3.43).

For each open set $\Omega' \subset\subset \Omega$, we employ the regularization methods to obtain functions $\boldsymbol{A}_\varepsilon$ and $\boldsymbol{B}_\varepsilon$ that converge uniformly to $\boldsymbol{A}$ and $\boldsymbol{B}$ on $\Omega' \times \mathbf{R} \times \mathbf{R}^n$ and satisfy the structure conditions (3.46). In order to establish existence of solutions $u \in \mathcal{C}^{1,\alpha}_{\text{loc}}(\Omega) \cap \mathcal{C}^0(\overline{\Omega})$ of the equation

$$\operatorname{div} \boldsymbol{A}(x, u, \nabla u) = \boldsymbol{B}(x, u, \nabla u),$$

with specified boundary data $f \in \mathcal{C}^0(\partial \Omega)$, we will investigate solutions solutions of the equation

$$\operatorname{div} \boldsymbol{A}_\varepsilon = \boldsymbol{B}_\varepsilon \tag{6.54}$$

on Ω'. We state the following Theorem without proof. The result can be found in [**GT**], Theorem 15.11. Notice that the structure (3.46) is uniformly elliptic (with ellipticity constant depending on ε).

6.19 THEOREM. *Suppose $\Omega' \subset\subset \Omega$ is a bounded domain with $\mathcal{C}^{1,\alpha}$ boundary and let $f \in \mathcal{C}^\infty(\mathbf{R}^n)$. Then there exists a solution $u \in \mathcal{C}^{2,\beta}(\overline{\Omega}')$ of (3.47) such that $u = f$ on $\partial \Omega'$.*

6.20 THEOREM. *Suppose $\Omega' \subset\subset \Omega$ is a domain with $\mathcal{C}^{1,\alpha}$ boundary and let $f \in \mathcal{C}^{1,\alpha}(\mathbf{R}^n)$. Then there exists a solution $u \in \mathcal{C}^{1,\beta}(\overline{\Omega}')$ of (3.42) with structure (3.43) such that $u = f$ on $\partial \Omega$ where $\beta = \beta(\alpha, \lambda, \Lambda, n, \Omega)$.*

PROOF. By Theorem 6.19, there is a sequence u_k of solutions of (6.54) with $\varepsilon = \varepsilon_k \to 0$. By Theorem 3.20, there is an exponent $\beta \in (0, 1)$ such that $\{u_k\}$ is bounded in $\mathcal{C}^{1,\beta}(\overline{\Omega}')$. Using Arzela-Ascoli compactness theorem, we infer that, passing to a subsequence, there are continuous functions $u, g_1, \ldots, g_n$ on $\overline{\Omega}'$ such that $u_k \to u$ and $D_i u_k \to g_i$ uniformly on $\overline{\Omega}$. Obviously, $g_i = D_i u$ for $i = 1, \ldots, n$, the limit function u belongs to $\mathcal{C}^{1,\beta}(\overline{\Omega}')$ and satisfies (3.42). □

6.21 THEOREM. *Suppose $\Omega \subset \mathbf{R}^n$ is bounded, but otherwise arbitrary domain and let $f: \partial\Omega \to \mathbf{R}$ be continuous. Then there exists a solution $u \in \mathcal{C}^{1,\alpha}_{\text{loc}}(\Omega)$ of (3.42) such that $u = f$ weakly at each point of $\partial\Omega$.*

PROOF. Let $\Omega_1 \subset\subset \Omega_2 \subset\subset \cdots \subset\subset \Omega_k \subset\subset \cdots \subset\subset \Omega$ be a sequence of smooth domains such that $\bigcup_k \Omega_k = \Omega$. Since $f: \partial\Omega \to \mathbf{R}$ is continuous, we may extend f continuously to $\mathbf{R}^n$. Using a partition of unity of Ω and mollification as in the proof of Theorem 1.45, we modify the extension so that it is $\mathcal{C}^\infty$-smooth in Ω; this smooth extension will also be denoted by f. By the previous result, for any k there exists $u_k \in \mathcal{C}^{1,\alpha}(\overline{\Omega}_k)$ solving (3.1) in Ω_k. According to Theorem 3.12, the structure (3.43) allows the following maximum principle:

$$\sup_{\Omega_k} |u_k| \leq C \sup_{\partial\Omega_k} |f| \leq C\boldsymbol{k} + \sup_{\Omega} |f|,$$

where C depends on the structure and $|\Omega_k|$. Using a standard energy estimate (like Theorem 4.4 or (3.14)) and the interior regularity result of Theorem 3.19, the sequence $\{u_j\}_{j>k}$ is bounded in $\mathcal{C}^{1,\alpha}(\overline{\Omega}_k)$ for each k. Using the argument of the previous proof and a diagonalization process, we obtain a subsequence whose limit u exists in $\mathcal{C}^{1,\alpha}(\Omega)$ and solves (3.42) on Ω. Now we will investigate the boundary behavior of u. Select $x_0 \in \partial\Omega$ and let $l < f(x_0)$. Choose $r > 0$ such that $l < f(x)$ for all $x \in B(x_0, r) \cap \overline{\Omega}$. Then $(u_k - l)^- \eta \in W^{1,p}_0(\Omega_k)$ whenever $\eta \in \mathcal{C}^\infty_c(B(x_0, r))$. We set

$$v_k = \begin{cases} \min\{u_k, l\} & \text{in } \Omega_k, \\ l & \text{outside } \Omega_k. \end{cases}$$

We may assume that $u \geq 0$. By Lemma 4.14, v_k is a supersolution up to level l on $B(x_0, r)$. Theorem 4.7 yields an energy estimate independent of k. Hence for η as above there is a weak limit of $(l - v_k)\eta$ in $W^{1,p}_0(\Omega)$ which cannot be anything else than $(l-u)^+ \eta$ extended by zero outside Ω. This shows that $u \geq l$ weakly at x_0, and consequently $u \geq f(x_0)$ weakly at x_0. Similarly we obtain that $u \leq f(x_0)$ weakly at x_0. $\square$

In view of Theorem 4.16 and the Kellogg property, Corollary 2.142, p. 146 we have the following corollary.

6.22 COROLLARY. *With the hypotheses of the previous theorem, there exists a solution $u \in \mathcal{C}^{1,\alpha}_{\text{loc}}(\Omega)$ of (3.42) such that, if we define $u = f$ on $\partial\Omega$,*
   (i) *u assumes the boundary values f continuously at each point $x_0 \in \partial\Omega$ at which $\mathbf{R}^n \setminus \Omega$ is not p-thin,*
   (ii) *u assumes the boundary values f continuously at p-quasi every point in $\partial\Omega$,*
   (iii) *$u \in \mathcal{C}(\overline{\Omega})$ provided that $\mathbf{R}^n \setminus \Omega$ is not p-thin at any $x_0 \in \partial\Omega$.*

6.3 Historical notes

During the first part of this century, the existence theory for solutions of quasilinear elliptic equations was restricted mainly to Euler-Lagrange equations. This method relies upon finding an extremal of a variation functional in an appropriate function space. Based on the pioneering work of De Giorgi and Nash, a method was developed to attack the existence problem for equations not related to variational problems. The work of De Giorgi and Nash provided $\mathcal{C}^{0,\alpha}$ *a priori* estimates which yielded existence theorems based on the Leray-Schauder fix point theory. However, in some situations, the assumptions needed to derive the $\mathcal{C}^{0,\alpha}$ estimates were too restrictive. As a way to overcome these restrictions, the theory of monotone operators was initiated with the work of Browder [**Bro**], Minty [**Min**], Vishik [**Vis**] and Leray–Lions [**LL**]. This technique was developed further in the book by Lions [**Li**].

The method used to prove Theorem 6.5 appears in [**MZ3**] and is based on an idea of Maeda [**Mae**].

References

[Ad1] Adams, D. R., *A trace inequality for generalized potentials*, Stud. Math. **48** (1973), 99–105.

[Ad2] _____, *L^p potential theory techniques and nonlinear PDE*, Potential Theory (Ed. M. Kishi), 1–15, Walter de Gruyter, Berlin 1992.

[Ad3] _____, *A sharp inequality of J. Moser for higher order derivatives*, Ann. Math. **128** (1988), 385–398.

[AH] Adams, D. R. and Hedberg, L. I., *Function spaces and potential theory*, Springer Verlag, 1996.

[AL] Adams, D. R. and Lewis, J. L., *Fine and quasi connectedness in nonlinear potential theory*, Ann. Inst. Fourier, Grenoble **35,1** (1985), 57–73.

[AM1] Adams, D. R. and Meyers, N. G., *Bessel potentials. Inclusion relations among classes of exceptional sets*, Bull. Amer. Math. Soc. **77** (1971), 968–970.

[AM2] _____, *Bessel potentials. Inclusion relations among classes of exceptional sets.*, Indiana Univ. Math. J. **22** (1972/73), 873–905.

[AM3] _____, *Thinness and Wiener criteria for non-linear potentials*, Indiana Univ. Math. J. **22** (1972), 169–197.

[AS1] Aronszajn, N. and Smith, K. T., *Functional spaces and functional completion*, Ann. Inst. Fourier (Grenoble) **6** (1956), 125–185.

[AS2] _____, *Theory of Bessel potentials I*, Ann. Inst. Fourier (Grenoble) **11** (1961), 385–475.

[AMS] Aronszajn, N., Mulla, N. and Szeptycki, P., *On spaces of potentials connected with L^p-classes*, Ann. Inst. Fourier **13** (1963), 211–306.

[Ba] Bagby, T., *Quasi topologies and rational approximation*, J. Funct. Anal. **10** (1972), 259–268.

[BZ] Bagby, T. and Ziemer, W., *Pointwise differentiability and absolute continuity*, Trans. Amer. Math. Soc. **194** (1974), 129–148.

[Bau] Bauman, P., *A Wiener test for nondivergence structure, second-order elliptic equations*, Indiana Univ. Math. J. **34** (1985), 825–844.

[BeV] Beirao da Veiga, H., *Sulla holderianita delle soluzioni di alcune disequazioni variazionali con condizioni unilatere al bordo*, Annali Mat. Pura Appl. **84** (1969), 73–112.

[B+5] Bénilan, P., Boccardo, L., Gallouët, Th., Gariepy, R., Pierre, M. and Vazquez, J.L., *An L^1-theory of existence and uniqueness of solutions of nonlinear elliptic equations.*, Ann. Scuola Norm. Sup. Pisa, Serie IV, **22,2** (1995), 241–273.

[Be1] Besicovitch, A. S., *A general form of the covering principle and relative differentiation of additive functions*, Proc. Camb. Phil. Soc. **41** (1945), 103–110.

[Be2] _____, *A general form of the covering principle and relative differentiation of additive functions, II*, Proc. Camb. Phil. Soc. **42** (1946), 1–10.

[BMo] Biroli, M. and Mosco, U., *Wiener criterion and potential estimates for obstacle problems relative to degenerate elliptic operators*, Ann. Mat. Pura Appl. **159** (1991), 255–281.

[BH] Bliedtner, J. and Hansen, W., *Balayage spaces – an analytic and probabilistic approach to balayage*, Universitext, Springer Verlag, 1986.

[BG1] Boccardo, L. and Gallouët, T., *Non-linear elliptic and parabolic equations involving measure data*, J. Funct. Anal. **87** (1989), 149–169.

[BG2] _____, *Non-linear elliptic equations with right hand side measures*, Comm. Partial Diff. Equations **17** (1992), 641–655.

[BMu] Boccardo, L. and Murat, F., *A property of nonlinear elliptic equations with the source term a measure*, Publications du Laboratoire d'Analyse Numérique, Université Pierre et Marie Curie, Preprint (1993.).

[Bo] Bojarski, B., *Remarks on some geometric properties of Sobolev mappings*, Functional Analysis & Related Topics (Shozo Koshi, ed.), World Scientific, 1991.

[BH] Bojarski, B. and Hajłasz, P., *Pointwise inequalities for Sobolev functions and some applications*, Studia Math. **106** (1993), 77–93.

[Br1] Brelot, M., *Familles de Perron et problème de Dirichlet*, Acta Sci. Math. (Szeged) **9** (1939), 133–153.

[Br2] _____, *Sur les ensembles effilés*, Bull Soc. Math. France **68** (1944), 12–36.

[Br3] _____, *Lectures on potential theory*, Tata Institute of Fundamental Research, Bombay, 1960.

[Br4] _____, *On Topologies and Boundaries in Potential Theory*, vol. 175, Springer Lecture Notes in Math., 1971.

[Bro] Browder, F. E., *Variational boundary value problems for quasilinear elliptic equations of arbitrary order*, Proc. Nat. Acad. Sci. USA **50** (1963), 31–37.

[Cac1] Caccioppoli, R., *Sulle equazioni ellittiche non lineari a derivate parziali*, Rend. Accad. Naz. Lincei **18** (1933), 103–106.

[Cac2] _____, *Sulle equazioni ellittiche a derivate parziali con n variabili indipendenti*, Rend. Accad. Naz. Lincei **19** (1934), 83–89.

[Ca] Calderón, A. P., *Lebesgue spaces of differentiable functions and distributions*, Proc. Symp. Pure Math. **4** (1961), 33–49.

[CFR] Calderón, C., Fabes, E. and Riviere, N., *Maximal smoothing operators*, Ind. Univ. Math. J. **23** (1973), 889–898.

[CZ] Calderón, A. P. and Zygmund, A., *Local properties of solutions of elliptic partial differential equations*, Studia Math. **20** (1961), 171–225.

[Cal] Calkin, J. W., *Functions of several variables and absolute continuity, I*, Duke Math. J. **6** (1940), 170–185.

[Cam1] Campanato, S., *Proprietà di hölderianità di alcune spaci di funzioni*, Ann. Sci. Norm. Sup. Pisa **(III)17** (1963), 175–188.

[Cam2] _____, *Proprietà di una famiglia di spaci funzionali*, Ann. Sci. Norm. Sup. Pisa **(III)18** (1964), 137–160.

[Car] Carleson, L., *On the connection between Hausdorff measures and capacity*, Ark. Mat. **3** (1958), 403–406.

[Cart] Cartan, H., *Théorie général du balayage en potentiel newtonien*, Ann. Univ. Grenoble Math. Phys **22** (1946), 221–280.

[Ce] Cesari, L., *Sulle funzioni assolutamente continue in due variabili*, Annali di Pisa **10** (1941), 91–101.

[Chi] Chiarenza, F., *Regularity for solutions of quasilinear elliptic equations under minimal assumptions*, Potential Analysis **4** (1995), 325–334.

[CL] Choe, H. J. and Lewis, J. L., *On the obstacle problem for quasilinear elliptic equations of p Laplacian type*, SIAM J. Math. Anal. **22** (1991), 623–638.

[Ch1] Choquet, G., *Theory of capacities*, Ann. Inst. Fourier **5** (1955), 131–295.

[Ch2] _____, *Sur les points d'effilément d'un ensemble. Application à l'étude de la capacité*, Ann. Inst. Fourier **9** (1959), 91–101.

[DD] Dal Maso, G. and Defranceschi, A., *A Kellogg property for μ-capacities*, Boll. Un. Mat. Ital. **2** (1988), 127–135.

[DMS] Dal Maso, G., and Skrypnik, I. V., *Capacity theory for monotone operators*, Preprint SISSA 6/95.

[DMV] Dal Maso, G., Mosco, U. and Vivaldi, M. A., *A poinwise regularity theory for the two-obstacle problem.*, Acta Math. **63** (1989), 57–107.

[DeG] De Giorgi, E., *Sulla differenziabilità e lánaliticità delle estremali degli integrali multipli regolari*, Memorie della Reale Accademia delle Science di Torino, Classe di Science Fisiche, Matematiche e Naturali **3** (1957), 25–43.

[De] Deny, J., *Les potentiels d'énergie finie*, Acta Math. **82** (1950), 107–183.

[DL] Deny, J. and Lions, J. L., *Les espaces du type de Beppo Levi*, Ann. Inst. Fourier, Grenoble **5** (1953–54), 305–370.

REFERENCES

[DiB]   DiBenedetto, E., $C^{1+\alpha}$ local regularity of weak solutions of degenerate elliptic equations, Nonlinear Anal. **7** (1983), 827–850.

[DM]   DiBenedetto, E. and Manfredi, J., *On the higher integrability of the gradient of weak solutions of certain degenerate elliptic systems*, Amer. J. Math. **115** (1993), 1107–1134.

[DT]   DiBenedetto, E. and Trudinger, N., *Harnack inequalities for quasiminima of variational integrals*, Ann. Inst. H. Poincaré Nonlin. Analysis **1** (1984), 295–308.

[Do]   Doob, J. L., *Application to analysis of a topological definition of a smallness of a set*, Bull. Amer. Math. Soc. **72** (1966), 579–600.

[Er]   Erohin, V., *The connection between metric dimension and harmonic capacity*, Usp. mat. Nauk. **13** (1958), 81–88.

[Evgc] Evans, G. C., *Application of Poincaré's sweeping-out process*, Proc. Nat. Acad. Sci. **37** (1935), 226–253.

[Ev]   Evans, L. C., *A new proof of local $C^{1,\alpha}$ regularity for solutions of certain degenerate elliptic p.d.e.*, J. Differential Equations **45** (1982), 356–373.

[EJ1]   Evans, L. C. and Jensen, R., *A boundary gradient estimate for harmonic functions and applications*, Nonlinear partial differential equations and their applications, Collége de France Seminar, vol. I, 1978/1979, pp. 160–176; Res. Notes in Math., vol. 53, Pitman, Boston, Mass.-London,, 1981.

[EJ2]   _____, *Correction to: A boundary gradient estimate for harmonic functions and applications*, In: Nonlinear partial differential equations and their applications, Collége de France Seminar I (1978/1979), 160–176; (1981), Pitman, Boston, Mass..

[EG]   Evans, L. C. and Gariepy, R. F., *Measure theory and fine properties of functions*, CRC Press, Boca Raton, Florida, 1992.

[FJK]   Fabes, E., Jerrison, D. and Kenig, C., *The Wiener test for degenerate elliptic equations*, Ann. Inst. Fourier (Grenoble) **32** (1982), 151–182.

[FZ]   Federer, H. and Ziemer, W. P., *The Lebesgue set of a function whose partial derivatives are p-th power summable*, Indiana Univ. Math. J. **22** (1972), 139–158.

[FP1]   Feyel, D. and de La Pradelle, A., *Espaces de Sobolev sur les ouverts fin*, C. R. Acd. Sci. Paris Ser. A **280** (1975), 1125–1127.

[FP2]   _____, *Le role des espaces de Sobolev en topologie fine*; Seminaire de Theorie du Potentiel Paris, Lecture Notes in Math 563, Springer, 1976, pp. 62–99.

[Fr]   Frehse, J., *Capacity methods in the theory of partial differential equations*, Jber. Deutsch. Math. Verein **84** (1982), 1–44.

[FM1]   Frehse, J. and Mosco, U., *Variational inequalities with one-sided irregular obstacles*, Manuscripta Math **28** (1979), 219–233.

[FM2]   _____, *Irregular obstacles and quasi-variational inequalities of stocastic impulse control*, Ann. Sc. Norm., Pisa **9** (1982), 105–157.

[Fro]   Frostman, O., *Potentiel d'équilibre et capacité des ensembles avec queques applicationa à la théorie des fonctions*, Medd. Lunds Univ. Mat. Sem. **3** (1935), 1–118.

[Fuc]   Fuchs, M., *Hölder continuity of the gradient for degenerate variational inequalities*, Nonlinear Anal. **15** (1990), 85–100.

[Fu1]   Fuglede, B., *Le théorème du minimax et la théorie fine du potentiel*, Ann. Inst. Fourier Grenoble **15.1** (1965), 65–88.

[Fu2]   _____, *Quasi topology and fine topology*, Séminaire de Théorie du Potentiel (Brelot–Choquet–Deny) **10** (1965–1966) no. 12.

[Fu3]   _____, *Application du thèoréme minimax à l'étude de diverses capacités*, C. R. Acad. Sci Paris Ser A–B **266** (1966), 921–923..

[Fu4]   _____, *The quasi topology asociated with a countably subadditive set function.*, Ann. Inst. Fourier Grenoble **21** (1971), 123–169.

[Fu5]   _____, *Finely Harmonic Functions*; Lecture Notes in Math. 289, Springer, 1972.

[Fu6]   _____, *Fonctions BLD et fonctions finement surharmoniques*; Seminaire de Theorie du Potentiel Paris, Lecture Notes in Math 906, Springer, 1982, pp. 126–157.

[Ga]   Gagliardo, E., *Proprieta di alcune classi di funzioni in piu variabili*, Ricerche di Mat. Napli **7** (1958), 102–137.

[GZ1]   Gariepy, R. and Ziemer, W. P., *Behavior at the boundary of solutions of quasilinear elliptic equations*, Arch. Rational Mech. Anal. **56** (1974/75), 372–384.

[GZ2]   _____, *A gradient estimate at the boundary for solutions of quasilinear elliptic equations*, Bull. Amer. Math. Soc. **82** (1976), 629–631.

[GZ3] _____, *A regularity condition at the boundary for solutions of quasilinear elliptic equations*, Arc. Rat. Mech. Anal. **67** (1977), 25–39.

[GMV] Gianazza, U., Marchi, S. and Villani, V., *Interior regularity for solutions to some degenerate quasilinear obstacle problems*, Preprint 1995.

[GT] Gilbarg, D. and Trudinger, N., *Elliptic partial differential equations of the second order*, Second Edition, Springer, Berlin-Heidelberg, 1983.

[GR1] Gol'dshtein, V. M. and Reshetnyak, Y. G., *Embedding and extension theorems, and capacity (in Russian)*, Novosibirsk. Gos. Univ., Novosibirsk, 1982.

[GR2] _____, *Quasiconformal mappings and Sobolev spaces (English translation)*, Kluwer Acad. Publ., Dordrecht-Boston-London, 1990.

[GLM] Granlund, S. P., Lindqvist, P. and Martio, O., *Conformally invariant variational integrals*, Trans. Amer. Math. Soc. **277** (1983), 43–73.

[GIS] Greco, L., Iwaniec, T. and Sbordone, C., *Inverting the p-harmonic operator*, Preprint 1996.

[Ha] Hajłasz, P., *Sobolev spaces on an arbitrary metric space*, Potential Analysis **5** (1996), 403–415.

[HL1] Hardy, G. G. and J. E. Littlewood, *A maximal theorem with function-theoretic applications*, Acta Math. **54** (1930), 81–86.

[HL2] _____, *Some properties of fractional integrals, I*, Math. Zeit. **27** (1928), 565 – 606.

[HL3] _____, *Some properties of fractional integrals, II*, Math. Zeit. **34** (1932), 403 – 439.

[HM1] Havin, V. P. and Maz'ya, V. G., *A nonlinear analogue of the Newtonian potential and metric properties of the (p,l)-capacity*, Dokl. Adak. Nauk SSSR **194** (1970), 770–773 (Russian); Soviet Math. Dokl. **11** (1970), 1294–1298.

[HM2] _____, *Non-linear potential theory*, Uspekhi Mat. Nauk **27** (1972), 67–148 (Russian); Russian Math. Surveys **27** (1972), 71–148.

[He1] Hedberg, L. I., *Nonlinear potentials and approximation in the mean by analytic functions*, Math. Z. **129** (1972), 299–319.

[He2] _____, *On certain convolution inequalities*, Proc. Amer. Math. Soc. **36** (1972), 505 – 510.

[He3] _____, *Spectral synthesis in Sobolev spaces, and uniqueness of solutions of the Dirichlet problem*, Acta Math. **147** (1981), 237–264.

[HW] Hedberg, L. I. and Wolff, T. H., *Thin sets in nonlinear potential theory*, Ann. Inst. Fourier **33** (1983), 161–187.

[HK1] Heinonen, J. and Kilpeläinen, T., *A-superharmonic functions and supersolutions of degenerate elliptic equations*, Ark. Mat. **26** (1988), 87–105.

[HK2] _____, *Polar sets for supersolutions of degenerate elliptic equations*, Math. Scand. **63** (1988), 136–150.

[HK3] _____, *On the Wiener criterion and quasilinear obstacle problems*, Trans. Amer. Math. Soc. **310** (1988), 239–255.

[HKMy] Heinonen, J., Kilpeläinen, T. and Malý, J., *Connectedness in fine topologies*, Ann. Acad. Sci. Fenn. Ser. A I Math. **15,1** (1990), 107–123.

[HKM1] Heinonen, J., Kilpeläinen, T. and Martio, O., *Fine topology and quasilinear elliptic equations*, Ann. Inst. Fourier **39** (1989), 293–318.

[HKM2] _____, *Nonlinear potential theory of degenerate elliptic equations.*, Oxford University Press, Oxford,, 1993.

[Hel] Helms, L. L., *Introduction to Potential Theory*, Pure and Applied Mthematics XXII, Willey–Interscience, New York, 1969.

[HMT] Hempel, J., Morris, G. R., and Trudinger, N. S., *On the sharpness of a limiting case of the Sobolev imbedding theorem*, Bull. Austrailian Math. Soc. **3** (1970), 369–373.

[IM] Iwaniec, T. and Manfredi, J., *Regularity of p-harmonic functions on the plane*, Rev. Mat. Iberoamericana **5** (1989), 1–19.

[IS] Iwaniec, T. and Sbordone, C., *Weak minima of variational integrals*, J. Reine Angew. Math. **454** (1994), 143–161.

[JN] John, F. and Nirenberg, L., *On functions of bounded mean oscillation*, Comm. Pure Appl. Math. **14** (1961), 415–426.

[Jo] Jones, P., *Quasiconformal mappings and extendability of functions in Sobolev spaces*, Acta. Math. **147** (1981), 71–88.

REFERENCES

[Ka] Kametani, S., *On some properties of Hausdorff's measure and the concept of capacity in generalized potentials*, Proc. Imp. Acad. Tokyo **18** (1942), 617–625.

[Kel1] Kellogg, O. D., *Unicité des fonctions harmoniques*, C. R. Acad. Sci. Paris **187** (1928), 526–527.

[Kel2] _____, *Foundations of Potential Theory*, J. Springer, 1929.

[Ke] Kerimov, T. M., *Behavior of the solution of the Zaremba problem near a singular point of the boundary*, Mat. Zametki **38** (1985), 566–575, 635. (Russian)

[Kic] Kichenassamy, S., *Quasilinear problems with singularities*, Manuscripta Math **57** (1987), 281–313.

[Ki1] Kilpeläinen, T., *Thinness and polarity in a nonlinear potential theory*; Proceedings of the summer school in potential theory at Jyväskylä (Jyväskylä), vol. 34, 1987, pp. 13–17.

[Ki2] _____, *Potential theory for supersolutions of degenerate elliptic equations*, Indiana Univ. Math. J. **38** (1989,), 253–275.

[Ki3] _____, *Hölder continuity of solutions to quasilinear ellliptic equations involving measures*, Potential Anal. **12** (1996), 461–475.

[KM1] Kilpeläinen, T. and Malý, J., *Generalized Dirichlet problem in nonlinear potential theory*, Manuscripta Math. **66** (1989), 25–44.

[KM2] _____, *Degenerate elliptic equations with measure data and nonlinear potentials*, Ann. Scuola Norm. Sup. Pisa. Cl. Science, Ser. IV **19** (1992), 591–613.

[KM3] _____, *Supersolutions to degenerate elliptic equations on quasi open sets*, Comm. Partial Diff. Equations **17** (1992), 371–405.

[KM4] _____, *The Wiener test and potential estimates for quasilinear elliptic equations*, Acta Math. **172** (1994), 137–161.

[KZ] Kilpeläinen, T. and Ziemer, W. P., *Pointwise regularity of solutions to nonlinear double obstacle problems*, Arkiv Mat. **29** (1991), 83–106.

[KX] Kilpeläinen, T. and Xu, X., *On the uniqueness problem for quasilinear elliptic equations involving measures*, Preprint 180, University of Jyväskylä 1995.

[KS] Kinderlehrer, D. and Stampacchia, G., *An introduction to variational inequalities and their applications*, Academic Press, New York, 1980.

[Ko] Kondrachov, V. I., *Sur certaines propriétés fonctions dans l'espace L^p*, C. R. (Doklady) Acad. Sci. URSS. (N.S.) **48** (1945), 535 – 538.

[LU1] Ladyzhenskaya, O. A. and Ural'tseva, N. N., *Quasilinear elliptic equations and variational problems with several independent variables*, Uspehi Mat. Nauk **16** (1961), 19–90 (Russian); English Translation: Russian Math. Surveys **16** (1961), 17–92.

[LU2] _____, *Linear and Quasilinear Elliptic Equations*, Academic Press, New York, 1968.

[La1] Latvala, V., *Finely superharmonic functions of degenerate elliptic equations*, Ann. Acad. Sci. Fenn, ser. A I. Math., Dissertationes 96 (1994) 1–52.

[La2] _____, *Finely $\mathcal{A}$-superharmonic functions*, Rev. Roumaine Math. Pures Appl. **39** (1994), 501–508.

[Le1] Lebesgue, H., *Lecons sur l'intégration el la recherche des fonctions primitives*, Paris (1904).

[Le2] _____, *Sur l'integration des fonctions discontinues*, Ann. Ecole. Norm. **27** (1910), 361–450.

[Le3] _____, *Sur des cas d'impossibilité du problème de Dirichlet*, C. R. Soc. Math. France (1913), 17.

[Le4] _____, *Condition de régularité, condition d'irrégularité, conditions d'impossibilité dans le problème de Dirichlet*, C. R. Acad. Sci. Paris **178** (1924), 349–354.

[LL] Leray, J. and Lions, J. L., *Quelques résultas de Višik sur les problèmes elliptiques non linéaires par les méthodes de Minty–Browder*, Bull. Soc. Math. France **93** (1965), 97–107.

[BL] Levi, B., *Sul prinzipio di Dirichlet*, Rend. Circ. Mat. Palermo **22** (1906), 293–359.

[Lewl] Lewis, L. G., *Quasiconformal mappings and Royden algebras in space*, Trans. Amer. Math. Soc. **158** (1971), 481–492.

[Lewj] Lewis, J., *Regularity of the derivatives of solutions to certain elliptic equations*, Ind. Univ. Math. J. **32** (1983), 849–858.

[Li1] Lieberman, G. M., *Boundary regularity for solutions of degenerate elliptic equations*, Nonlinear Analysis, TMA **12** (1988), 1203–1219.

[Li2] _____, *Regularity of solutions to some degenerate double obstacle problems*, Indiana Univ. Math. J. **40** (1991), 1009–1028.

[Li3] _____, *Sharp forms of estimates for subsolutions and supersolutions of quasilinear elliptic equations involving measures*, Comm. Partial Differential Equations **18** (1993), 1191–1212.

[Lin1] Lindqvist, P., *On the definition and properties of p–superharmonic functions*, J. Reine Angew. Math. **365** (1986), 67–79.

[Lin2] _____, *Regularity for the gradient of the solution to a nonlinear obstacle problem with degenerate ellipticity*, J. Nonlinear Analysis, TMA, **12** (1988), 1245–1255.

[Lin3] _____, *On nonlinear Rayleigh quotients*, Potential Anal. **2** (1993), 199–218.

[LM] Lindqvist, P. and Martio, O., *Two theorems of N. Wiener for solutions of quasilinear elliptic equations*, Acta Math. **155** (1985), 153–171.

[Li] Lions, J. L., *Quelques méthodes de résolution des problèmes aux limites nonlinéaires*, Dunod Gautheire–Villans, 1969.

[LMu] Lions, P. L. and Murat, F., *Sur les solutions renormalisées d'équations elliptiques non linéaires*, C. R. Acad. Sci. Paris (to appear).

[Lit] Littman, W., *Polar sets and removable singularities of partial differential equations*, Ark. Math. **7,1** (1967), 1–9.

[LSW] Littman, W., Stampacchia, G. and Weinberger, H. F., *Regular points for elliptic equations with discontinuous coefficients*, Ann. Scuola Norm. Sup. Pisa. Serie III. **17** (1963), 43–77.

[Liu] Liu, Fon-Che, *A Lusin type property of Sobolev functions*, Ind. Univ. Math. J. **26** (1977), 645 – 651.

[Lo] Loewner, C., *On the conformal capacity in space*, J. Math. Mech. **8** (1959), 411–414.

[LMy] Lukeš, J. and Malý, J., *Thinness, Lebesgue density and fine topologies (An interplay between real analysis and potential theory)*, Proceedings of the Summer School on Potential Theory, Joensuu 1990. University of Joensuu, Publication in Sciences 26, Joensuu 1992.

[LMZ] Lukeš, J, Malý, J. and Zajíček, L., *Fine topology methods in real analysis and potential theory*, Lecture Notes in Math. 1189, Springer, 1986.

[Mae] Maeda, F., *A convergence property for solutions of certain quasi-linear elliptic equations*, Lecture Notes, Springer **794** (1980), 547–553.

[Mal1] Malý, J., *Hölder type quasicontinuity*, Potential Analysis **2** (1993), 249–254.

[Mal2] _____, *Nonlinear potentials and quasilinear PDE's*, Potential Theory ICPT 94 (Král, J., Lukeš, J., Netuka, I. and Veselý, J., eds.), Walter de Gruyter, Berlin, 1996.

[Mal3] _____, *Potential estimates and Wiener criteria for quasilinear elliptic equations*, XVIth Rolf Nevanlinna Colloquium (Laine,I. and Martio, O., eds.), Walter de Gruyter, Berlin, 1996.

[Mal4] _____, *Pointwise estimates of nonnegative subsolutions of quasilinear elliptic equations at irregular boundary points.*, Comment. Math. Univ. Carolinae **37,1** (1996), 23–42.

[Man1] Manfredi, J., *Regularity of the gradient for a class of nonlinear possibly degenerate elliptic equations*, Ph.D. thesis, Washington University (1986).

[Man2] _____, *Regularity for minima of functionals with p-growth*, J. Diff. Equations **76** (1988), 203–212.

[Ma1] Marchi, S., *Wiener estimates at boundary points for degenerate quasilinear elliptic equations*, Istit. Lombardo Accad. Sci. Lett. Rend. A **120** (1986), 17–33.

[Ma2] _____, *Boundary regularity for parabolic quasiminima*, Ann. Mat. Pura Appl. **166** (1994), 17–26.

[Mar] Marcinkiewicz, J., *Sur l'interpolation d'opérateurs*, C.R. Acad. Sci. Paris **208** (1939), 1271–1273.

[MM] Marcus, M. and Mizel, V. J., *Complete characterization of functions which act, via superposition, on Sobolev spaces*, Trans. Amer. Math. Soc. **251** (1979), 187–218.

[Mart] Martio, O., *Capacity and measure densities*, Ann. Acad. Scient. Fennicae. Series A I. Mathematica 4 (1978–79), 109–118.

[Maz1] Maz'ya, V. G., *The p-conductivity and theorems on imbedding certain function spaces into a C-space*, Dokl. Akad. Nauk SSSR **140** (1961), 299–302 (Russian). English translation: Soviet Math. **2** (1961), 1200–1203.

[Maz2] _____, *Regularity at the boundary of solutions of elliptic equations, and conformal mappings*, Dokl. Akad. Nauk SSSR **150** (1963), 1297–1300 (Russian); English transl. in Soviet Math. Dokl. **4,5** (1963), 1547–1551.

[Maz3] _____, *On modulus of continuity of the solutions of the Dirichlet problem near an irregular boundary*, Problems Math. Anal., Izdat. leningrad Univ., Leningrad, 1966, pp. 45–58. (Russian)

[Maz4] _____, *Behavior, near the boundary, of solutions of the Dirichlet problem for a second-order elliptic equation in a divergent form*, Math. Zametki **2,2** (1967), 209–220 (Russian); English transl. in Math. Notes **2** (1967), 610–617.

[Maz5] _____, *On the continuity at a boundary point of solutions of quasi-linear elliptic equations*, Vestnik Leningrad. Univ. **25** (1970), 42–55 (Russian); English translation:, Vestnik Leningrad Univ. Math. **3** (1976), 225–242.

[Maz6] _____, *Sobolev spaces*, Springer, Berlin-New York, 1985.

[MaSh] Maz'ya, V. G. and Shaposhnikova, T. O., *The theory of multipliers in spaces of differentiable functions,*, Pitman, Boston,, 1985.

[Me1] Meyers, N. G., *A theory of capacities for potentials of functions in Lebesgue classes*, Math. Scand. **26** (1970), 255–292.

[Me2] _____, *Taylor expansion of Bessel potentials*, Ind. U. Math. J. **23** (1974), 1043–1049.

[Me3] _____, *Continuity properties of potentials*, Duke Math. J. **42** (1975), 157–166.

[Me4] _____, *Integral inequalities of Poincaré and Wirtinger Type*, Arch. Rat. Mech. Analysis **68** (1978), 113–120.

[MS] Meyers, N. G. and Serrin, J., $H = W$, Proc. Nat. Acad. Sci. U.S.A. **51** (1964), 1055–1056.

[MZ1] Michael, J. H. and Ziemer, W. P., *A Lusin type approximation of Sobolev functions by smooth functions*, Contemporary Mathematics, AMS, **42** (1985), 135–167.

[MZ2] _____, *Interior regularity for solutions to obstacle problems*, Nonlin. Anal. TMA **10** (1986), 1427–1448.

[MZ3] _____, *Existence of solutions to obstacle problems*, Nonlin. Anal. TMA **17** (1991), 45–71.

[Mik] Mikkonen, P., *On the Wolff potential and quasilinear elliptic equations involving measures*, Ann. Acad. Sci. Fenn. Math. Dissertationes **104** (1996), 1–71.

[Mi] Miller, K., *Exceptional boundary points for the nondivergence equation which are regular for the Laplace equation and vice-versa*, Ann. Scuola Norm. Sup. Pisa Cl. Sci. **22** (1968), 315–330.

[Min] Minty, G., *Monotone (nonlinear) operators in Hilbert space*, Duke Math. J. **29** (1962), 341–346.

[Mo1] Morrey, C. B.,Jr., *On the solutions of quasi-linear elliptic partial differential equations*, Trans. Amer. Math. Soc. **43** (1938), 126–166.

[Mo2] _____, *Functions of several variables and absolute continuity, II,*, Duke Math. J. **6** (1940), 187–215.

[Mo3] _____, *Second order elliptic equations in several variables and Hölder continuity*, Math. Z. **72** (1945), 146–164.

[Mo4] _____, *Multiple integrals in the Calculus of Variations*, Springer, Berlin, 1966.

[Mor1] Morse, A. P., *The behavior of a function on its critical set*, Ann. Math. **40** (1939), 62–70.

[Mor2] _____, *Perfect blankets*, Trans. Amer. Math. Soc. **6** (1947), 418–442.

[Mosc] Mosco, U., *Wiener criterion and potential estimates for the obstacle problem*, Indiana Univ. Math. J. **36** (1987), 455–494.

[Mos1] Moser, J., *A new proof of De Giorgi's theorem concerning the regularity problem for elliptic differential equations*, Comm. Pure Appl. Math. **13** (1960), 457–468.

[Mos2] _____, *On Harnack's theorem for elliptic differential equations*, Comm. Pure Appl. Math. **14** (1961), 577–591.

[Mos3] _____, *A sharp form of an inequality by N. Trudinger*, Ind. Univ. Math. J. **20** (1971), 1077–1092.

[Mu] Mu, Jun, *Higher regularity of the solution to the p-Laplacian obstacle problem*, J. Differential Equations **95** (1992), 370–384.

[MuZ] Mu, Jun and Ziemer, W. P., *Smooth regularity of solutions of double obstacle problems involving degenerate elliptic equations*, Comm. Partial Diff. Equations **16** (1991), 821–843.

[Na] Nash, J., *Continuity of solutions of parabolic and elliptic equations*, Amer. J. Math. **80** (1958), 931–954.

[Nik] Nikodým, O., *Sur une classe de fonctions considérées dans le probléme de Dirichlet*, Fund. Math. **21** (1933), 129–150.

[Ni1]   Nirenberg, L., *Remarks on strongly elliptic partial differential equations*, Comm. Pure Apppl. math. **8** (1955), 649–675.

[Ni2]   _____, *On elliptic partial differential equations*, Ann. Scuola Norm. Pisa(III) **13** (1959), 1–48.

[Nor]   Norando, T., $C^{1,\alpha}$ *local regularity for a class of nonlinear possibility degenerate elliptic variation inequalities*, Boll. Un. Ital. Mat. **5** (1986), 281–292.

[No]   Novruzov, A. A., *On the regularity of boundary points for second-order degenerate linear elliptic equations*, Mat. Zametki **30** (1981), 353–362. (Russian)

[Os]   Osgood, W. F., *On the existence of the Green's function for the most general simply connected plane region*, Trans Amer. Math. Soc. **1** (1900), 310–314.

[Pe]   Perron, O., *Eine neue behandlung der ersten Randwertaufgabe für* $\Delta u = 0$, Math. Z. **18** (1923), 42–54.

[Po]   Poincaré, H., *Sur les équations de la Physique mathématique*, R. C. Palermo **8** (1894), 57–155.

[Ra]   Rademacher, H., *Über partielle und totale Differenzierbarkeit, I*, Math. Ann. **79** (1919), 340–359.

[R1]   Rakotoson, J. M., *Equivalence between the growth of* $\int_{B(x,r)} |\nabla u|^p dy$ *and T in the equation* $P[u] = T$, J. Diff. Eq. 86 (1990), 102–122.

[R2]   _____, *Quasilinear elliptic problems with measures as data*, Diff. and Integral Equations **4** (1991), 449–457.

[R3]   _____, *Generalized solutions in a new type of sets for problems with measures as data*, Diff. and Integral Equations **6** (1993), 27–36.

[RZ]   Rakotoson, J. M. and Ziemer, W. P., *Local behavior of solutions of quasilinear elliptic equations with general structure*, Trans. Amer. Math. Soc. **319** (1990), 747–764.

[Re]   Rellich, R., *Ein Satz über mittlere Konvergenz*, Nachr. Akad. Wiss. Göttingen Math.–Phys. **K1** (1930), 30–35.

[Re1]   Reshetnyak, Y. G., *On the concept of capacity in the theory of functions with generalized derivatives*, Sibirsk. Mat. Zh. **10** (1969), 1109–1138 (Russian); Siberian Math. J. **10** (1969), 818–842.

[Re2]   _____, *On the boundary behavior of functions with generalized derivatives*, Sibirsk. Mat. Zh. **13** (1972), 411–419 (Russian); Siberian Math. J. **13** (1972), 285–290.

[Rif1] Riesz, F., *Sur les functions subharmoniques et leur raport à la théorie du potentiel*, Acta Math. **48** (1926), 329–343.

[Rif2] _____, *Sur les functions subharmoniques et leur raport à la théorie du potentiel*, Acta Math. **54** (1930), 321–360.

[Rim]   Riesz, M., *L'intégral di Riemann-Liouville et la probléme de Cauchy*, Acta Math. **81** (1949), 1–223.

[Roy]   Royden, H., *On the ideal boundary of an open Riemann surface*, Ann. Math. Studies **30** (1953), 107–109.

[Se1]   Serrin, J., *Local behavior of solutions of quasi-linear equations*, Acta Math. **111** (1964), 247–302.

[Se2]   _____, *Pathological solutions of elliptic differential equations*, Ann. Scuola Norm. Sup. Pisa (1964), 385–387.

[Se3]   _____, *Isolated singularities of solutions of quasi-linear equations*, Acta Math. **113** (1965), 219–240.

[Se4]   _____, *Removable singularities of solutions of elliptic equations, II,*, Arch. Rational Mech. Anal. **20** (1965), 163–169.

[Sk1]   Skrypnik, I. V., *A criterion for regularity of a boundary point for quasilinear elliptic equations*, Dokl. Akad. Nauk SSSR **274**(5) (1984), 1040–1044.

[Sk2]   _____, *Nonlinear elliptic boundary value problems*, Teubner Verlag, Liepzig, 1986.

[So1]   Sobolev, S. L., *On some estimates relating to families of functions having derivatives that are square integrable*, Douk. Adak. Nauk SSSR, **1** (1936), 267–270. (Russian)

[So2]   _____, *On a theorem of functional analysis*, Mat. Sb. **46** (1938), 471–497 (Russian); English translation: Am. Math. Soc. Translations **34** (1963), 39–68.

[So3]   _____, *Applications of functional analysis in mathematical physics*, Leningrad: Izd. LGU im. A.A. Zdanova (1950) (Russian); English Translation: Amer. Math. Soc. Translations, **7** (1963).

[Sta]   Stampacchia, G., *Problemi atl contorno ellittici, con dati discontinui, dotati di soluzioni hölderiane*, Ann. Mat. Pura Appl. **51** (1960), 1–37.
[St1]   Stein, E. M., *The characterization of functions arising as potentials, I*, Bull. Amer. Math. Soc. **67** (1961), 102–104.
[St2]   _____, *The characterization of functions arising as potentials, II*, Bull. Amer. Math. Soc. **68** (1962), 577–582.
[St3]   _____, *Singular integrals and differentiability properties of functions*, Princeton University Press, 1970.
[Str]   Strichartz, R., *A note on Trudinger's extension of Sobolev's inequalities*, Ind. Univ. Math. J. **21** (1972), 841–842.
[Ta]   Taylor, S. J., *On the connection between Hausdorff measures and generalized capacity*, Proc. Camb. Phil. Soc. **57** (1961), 524–531.
[To]   Tolksdorf, P., *Regularity for a more general class of quasilinear elliptic equations*, J. Diff. Equations **51** (1984), 126–150.
[Ton1] Tonelli, L., *Sulla quadratura delle superficie*, Atti Reale Accad. Lincei **6** (1926), 633–638.
[Ton2] _____, *L'estremo assuluto degli integrali doppi*, Ann. Scuola Norm. Sup. Pisa **2** (1933), 89–130.
[T1]   Trudinger, N. S., *On Harnack type inequalities and their application to quasilinear elliptic equations*, Comm. Pure Appl. Math. **20** (1967), 721–747.
[T2]   _____, *On imbeddings into Orlicz spaces and some applications*, J. Math. Mech. **17** (1967), 473–483.
[Uh]   Uhlenbeck, K., *Regularity for a class of nonlinear elliptic systems*, Acta. Math. **111** (1964), 219–240.
[Ur]   Ural'tseva, N., *Degenerate quasilinear elliptic systems*, Zap. Naucn. Sem. Leningrad Otdel. Mat. Inst. Steklov (Russian) **7** (1968), 184–222.
[Vis]   Vishik, M. I., *On the solvability of the first bouundary value problem for nonlinear elliptic systems of differential equations*, Dokl. A. N. S. **1** (1960), 1126–1129.
[Vi]   Vitali, G., *Sui gruppi di punti e sulle funzioni di variabili reali*, Atti Accad. Sci. Torino **43** (1908), 75–92.
[We]   Weyl, H., *The method of orthogonal projections in potential theory*, Duke Math. J. **7** (1940), 411–444.
[Wie1] Wiener, N., *The Dirichlet problem*, J. Math. and Physics **3** (1924), 127–146.
[Wie2] _____, *Certain notions in potential theory*, J. Math. Phys. **3** (1924), 24–51.
[Wie3] _____, *The ergodic theorem*, Duke Math. J. **5** (1939), 1–18.
[Yo]   Yosida, K., *Functional analysis*, Springer, 5th edition, 1978.
[Za]   Zaremba, *Sur le principe de Dirichlet*, Acta Math. **34** (1911), 293–316.
[Z1]   Ziemer, W. P., *Extremal length and conformal capacity*, Trans. Amer. Math. Soc. **126** (1967), 460–473.
[Z2]   _____, *Extremal length and p-capacity*, Michigan Math. J. **16** (1969), 43–51.
[Z3]   _____, *Extremal length as a capacity*, Michigan Math. J. **17** (1970), 117–128.
[Z4]   _____, *Mean values of subsolutions of elliptic and parabolic equations*, Trans. Amer. Math. Soc. **279** (1983), 555–566.
[Z5]   _____, *Weakly differentiable functions: Sobolev spaces and functions of bounded variation*, Springer New York, 1989.

Index

(ε, δ) domain, **34**
absolutely continuous representative, 22, 61
Adams, x–xii, 61–62, 168, 158–159, 182, 252
Adams' theorem, x, **53**, **56**, 168, 163
approximate
  derivative, **43**, 143, 148, 186, 225
approximately
  continuous, **89**
  differentiable, **43**, 143, 149
approximation, 150
  by smooth functions, 22–27, 152, 46, 48, 96
  Lusin-type, 40–43
Aronszajn, 62, 157
Bagby, 157–159
Bagby's theorem, 75, **147**, 156
barrier, **108**–109
Bauman, 230
Beirao da Veiga, 251
Bénilan, 231–232
Besicovitch, 61
Besicovitch covering lemma, x, **7**, 61
Bessel kernel, **59**
Bessel potentials, **59**–60
Biroli, 251
Bliedtner, 231
Boccardo, 231–232
Bojarski, 62
boot-strap argument, 118
boundary
  regularity, **108**–109, **141**–142, 197–225
  value, weakly assumed, **135**–139, **197**–204
Brelot, 66, 158–159
Browder, 271

Caccioppoli, 182
Caccioppoli's inequality, 118, 123, 124, 187
Calderón, 34, 58, 62, 157
Calkin, 61
Campanato 62
Campanato's theorem, **31**
capacity, x–xi, **63**–92, 139–157
  classical, 102
  $(p;r)$-capacity, **63**–92, 139–157
  Choquet capacity, **66**
  of a ball, 68
capacitability, **66**
capacitary distribution, **80**–84
capacitary extremal, **78**–84
Carleson, 157
Cartan, 158
Cesari, 62
chain rule, **44**
Chiarenza, 183
Choe, 252
Choquet, 66, 157–159
Choquet
  capacity, **66**
  property, **145**, 159
coefficients, 111, 115, 118, 162, 180
coercive, **110**–114
compact imbedding, 35, 37
comparison principle, 136
condition (H), 187
convex functional, 110–115
convolution kernel, 5
covering, 7–9, 147
Dal Maso, 157–159, 252
De Giorgi, ix–xi, 109, 118, 158, 182, 270
  De Giorgi method, ix, **118**-**122**
de La Pradelle, 159
Defranceschi, 159

density point, **14**, 43, 86, 89, 143
density of measures, 10
dense subspace, 21–27, 152, 46, 48
Deny, 61, 157–159
diagonalization, **49**
DiBenedetto, 181, 183
difference quotient, 27, 116
differentiable
  almost everywhere 43,
  approximately, **43**, 143
differentiable structure, x–xii
Dirichlet, 157
Dirichlet
  integral, 93, 115
  problem, **105**–109, 113, **135**–142,
    fine, 165, 170–174, 197–204, 210–225
distribution function, 3, **51**
distributional
  derivative, 21
  solution (supersolution, subsolution)
    of the Laplace equation, **92**
double obstacle problem, xi, 233–238,
  238–251
elliptic equations, 111–115, 118, 162,
  180
energy, **79**–84
  estimate, 118, 123, 124, 187, 193, 205
  minimizer, **79**–84
exponential integrability, 19
extension domain, 27, **34**, 37
Doob, 159
Erohin, 157
Euler, 157
Evans, G.C., 158–159
Evans, L.C., x–xii, 181, 230–231
Fabes, 157, 230
Federer, 157
Feyel, 158
fine Sobolev spaces, xi, 148–157, **161**
fine topology, x, **143**–148
p-finely continuous, **130**, 139, **143**–148
  representative, 130, 194
p-finely open set, **143**, 146, 148–157,
  161
Fourier, 157
Fourier transform, **58**
Frehse, 157, 251–252
Frostman, 158

Fuchs, 252
Fuglede, 157–159, 231
fundamental harmonic function, **99**–101
Gagliardo, 62
Gallouët, 231–232
Gariepy, x–xii, 230, 230–232
Gauss, 157
generalized derivative, **21**
Gianazza, 252
Gilbarg, xii
Gol'dshtein, 35
Granlund, 158
Greco, 232
Green, 157
Green
  domain, **47**–50, 63, 72, 98–101
  function, 98–101
  potential, 98–101
Gâteaux derivative, **110**–115
Hajłasz, 62
Hansen, 231
Hardt, 61
Hardy, 61
harmonic function **92**–109, 204–210
Harnack
  inequality, 178
  weak, **125**, 174, 190
Hausdorff measure 10, 86–89
Havin, 157–159
Hedberg, xii, 61, 158–159
Heinonen, xi–xii, 158–159, 252
Helms, 158
Hempel, 61
higher regularity, 97, 115, 180–182, 238–251
Hilbert's twentieth problem, ix
Hölder's inequality, **2**
Hölder continuous, 30–31, 37, 118, 122,
  126, 195, 202
  derivatives
imbeddings, 31–40, 56
interior regularity, 97, 109, 115, 118,
  122, 180–182, 185–197, 233–251
interpolation, 3
  Marcinkiewicz interpolation theorem,
    51, 62
irregular boundary point, 224
irregular obstacle, x
Iwaniec, 181, 231–232

Jensen, x, 230–231
Jerrison, 230
John, 61, 182
John-Nirenberg inequality, **20**, **40**, 174
Jones, 34
Kametani, 157
Kellogg, 157–159
Kellogg property, xii, **146**, 270
Kenig, 230
Kerimov, 158
Khavin, see Havin
Kichenassamy, 232
Kilpeläinen, xi–xii, 158–159, 230–232, 252
Kinderlehrer, xii, 251
Kondrachov, 62
Kondrachov-Rellich imbedding theorem, **35**
Ladyzhenskaya, ix–xii, 182
Latvala, 159, 231
Laplace equation, x, **92**–109, 204–210
p-Laplace equation, **115**, 143–146
de La Pradelle, 159
Lebesgue, 61, 158
Lebesgue
　density, **14**, 43, 86, 89, 143
　point, **5**, 13, 89–92, 194
Leray, 271
Lewis, 159, 181, 252
Lieberman, 182, 231, 252
Lindqvist, 158, 230–231
Lions, J.L., 61, 157, 271
Lions, P.L., 232
Lipschitz
　domain, **34**
　function, 40–44
Littman, ix, 157, 230–231
Liu, 62
local
　minimizer, **111**
　regularity, 97, 109, 115, 118, 122, 118, 122, 180–182, 185–197, 233–251
Loewner, 157
lower function, 106
lower semicontinuous representative, 130, 194
Lukeš, 159, 231
Lusin-type approximation, 40–43
Maeda, 271

Malý, 157–159, 230–231, 252
Manfredi, 181
Marchi, 252
Marcinkiewicz, 62
Marcinkiewicz interpolation theorem, **51**
Marcus, 62
Martio, x–xii, 157–158, 230
maximal function, x, **11**–14, 17–18, 41, 43, 89–92
maximum principle, 170–174
Maz'ya, x–xii, 77, 157–159, 182, 230
Maz'ya-Shaposhnikova multiplier theorem, 77, 163
mean valued, 93
measure data, 131–135, 210–223, 225–230
Meyers, 61, 157–159
Michael, 62, 251, 271
Mikkonen, 159
Miller, 230
minimizers, 110–115
minimum principle, 106
　strong, 106
Minkowski's inequality, 2
Minty, 271
Mizel, 62
modulus of continuity, 195–197, 202–204
mollification, **5**–6, 22, 30
mollifier, **5**–6, 22, 30
Morrey, xii, 61–62, 182
Morrey
　lemma, **30**
　space, **15**, 20, 30, 40, 162, 163
Morris, 61
Morse, 61
Mosco, 251–252
Moser, ix–xi
Moser
　iteration technique, ix, 122–126, 174–178
Mu, 252
Mulla, 157
multiplier space, x, **77**, 162, 163
Murat, 231–232
Nash, ix–xi, 109, 110, 182, 270
necessity
　of classical Wiener condition, 108, 207–210

condition for boundary regularity, 108, 142, 207–210, 224–225
Newton, 157
newtonian potential, 99, 159
Nikodým, 61
Nirenberg, 61–62, ,158, 182
nonlinear potential, 132–135
Norando, 252
norm
  Sobolev norm, **21**
  multiplier norm, **77**, 162
  Morrey space norm, **15**, 162
obstacle, 233
obstacle problem, 233–251, 253–269
obstacles, irregular, 233
$(p;r)$-capacity, **63**–92, 139–157
$(p;r)$-energy, **79**–84
p-finely continuous, **130**, 139, **143**–148
  representative, 130, 194
p-finely open set, **143**, 146, 148–157, 161
p-fine topology, x, **143**–148
p-Laplace equation, **115**, 143–146
p-polar, **71**, 72, 73, 81, 143–148
p-q.e., **73**
p-quasicontinuous, **73**–76, 146, 147, 156
  representative, 74, 75, 79, 90, 149
p-quasi every, **73**
p-quasi everywhere, **73**
p-thin, **84**, 139–142, 143–148, 199–204
partition of unity, 25, 150
p-quasiopen, 146
Perron, 158
Perron-Wiener-Brelot solution, **106**–109
Pierre, 231–232
Poincaré, 157
Poincaré's inequality, 30, 33, 39
Poisson, 157
polar sets, **71**, 72, 73, 81, 143–148, 178
pseudomonotone operator, xi, 262–263
PWB solution, **106**–109
q.e., **73**, **102**
quasi-Lindelöf property, **147**
quasicontinuous, **73**–76, 146, 147, 156
  representative, 74, 75, 79, 90, 149
quasi everywhere, 73
quasiopen set, 146
  p-quasiopen, 146
Rademacher, 62

Rademacher's theorem 44
Rakotoson, 231
reference domain, 67
regular boundary point, **108**–109, **141**–142, 197–200
regularity
  boundary, **108**–109, **141**–142, 197–225
  higher, 97, 115, 180–182, 238–251
  interior, 97, 109, 115, 118, 122, 118, 122, 180–182, 185–197, 233–251
regularization, **5**–6, 22, 30
regularizer, **5**–6, 22, 30
Rellich, 62
Rellich-Kondrachov imbedding theorem, 35
removability, 72
removable singularities, 126–128, 178–179
representative, 22, 90, 149
  p-finely continuous, 130, 194
  p-quasicontinuous, 74, 75, 79, 90, 149
  absolutely continuous, 22
  lower semicontinuous, 130, 194
  harmonic, 95
  quasicontinuous, 97
  superharmonic, 95
rescaling, 163
Reshetnyak, 35, 157
resolutive, **106**–108
Riemann, 157
Riesz, F., 158
Riesz, M., 61
Riesz
  kernel, 14
  potential, x, **14**–21
  transform, 58–60
Riviere, 157
Royden, 159
Royden algebra, 159
Sbordone, 231–232
Schauder, 118
separation, 143
Serrin, ix–xi, 61, 182–183, 231
singularity
  removable, 126–128, 178–179
Shaposhnikova, x, 157, 182
Skrypnik, 157, 230
Smith, 62, 157

Sobolev, 61–62
Sobolev
  imbedding theorem, **31**–40
  inequality, **31**–40
  norm, 21
  space, x–xii, **21**-60
    fine, **148**–157
Sobolev-Poincaré inequality, 38
solution
  distributional of the Laplace equation, 92–98
  of the Dirichlet problem, **105**–109, 113, **135**–142,
  of the obstacle problem, PWB, **106**–109
  weak, **111**–142, **161**–270
  of the Laplace equation, **93**
Stampacchia, ix–xii, 182, 230–231, 251
Stein, 34, 51, 59, 58
Strichartz, 61
strong minimum principle, 106
strong type operator, **52**
structure, 111–115, 115, 118, 162, 180, 186
  differentiable, 180–182, 238–251, 269–270
  general, 162, 253
subharmonic, **93**
submeanvalued, 93
subsolution
  distributional of the Laplace equation, 92–98
  weak, **111**–142, 161–179, 185–230, 253–269
  of the Laplace equation, **93**
sufficiency
  of classical Wiener condition, 108, 204–210
  of condition for boundary regularity, 108, 142, 197–210
superharmonic, **93**–109, **205**
superharmonic representative, 95
supersolution
  distributional of the Laplace equation, 92–98
  weak, **111**–142, 161–179, 185–230, 253–269
  of the Laplace equation, **93**
  up to all levels, **185**–197, 225–230
  up to level l, **185**–197, 198, 235, 270
  supremum estimate, 167–174
Suslin
  set, **66**
Szeptycki, 157
Taylor, 157
test function, **111**, **161**, 165
thin
  set thin at a point, **84**, 102, 139–142, 143–148, 199–204, 206
Tolksdorf, 181
Tonelli, 61–62
Trudinger, ix–xii, 61, 182–183
$\mathcal{U}$-potential, **79**–84
Uhlenbeck, 181
upper function, **106**
upper obstacle problem, 185
upper solution, **106**
Ural'tseva, ix–xii, 181, 182
variational inequality, 233–251, 253–269
Vazquez, 231–232
Villani, 252
Vishik, 271
Vitali, 61
Vivaldi, 252
weak
  boundary value, **135**–139, **197**–204
  derivative 21
  Harnack inequality, ix, **125**, 174, 190
  limits at the boundary, 197–210
  solution, **111**–142, **161**–270
  subsolution, **111**–142, 161–179, 185–230, 253–269, **161**–179, 185–230
  supersolution **111**–142, 161–179, 185–230, 253–269
weak type operator, 51
Weierstrass, 157
Weinberger, ix, 230–231
Weyl, 158
Wiener, ix, 61, 158
Wiener condition, ix–xii, 108, 142, 197–210, 224–225
  classical, 108, 204–210
  necessity, 207–210, 224–225
  sufficiency, 197–210
Wiener criterion, 108, 142, 197–210, 224–225
Wolff, 158–159,
Wolff potential, 158, 225, 132–135

Young's inequality, 1
Zajíček, 231
Zaremba, 158

Ziemer, x–xii, 61, 157–159, 230–231, 251–252, 271
Zygmund, 58, 62

Notation Index

Binary operations

$f * g$	xiv
$f \circ g$	xiv
$a \cdot b$	xiv
$f \wedge g$	242

Capacity and energy

$\mathbf{c}_p(E)$	63
$\mathbf{C}_p(E)$	63
$\gamma(E)$, $\gamma_{p;r}(E;\mathcal{O})$	63
$\mathcal{E}(\mu)$	79
$\mathcal{E}_{p;r}(\mu,\mathcal{O})$	79

Convolution kernels

ϕ, ϕ_ε	5
g_α	59
I_α	14

Constants and numbers

p'	xiv
p^*	xiv
$\boldsymbol{\alpha}(n)$	xiii
$\boldsymbol{\rho}_\alpha$	14, 58
ε_1	244
ϵ	162
$\mathfrak{S}$	187

Differentiation

D_i	xiv
D^α	xiii
∇u, $\nabla^k u$	21
Du	148
$\partial J(u)$	110

Function spaces

$\mathcal{C}(E)$	xiii
$\mathcal{C}_0(E)$	xiii
$\mathcal{C}^k(\Omega)$	xiii
$\mathcal{C}^k_c(\Omega)$	xiii
$\mathcal{C}^{0,\alpha}(\Omega)$	xiii
$\mathcal{C}^{k,\alpha}(\Omega)$	xiii
L^p	xiii
$L^p_{\text{loc}}(\Omega)$	xiii
$\mathcal{M}^p$	15
$L^{\alpha,p}$	60
$W^{k,p}$	21, 148
$W^{k,p}_0$	21, 149
$W^{k,p}_c$	21, 149
$W^{k,p}_{\text{loc}}$	21, 161
$Y^{1,p}$	46, 47
$Y^{1,p}_0$	46, 47
$R^p_c(\Omega)$	149
$R^p(\Omega)$	149
$W^{1,p}_i(\Omega)$	149
$W^{1,p}_{cc}(\Omega)$	161
$W^{1,p}_{p\text{-loc}}(\Omega)$	149
$\mathcal{U}$, $\mathcal{U}_{p;r}(\mathcal{O})$	63
$M(X,Y)$	77
$M(\mathcal{U}, L^p(\mathcal{O}))$	77
$M(W^{1,p}(\mathbf{R}^n), L^p(\mathbf{R}^n))$	162

Functions

$u \mathbin{\llcorner} E$	xiii
χ_E	xiii
$g_{z,r}(x)$	98
$g_z(x)$	98
$\Gamma_z(x)$	99
$\Gamma(r)$	99
$G(x,z)$	98
H_f	106
$\overline{H}_f$	106
$\underline{H}_f$	106
u^+	xiv
u^-	xiv
u_ε	5
u_E	82, 102
$\lambda_{f;\nu}$	51
h_k	257

Integrals and mean values

$\int_E u(x)\,dx$	xiv
$\fint_E u\,dx$	xiv
$\overline{u}_E$	xiv
$\overline{u}_r(x)$	93
$\overline{v}(a,r)$	87
$(\nabla u)_{x_0,\rho}$	247
$\Phi(R)$	249
$D(x_0,r)$	201
$D(r)$	201

Measures

$\|E\|$	xiii
H^k	xiii
$\lambda_{z,r}$	98
δ_z	xiii
μ_E	82, 102

Norms

$\|u\|_p$	xiii
$\|u\|_{k,p}$	21
$\|u\|_{k,p;\Omega}$	21
$\|u\|_\mathcal{U}$	63
$\|u\|_{M(X,Y)}$	77
$\|u\|_{M(p,p);\Omega}$	162
$\|u\|_{1+\alpha}$	182

Operators, transforms

$\hat{f}$	58
$I_\alpha f$	14
$M(f)$	11
$R_i(f)$	58
Δu	92
$\Delta_p u$	115
$P\mu$	98
$G\mu(x)$	99
T	227
$\mathbf{W}_p^\mu(x)$	225
M_Ω	254

Relations

$U \subset\subset \Omega$	xiii, 161

Sets

$b_p F$	84
$B(x,r)$	xiii
$K(x,r)$	xiii
$\mathbf{R}^n$	xiii
E°	xiii
$\overline{E}$	xiii
$\operatorname{spt} u$	xiv
∂E	xiii
$\mathcal{O}$	63
$\{u > \alpha\}$	xiii
Ω_r	93

Structure and coefficients

$\mathbf{F}(x,\zeta,\xi)$	111
$A(x,\zeta,\xi)$	111, 162
$B(x,\zeta,\xi)$	111, 162
$F(x,\xi)$	111
$A(x,\xi)$	111
$A_0(h)$	239
$\mathbf{J}, \mathbf{J}[u,\Omega]$	111
a, c	111
α, β	111
a_i, b_i, c_i	162, 253
ϵ	162, 210
$\overline{a}, \overline{b}$	163, 186
f	189,
$k(r)$	162,
$h(r)$	186
$\kappa(r)$	194, 186
$a(r)$	210, 210
$b(r)$	210
$d(r)$	214
λ	180, 239
Λ	180, 239
Λ_1	239

Other notation

$\operatorname{dist}(A,B)$	xiii
$\operatorname{diam}(A)$	xiii
$\operatorname{osc}_A u$	xiii
X^*	xiii
$\langle \mu, v \rangle$	79
$\overline{u}$	187
u_l	187, 200
m, l	200
$m_l(r)$	200
$\operatorname{w}\liminf_{x \to x_0} u(x)$	197
$\operatorname{w}\limsup_{x \to x_0} u(x)$	197
$\operatorname{w}\lim_{x \to x_0} u(x)$	197
$\omega^-(r)$	201
$\omega^+(r)$	201
$\omega(r)$	201
q, τ, γ	210
$\mathcal{Y}(E), \mathcal{Y}_{p;r}(E;\mathcal{O})$	63
$\overline{\mathcal{Y}}(E)$	74
ψ_1, ψ_2	234

NOTATION INDEX

$\mathcal{K}$	234	$\overline{\psi}_1(x_0)$	234
$\mathcal{T}$	234	$\underline{\psi}_1(x_0)$	234
$p\text{-}\sup_{B(x_0,r)} \psi$	234	$M_1(r)$	236
		$M_2(r)$	236
$p\text{-}\inf_{B(x_0,r)} \psi$	234	E_1^λ	236
		E_2^λ	236
$\overline{\psi}_1(r)$	234	$A_i^\lambda(s)$	236
$\underline{\psi}_1(r)$	234	$S_i(t,r)$	236

图字：01-2019-0853 号

Fine Regularity of Solutions of Elliptic Partial Differential Equations, by Jan Malý, William P. Ziemer,
first published by the American Mathematical Society.
Copyright © 1997 by the American Mathematical Society. All rights reserved.
This present reprint edition is published by Higher Education Press Limited Company under authority
of the American Mathematical Society and is published under license.
Special Edition for People's Republic of China Distribution Only. This edition has been authorized by
the American Mathematical Society for sale in People's Republic of China only, and is not for export therefrom.

本书原版最初由美国数学会于 1997 年出版，原书名为 *Fine Regularity of Solutions of Elliptic Partial Differential Equations*，作者为 Jan Malý, William P. Ziemer。

美国数学会保留原书所有版权。

原书版权声明：Copyright © 1997 by the American Mathematical Society。

本影印版由高等教育出版社有限公司经美国数学会独家授权出版。

本版只限于中华人民共和国境内发行。本版经由美国数学会授权仅在中华人民共和国境内销售，不得出口。

椭圆偏微分方程的解的
精细正则性
Tuoyuan Pianweifen Fangcheng
de Jie de Jingxi Zhengzexing

图书在版编目 (CIP) 数据

椭圆偏微分方程的解的精细正则性 = Fine Regularity of Solutions of Elliptic Partial Differential Equations : 英文 / (捷克) 简·马利 (Jan Maly), (美) 威廉姆·P. 齐默 (William P. Ziemer) 著 . — 影印本 . — 北京：高等教育出版社，2019.5
ISBN 978-7-04-051724-8

Ⅰ．①椭⋯　Ⅱ．①简⋯　②威⋯　Ⅲ．①二阶—椭圆型方程—偏微分方程—正则性—研究—英文　Ⅳ．① O175.23

中国版本图书馆 CIP 数据核字 (2019) 第 067185 号

策划编辑　李　鹏　　责任编辑　李　鹏
封面设计　张申申　　责任印制　赵义民

出版发行　高等教育出版社
社　址　北京市西城区德外大街 4 号
邮政编码　100120
购书热线　010-58581118
咨询电话　400-810-0598
网　址　http://www.hep.edu.cn
　　　　　http://www.hep.com.cn
网上订购　http://www.hepmall.com.cn
　　　　　http://www.hepmall.com
　　　　　http://www.hepmall.cn
印　刷　北京中科印刷有限公司

开　本　787mm×1092 mm　1/16
印　张　19.75
字　数　500 千字
版　次　2019 年 5 月第 1 版
印　次　2019 年 5 月第 1 次印刷
定　价　135.00 元

本书如有缺页、倒页、脱页等质量问题，请到所购图书销售部门联系调换
版权所有　侵权必究
[物　料　号 51724-00]

郑重声明

高等教育出版社依法对本书享有专有出版权。任何未经许可的复制、销售行为均违反《中华人民共和国著作权法》，其行为人将承担相应的民事责任和行政责任；构成犯罪的，将被依法追究刑事责任。为了维护市场秩序，保护读者的合法权益，避免读者误用盗版书造成不良后果，我社将配合行政执法部门和司法机关对违法犯罪的单位和个人进行严厉打击。社会各界人士如发现上述侵权行为，希望及时举报，本社将奖励举报有功人员。

反盗版举报电话　　(010) 58581999　58582371　58582488
反盗版举报传真　　(010) 82086060
反盗版举报邮箱　　dd@hep.com.cn
通信地址　　　　　北京市西城区德外大街 4 号
　　　　　　　　　高等教育出版社法律事务与版权管理部
邮政编码　　　　　100120

美国数学会经典影印系列

1. **Lars V. Ahlfors**, Lectures on Quasiconformal Mappings, Second Edition
2. **Dmitri Burago**, **Yuri Burago**, **Sergei Ivanov**, A Course in Metric Geometry
3. **Tobias Holck Colding**, **William P. Minicozzi II**, A Course in Minimal Surfaces
4. **Javier Duoandikoetxea**, Fourier Analysis
5. **John P. D'Angelo**, An Introduction to Complex Analysis and Geometry
6. **Y. Eliashberg**, **N. Mishachev**, Introduction to the *h*-Principle
7. **Lawrence C. Evans**, Partial Differential Equations, Second Edition
8. **Robert E. Greene**, **Steven G. Krantz**, Function Theory of One Complex Variable, Third Edition
9. **Thomas A. Ivey**, **J. M. Landsberg**, Cartan for Beginners: Differential Geometry via Moving Frames and Exterior Differential Systems
10. **Jens Carsten Jantzen**, Representations of Algebraic Groups, Second Edition
11. **A. A. Kirillov**, Lectures on the Orbit Method
12. **Jean-Marie De Koninck**, **Armel Mercier**, 1001 Problems in Classical Number Theory
13. **Peter D. Lax**, **Lawrence Zalcman**, Complex Proofs of Real Theorems
14. **David A. Levin**, **Yuval Peres**, **Elizabeth L. Wilmer**, Markov Chains and Mixing Times
15. **Dusa McDuff**, **Dietmar Salamon**, *J*-holomorphic Curves and Symplectic Topology
16. **John von Neumann**, Invariant Measures
17. **R. Clark Robinson**, An Introduction to Dynamical Systems: Continuous and Discrete, Second Edition
18. **Terence Tao**, An Epsilon of Room, I: Real Analysis: pages from year three of a mathematical blog
19. **Terence Tao**, An Epsilon of Room, II: pages from year three of a mathematical blog
20. **Terence Tao**, An Introduction to Measure Theory
21. **Terence Tao**, Higher Order Fourier Analysis
22. **Terence Tao**, Poincaré's Legacies, Part I: pages from year two of a mathematical blog
23. **Terence Tao**, Poincaré's Legacies, Part II: pages from year two of a mathematical blog
24. **Cédric Villani**, Topics in Optimal Transportation
25. **R. J. Williams**, Introduction to the Mathematics of Finance
26. **T. Y. Lam**, Introduction to Quadratic Forms over Fields

27 **Jens Carsten Jantzen**, Lectures on Quantum Groups

28 **Henryk Iwaniec**, Topics in Classical Automorphic Forms

29 **Sigurdur Helgason**, Differential Geometry,
    Lie Groups, and Symmetric Spaces

30 **John B. Conway**, A Course in Operator Theory

31 **James E. Humphreys**, Representations of Semisimple Lie Algebras
    in the BGG Category O

32 **Nathanial P. Brown**, **Narutaka Ozawa**, C*-Algebras and
    Finite-Dimensional Approximations

33 **Hiraku Nakajima**, Lectures on Hilbert Schemes of Points on Surfaces

34 **S. P. Novikov**, **I. A. Taimanov**, **Translated by Dmitry Chibisov**,
    Modern Geometric Structures and Fields

35 **Luis Caffarelli**, **Sandro Salsa**, A Geometric Approach to
    Free Boundary Problems

36 **Paul H. Rabinowitz**, Minimax Methods in Critical Point Theory with
    Applications to Differential Equations

37 **Fan R. K. Chung**, Spectral Graph Theory

38 **Susan Montgomery**, Hopf Algebras and Their Actions on Rings

39 **C. T. C. Wall**, **Edited by A. A. Ranicki**, Surgery on Compact Manifolds,
    Second Edition

40 **Frank Sottile**, Real Solutions to Equations from Geometry

41 **Bernd Sturmfels**, Gröbner Bases and Convex Polytopes

42 **Terence Tao**, Nonlinear Dispersive Equations: Local and Global Analysis

43 **David A. Cox**, **John B. Little**, **Henry K. Schenck**, Toric Varieties

44 **Luca Capogna**, **Carlos E. Kenig**, **Loredana Lanzani**,
    Harmonic Measure: Geometric and Analytic Points of View

45 **Luis A. Caffarelli**, **Xavier Cabré**, Fully Nonlinear Elliptic Equations

46 **Teresa Crespo**, **Zbigniew Hajto**, Algebraic Groups and Differential Galois Theory

47 **Barbara Fantechi**, **Lothar Göttsche**, **Luc Illusie**, **Steven L. Kleiman**,
    Nitin Nitsure, **Angelo Vistoli**, Fundamental Algebraic Geometry:
    Grothendieck's FGA Explained

48 **Shinichi Mochizuki**, Foundations of p-adic Teichmüller Theory

49 **Manfred Leopold Einsiedler**, **David Alexandre Ellwood**, **Alex Eskin**,
    Dmitry Kleinbock, **Elon Lindenstrauss**, **Gregory Margulis**, **Stefano Marmi**,
    Jean-Christophe Yoccoz, Homogeneous Flows, Moduli Spaces and Arithmetic

50 **David A. Ellwood**, **Emma Previato**,
    Grassmannians, Moduli Spaces and Vector Bundles

51 **Jeffery McNeal**, **Mircea Mustaţă**, Analytic and Algebraic Geometry:
    Common Problems, Different Methods

52 **V. Kumar Murty**, Algebraic Curves and Cryptography

53 **James Arthur**, **James W. Cogdell**, **Steve Gelbart**, **David Goldberg**,
    Dinakar Ramakrishnan, **Jiu-Kang Yu**, On Certain L-Functions

54 **Rick Miranda,** Algebraic Curves and Riemann Surfaces

55 **Hershel M. Farkas, Irwin Kra,** Theta Constants, Riemann Surfaces and the Modular Group

56 **Fritz John,** Nonlinear Wave Equations, Formation of Singularities

57 **Henryk Iwaniec, Emmanuel Kowalski,** Analytic Number Theory

58 **Jan Malý, William P. Ziemer,** Fine Regularity of Solutions of Elliptic Partial Differential Equations

59 **Jin Hong, Seok-Jin Kang,** Introduction to Quantum Groups and Crystal Bases

60 **V. I. Arnold,** Topological Invariants of Plane Curves and Caustics

61 **Dusa McDuff, Dietmar Salamon,** *J*-Holomorphic Curves and Quantum Cohomology

62 **James Eells, Luc Lemaire,** Selected Topics in Harmonic Maps

63 **Yuri I. Manin,** Frobenius Manifolds, Quantum Cohomology, and Moduli Spaces

64 **Bernd Sturmfels,** Solving Systems of Polynomial Equations

65 **Liviu I. Nicolaescu,** Notes on Seiberg-Witten Theory